PIT BULL

Pit Bull

THE BATTLE OVER AN AMERICAN ICON

BRONWEN DICKEY

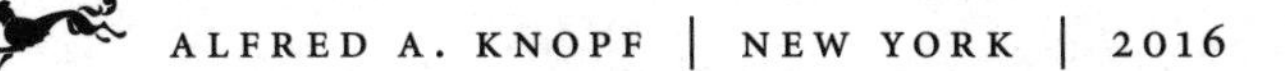

THIS IS A BORZOI BOOK
PUBLISHED BY ALFRED A. KNOPF

Published in the United States by Alfred A. Knopf,
a division of Penguin Random House LLC, New York,
and distributed in Canada by Random House of Canada,
a division of Penguin Random House Canada Limited, Toronto.

www.aaknopf.com

Knopf, Borzoi Books, and the colophon are registered trademarks
of Penguin Random House LLC.

Library of Congress Cataloging-in-Publication Data
Names: Dickey, Bronwen, author.
Title: Pit bull : the battle over an American icon / by Bronwen Dickey.
Description: First U.S. edition. | New York : Alfred A. Knopf, 2016. |
Includes bibliographical references and index.
ISBN 9780307961761 (hardcover) | ISBN 9780307961778 (ebook)
Identifiers: LCCN 2015033292
Subjects: LCSH: Pit bull terriers.
Classification: LCC SF429.P58 D53 2016 | DDC 636.755/9—dc23
LC record available at http://lccn.loc.gov/2015033292

Jacket design by Oliver Munday

Manufactured in the United States of America
Published May 13, 2016
Reprinted Three Times
Fifth Printing, November 2016

For Sean, who makes everything possible

Neither a man, nor a crowd, nor a nation
can be trusted to act humanely or to think sanely
under the influence of a great fear.

—BERTRAND RUSSELL

CONTENTS

ABBREVIATIONS

ABKC—American Bully Kennel Club
ADBA—American Dog Breeders Association
ADOA—American Dog Owners Association
AFF—Animal Farm Foundation
AKC—American Kennel Club
AmStaff—American Staffordshire terrier
APBT—American pit bull terrier
ASPCA—American Society for the Prevention of Cruelty to Animals
AVMA—American Veterinary Medical Association
BAD RAP—Bay Area Dog Owners Responsible About Pit Bulls
BSL—breed-specific legislation
DBIH—dog-bite-injury hospitalization
DBRF—dog-bite-related fatality
F1, F2—first-generation cross (purebred A × purebred B), second-generation cross
HSUS—Humane Society of the United States
PETA—People for the Ethical Treatment of Animals
PFL—Pets for Life
SNP—Single-nucleotide polymorphism
Stafford, Staffie Bull—Staffordshire bull terrier
UKC—United Kennel Club

BREED GLOSSARY

PIT BULL BREEDS

American pit bull terrier

American Staffordshire terrier

Staffordshire bull terrier

American bully

RELATED BREEDS

English bulldog

American bulldog

Boxer

French bulldog

English bull terrier

Boston terrier

Bullmastiff

Dogo Argentino

AUTHOR'S NOTE

This book is the culmination of seven years of research that took me across fifteen states, during which I interviewed more than 350 people, including pet owners, dog breeders, dog trainers, animal control officers, animal rescuers, conformation judges, shelter workers, cruelty investigators, animal advocates, law enforcement officers, and dog bite victims, as well as experts in veterinary medicine, animal behavior, canine genetics, constitutional and criminal law, public policy, risk analysis, and public health epidemiology. I also attended multiple conformation shows, animal welfare conferences, law enforcement training seminars, and community outreach events and read many thousands of pages of scientific journal articles, legal depositions, police and coroner's reports, and archival materials that spanned more than 250 years.

When it has been necessary for me to re-create past events, I have drawn most heavily from primary sources, including contemporaneous news coverage, letters, diaries, photographs, incident reports, and other public documents. If key participants were still alive, I contacted them for verification whenever possible. In some cases, multiple days' or weeks' worth of interviews and follow-ups have been condensed for easier reading, but quotations have been fact-checked with sources or checked against the transcripts of recorded conversations.

Many people on all sides of the pit bull debate will disagree with what I have written here. Breeders, in particular, will argue with me about which dogs "count" as pit bulls, as they do so often among themselves. I have included the breeds in this group that are most closely related and physically similar, with full understanding that some of those breeds have changed over time. While a growing number of animal advocates prefer the word "guardian" to "owner" when referring to humans and their canine companions, I have chosen to use the latter term because of its (current) legal accuracy. I do not claim that this book encapsulates everything there is to know about pit bulls, but I do hope that my efforts in the following pages will contribute to a much larger dialogue about human-animal relationships in America. Please be advised that some of the material and several of the images in this book are graphic in nature.

A few last names have been omitted to protect the privacy of certain people. Any factual errors are entirely my own.

PIT BULL

PROLOGUE

There are very few monsters who warrant the fear we have of them.

—ANDRÉ GIDE, *Les Nouvelles Nourritures*

The sky had just begun to darken when the Smoak family stopped for gas in Davidson County, Tennessee. It was the first day of 2003, and James Smoak; his wife of twelve years, Pamela; his seventeen-year-old stepson, Brandon; and the family's two dogs, Patton and Cassie, had just finished a vacation in Nashville. They were now headed home to Saluda, North Carolina, as part of a two-car caravan, with James's parents and the Smoaks' two younger children bringing up the rear.

Inserting a credit card with one hand, James needed the other to operate the gas pump, so he tossed his wallet onto the roof of his green station wagon as he filled the tank, then climbed back into the car without retrieving it. A few minutes later, when a patrol car flashed its lights and directed him to pull over onto the shoulder of I-40, Smoak reached into his pocket and realized his wallet was missing. Assuming this was a standard-issue traffic stop, James and Pamela were more annoyed than concerned. When three additional cruisers from the Tennessee Highway Patrol and the Cookeville, Tennessee, police department arrived, however, both grew nervous.

A patrolman shouted instructions at Smoak: He was to drop his keys out of the window, get out of the car with his hands raised, and walk backward toward the rear of the vehicle. Bewildered, James did as he'd been told. Once he reached the back of the station wagon and kneeled down as directed, an officer ordered him to put his hands behind his back. Another policeman trained an assault rifle on him, and a third cinched handcuffs around his wrists. Pamela and Brandon were then instructed to get out, kneel, and put their hands behind their backs to be cuffed as well.

One of the officers informed James that the local police dispatcher had been notified of a possible robbery in the area. James and his family were the primary—in fact, the only—suspects. Insisting that he was not involved and knew nothing about it, James stopped his protest midsentence when he noticed the wagon's open passenger door. Pamela had forgotten to nudge it closed when she got out. "Hey, wait!" James shouted at the officers. "I've got dogs in the car! *I don't want them to run into the road!*"

The officers' twenty-five-minute dash-cam recording of these events would later be submitted into evidence as part of a formal investigation into police procedure. Around the video's three-minute mark, Patton, whom the family described as a brindle "pit bull/boxer mix," bounds out of the wagon with his tail wagging. At fifty-five pounds, he is smaller than most Labradors. If Patton growls or barks, none of the officers' microphones pick it up. Then Officer Eric Hall of the Cookeville Police Department shoulders his 12-gauge shotgun. The family members, who are kneeling next to one another in the bright glare of the police car's lights, begin to scream, their voices merging into a single panicked chorus.

"No!" Pamela begs Hall on the recording. "*No! No . . . !*"

Hall fires his weapon at close range, killing the animal instantly.

James springs to his feet. "*You shot my dog!*" he bellows at the officers. "*Goddamnit!* Oh, my God! *Oh, my God!*" Two of the patrolmen tackle Smoak, pinning the man to the ground. They order a hysterical Pamela, who has also jumped up, back to her knees. The family's wailing sobs overwhelm the audio. Finally, the officers shove James, Pamela, and Brandon into two separate patrol cars and await the dispatcher's instructions.

Within the hour, the officers would learn over the radio that no robbery had occurred. A female driver in a passing car had seen dollar bills from James's wallet fluttering away from the Smoaks' station wagon and called the Tennessee Highway Patrol to report a possible holdup. After a volley of calls between the highway patrol and the police dispatchers, both departments sent officers on a felony traffic stop—the kind reserved for drug traffickers and those suspected of violent crimes—despite having few details.

Once the mix-up had been relayed to them, the policemen sheepishly uncuffed the Smoak family, apologized, and told them they were free to go. Brandon crawled over to Patton's lifeless body and cradled it as he cried. His stepfather then wrapped the dog in the only item he could use as a makeshift shroud, a plastic bag, and placed him in the back of the wagon. James Smoak later said that he was so emotionally distraught over the shooting that Pamela took him to the nearest hospital emergency department for severe chest pain. He would also undergo surgery for a knee injury he said was incurred when the officers threw their weight into him.

For Hall, a professed animal lover who holds a degree in wildlife law enforcement, the presence of a brindle "pit bull" in the Smoaks' car fit into the "dangerous criminal" narrative. Pit bulls, he believed, did not behave like other dogs. Though he admitted that the dog had been out of the car for only three seconds, this was the version of events Hall recounted in his internal affairs interview:

> OFFICER ERIC HALL: This . . . dog jumps out of the car, the suspect car, and I believe it's a . . . full-grown pit bull. I'm the officer furtherest [*sic*] to the right and it circles around, and I even see Officer McWhorter [a fellow patrolman] kind of track the dog with his weapon and then come back up, and that dog's coming right at me. I mean it just singled me out—here it comes charging and growling . . . definitely an aggressive posture and stance it was taking . . . trying to circle me . . . I yelled at the dog, and I yelled I know once real loud [for it] to get back. Scare it, something, get it, just get back off of me. And no effect at all. And it was trying to get round me. And it was probably two foot [*sic*] from the muzzle of my shotgun . . . and trying to, trying to get at me, and I fired one time at the dog . . . put the dog down.
> DETECTIVE SERGEANT BILL BOWMAN: When a pit bull is, is on the attack, what mannerisms do they have?
> HALL: . . . He was sidestepping, trying to get, trying to come at me . . . Growling, kind of hunched like he was going to spring up at me and so forth . . .
> BOWMAN: Did you feel that the dog was stalking you?
> HALL: Oh, yeah.

BOWMAN: . . . Okay. And your knowledge of pit bulls [is] that they are extremely dangerous . . . ?
HALL: Yes.
BOWMAN: . . . They have killed people before . . . ?
HALL: Yes.

Officer Mead McWhorter backed up his colleague: Shooting the dog was the only option. The "pit bull" had been large and menacing, "sixty-five or seventy-five pounds." In other words, "big enough to bite you." McWhorter explained, "There's two thousand pounds per square inch in a dog bite . . . I don't know that I would take the chance on it." The internal affairs board concurred with Hall and McWhorter, deciding that Hall "was placed in a position that a reasonable person would have believed that he was in imminent danger of serious bodily harm or death."

This was not the first time Hall had encountered dogs in the line of duty; before the Smoak incident, he had shot two others in "self-defense." When asked why he did not use Mace, a Taser, or a nightstick to protect himself, Hall replied, "Mace doesn't work on animals as well as it does on people, and both my hands were full with the shotgun. You can't try to kick a dog or be jumping around with a loaded shotgun in your hand."

The judge who presided over the Smoaks' lawsuit and the officers' subsequent appeals did not see it that way. In fact, she excoriated members of the Cookeville PD for "prowling" around the family and pointing both a shotgun and an assault rifle at them in the first place, rather than keeping their weapons in the customary "down ready" position. "Although this gesture was superficially menacing and surely intended as an intimidating show of force," the judge wrote in her formal opinion, "the way these two officers paced and shuffled behind the troopers, wavering guns gripped white-knuckled in their hands, exhibits such nervousness and fear that, even on video, the tension is palpable and the ensuing events are almost predictable."

The mere presence of a pit bull caused Eric Hall such anxiety that he abandoned his professional training, his good judgment, and his common sense in the line of duty. On the evening of January 1, 2003, he didn't see a family dog at all. He saw a monster.

CHAPTER 1

PARIAH DOGS

> The animal has secrets which, unlike the secrets of caves, mountains, seas, are specifically addressed to man.
>
> —JOHN BERGER, "Why Look at Animals?"

> All knowledge, the totality of all questions and answers, is contained in the dog.
>
> —FRANZ KAFKA, "Investigations of a Dog"

On a hot summer day a few years ago, my husband, Sean, brought home Nola, a slightly underweight thirty-eight-pound pit bull with a caramel-and-white coat, a pink nose, and eyes the color of honey. Pronounced cheek muscles and a cleft in the top of her head gave her face the shape of a small but eager heart. When Sean and I visited our city's animal shelter earlier that month, I had barely noticed her. "Wait a minute," he said as I cooed over a flashier and more extroverted candidate. "What about this one?" When she glued herself to Sean's legs in the shelter's play yard, we agreed to give her a home.

We did not need another dog. We had been married less than a year, and we still did not know how best to shape our independent selves to the contours of a shared life. Sean worked long hours at a local hospital, and I spent weeks away from home on reporting assignments. Oscar, our imperious black pug, had finally gotten to the age where we no longer worried about leaving him alone in the house, and not worrying was an activity I had come to enjoy. Any new addition would put a large speed bump in our otherwise smooth routine.

I'm still not quite sure why we did it, but I do remember an unnerving sense of recognition when she looked at me through her chain-link kennel. It was a feeling I had never experienced before and haven't since. I can't say that she stared at me, exactly, but she never looked away, either. Somewhere underneath that look was a spark of intense curiosity. It was a look with layers.

In his classic essay "Why Look at Animals?," the critic John Berger writes that the look between man and animal is one of enormous weight. It is a bridge between our species and theirs, one of the few that can be built between creatures that do not share a common language. "The animal scrutinises [man] across a narrow abyss of non-comprehension," Berger says. When man looks back, however, he "is always looking across ignorance and fear."

I did not feel fear in the presence of this particular dog, but hundreds of news reports about dog attacks told me that I had good reason to fear her kind. For almost forty years, the media have recycled a seductively simple narrative about "dangerous breeds" that relies on terrifying anecdotes and a vocal minority of people who have declared themselves "experts."

Those closest to me were quick to express their reservations about adopting a pit bull. "I wouldn't do it," one woman told me solemnly. "Those dogs will turn on you. And once the aggression switch is turned on, there's no turning it off." Another lowered her voice and said she "assumed certain things about people who own pit bulls." She did not need to elaborate; I understood that she assumed pit bull owners were either criminally inclined or otherwise morally suspect. But my friends who wrung their hands had never physically encountered these dogs; they were not even sure how to identify them. Even so, they believed they knew the pit bull's *story*, into which was written a long legacy of human bloodlust and betrayal, hard-line genetic determinism, and unprovoked animal rage.

Our popular culture is saturated with these stereotypes, and they are highly profitable for corporations and celebrities looking to cultivate an image of toughness. In 2003, Nike ignited a media firestorm when it released a television ad that cut images of chained pit bulls lunging at each other into footage of basketball face-offs between the NBA superstars Tony Parker, Jason Kidd, Gary Payton, and Steve Nash. The company was adamant that it never intended to encourage dogfighting, only to capture what it called "the edginess of urban basketball." Likewise, in 2012, when a reporter for *GQ* magazine asked the Cuban-American rapper Armando Pérez about the origins of his stage name, Pitbull, the artist replied, "In dogfights, pit bulls are too stupid to lose," a muddled way of saying that the dogs use their hearts more than their heads, and their hearts know only how to fight.

But images of pit bulls as snarling beasts are by no means limited to urban iconography. Take a walk through any American shopping center, and you are likely to find pit bull sunglasses, pit bull soccer cleats, pit bull tires, pit bull colognes, even pit bull energy drinks, wines, and hot sauces, all of which trade on the idea of a Nietzschean *Überhund* who "won't back down" or "won't let go." Some of our most common metaphors invoke the days of animal combat as well: The terms "top dog" and "underdog" both originated in the fighting pits. When the Republican vice presidential candidate Sarah Palin joked during her 2008 acceptance speech that the only difference between a hockey mom and a pit bull was lipstick, the audience laughed knowingly. The irony of this is that many who valorize the stereotype—who proudly identify themselves as pit bull politicians, pit bull lawyers, pit bull stock traders—vilify the actual dogs at the center of it.

But which dogs are those? The term "pit bull" is an elastic, imprecise, and subjective phrase. At its narrower end, it denotes one breed: the American pit bull terrier (APBT), which originated as a fighting dog in Massachusetts in 1889. A century later, American lawyers and politicians expanded the term to include the American Staffordshire terrier (AmStaff), the APBT's preppy kennel-club cousin, and the Staffordshire bull terrier (Stafford), a pint-sized variant of the nineteenth-century British fighting dog that is now primarily bred for conformation shows in its native United Kingdom. In 2013, one more breed was added to this already diverse group when the United Kennel Club (UKC) formally recognized the American bully, a heavier conformation and companion dog derived from the AmStaff in the 1990s. Dogs from these four breeds can weigh anywhere from twenty-five to one hundred pounds (with most about forty to fifty) and come in at least sixteen different coat colors and patterns.

Ask a hundred different Americans to define what a "pit bull" is, however, and you will get a hundred different answers. Over the past two decades, the category has swelled to include mixed-breed dogs that possess supposed "pit bull characteristics," such as blocky heads, white chest markings, or brindle coats. Those same characteristics can be found in more than twenty breeds of dog, and the latest genetic research indicates that many mixed-breed dogs identified as "pit mixes" actually aren't. "Pit bull," as it is most commonly used, has become a slap-dash shorthand for a general *shape* of dog—a medium-sized, smooth-coated mutt—or a "dog not otherwise specified." Most of the seventy-seven million dogs thought to be living in the United States are not registered with kennel clubs, and many are not even licensed, so it is impossible to know how many dogs from pit bull breeds live, die, or enter the American shelter system each year.

We do know that there are a lot of them. Between 1995 and 2005 (the most recent time frame for which data are available), a relatively small specialty registry called the American Dog Breeders Association (ADBA) registered more than 700,000 American pit bull terrier puppies. In that same decade, the American Kennel Club (AKC) registered a little more than 25,000 AmStaffs and Staffords. The American Bully Kennel Club (ABKC), another niche organization, currently registers almost 40,000 American bullies each year, as well as roughly 60,000 dogs from similar bulldog breeds. If the UKC were to make its statistics public, those numbers would add considerably to the total.

Combined, these figures still don't account for the untold thousands of purebred dogs whelped in backyards across the United States, or the much higher number of mixed-breed animals who may have one or more pit bull "characteristics." In 2011, when the first "mutt census" was conducted by Mars Veterinary, a research division of the consumer brand Mars Incorporated that sells dog DNA tests, the American Staffordshire terrier was the seventh most common breed identified in America's mixed-breed dog population, despite being seventieth in AKC popularity.

The number of Americans who self-identify as pit bull owners is increasing as well. Banfield pet hospitals, the largest chain of veterinary clinics in the country, reports that the number of their pit-bull-owning clients rose 47 percent between 2003 and 2013. A separate analysis of national veterinary records lists the American pit bull terrier among the top three most popular breeds in twenty-eight states and among the top five in thirty-four. Once again, we can't know which of these animals are pedigreed pit bulls and which are simply mixed breeds, but this does tell us that the "pit bull" label is prolific and that many pet owners are quite proud of it.

Contrary to the media narrative, only a tiny subset of American pit bulls will ever have any contact with the world of illegal dogfighting, which is a felony in all fifty states. Only a handful of dogs from specific bloodlines of one breed—the American pit bull terrier—are still selected and trained for that purpose. Cruelty investigators at the American Society for the Prevention of Cruelty to Animals (ASPCA) and the Humane Society of the United States (HSUS) report that even within this highly specialized subset only one dog per litter may show the necessary temperament and stamina for the grim task of mortal combat, which is on par with historical estimates, which place the number of purpose-bred APBTs matched in pit contests somewhere between 1 and 10 percent. Therefore, comparing the temperaments and behaviors of elite fighting dogs with those of all pit bulls is a bit like using the U.S. Navy SEALs as a benchmark for all American men. Fortunately for the canine victims involved, law enforcement officials are seeing the numbers of APBTs bred for fighting dwindle, thanks to increased awareness and tougher enforcement of cruelty laws. They insist that the overwhelming majority of pit bulls, like most dogs in America, live uneventful lives as family pets.

You would not know this from reading, watching, or listening to the news. Nor would you know that only about thirty-five Americans are killed by any type of dog each year, as opposed to the thirty-five thousand who die of accidental overdoses or the thirty-six thousand who perish in car accidents. Unlike overdoses and car accidents, however, exceedingly rare events like dog attacks terrify the most primitive, reptilian parts of our brains. And unlike falls or drownings, which also account for thousands of American deaths each year, dog bite deaths, especially when pit bulls are involved, allow the audience to choose sides when looking to place blame: *Are pit bulls inherently dangerous, or is it all in how you raise them? Which is more important, human rights or animal rights? Which is stronger, nature or nurture?* Like the existence of God, a woman's right to choose, or the ethics of capital punishment, such reductions are good at creating conflict (and conflict drives narrative, which in turn draws viewers and clicks), but they foreclose the possibility of solving the complicated problems of cruelty and violence.

Underneath America's need to define a singular "truth" about pit bulls is a much more revealing division: that pit bulls are not for people like "us"—the respectable and morally upstanding members of society; pit bulls belong to *them.*

For the better part of two hundred years, the history of bull-and-terrier dogs was illustrious, rather than infamous. Advertisers across the United

top left Landon Rives and dog, ca. 1900; *top right* family portrait, date and location unknown; *above* family portrait, date and location unknown; *left* girl poses with Pete the Pup from the *Our Gang* comedies at the Steel Pier in Atlantic City, 1937

Original caption, 1927: "Pal the Wonder Dog, not content with acting, picks up a megaphone to direct his co-star Mildred Davis during production of the Paramount picture *Too Many Crooks.* Davis is the wife of famous silent actor Harold Lloyd."

States clamored to use pit bulls in their campaigns during the 1920s, not because the dogs were believed to be menacing, but because they were thought to be so friendly and appealing to the "average Joe." They are the only dogs to have appeared on the cover of *Life* magazine three times, for example. The animals' widespread popularity among people of all ages, races, and classes owed much to their reputations as plucky, unfussy sidekicks and hardy all-purpose workers. More than that, however, "the dog with the patch over his eye" was seen as quintessentially *American:* good-natured, brave, resilient, and dependable. By World War I, pit bulls were so beloved as national symbols that we literally and figuratively wrapped them in the flag. We even called them "Yankee terriers."

Haphazardly classified under almost twenty other names over the years, bull-and-terrier dogs marched onto the field at the Battle of Gettysburg and sniffed out snipers at Normandy. They peeked out of covered wagons bound for California and stumped for women's suffrage. One greeted visitors at New York City's first pizzeria in 1907, while another lived in Teddy Roosevelt's White House. They also accompanied us into the brave new world of modern technology, listening to "his master's voice" on the recently invented gramophone and riding shotgun in the first cross-country road trip by automobile.

Cultural icons as diverse as Sir Walter Scott, William "Buffalo Bill" Cody, Anna Pavlova, Helen Keller, Jack Dempsey, Jack Johnson, Andy Devine, Roscoe "Fatty" Arbuckle, Gary Cooper, Douglas Fairbanks, James Thurber, Theodor "Dr. Seuss" Geisel, and Jimmy Carter proudly kept bull-and-terrier dogs as pets, and years before anyone heard of a German shepherd named Rin Tin Tin, pit bull actors ruled the silver screen. In fact, "Rinty" only appeared in 27 motion pictures, while a pit bull named Pal the Wonder Dog appeared in 224.

Then, in the 1970s, like a bright light snapping off, everything went terribly wrong. The crime of dogfighting exploded in the headlines, and the well-intentioned, well-publicized crusade to stamp out a barbaric but moribund form of animal torture unwittingly made it more popular. Once reporters and misinformed activists cast the dogs as willing participants in their own abuse, pit bulls were exiled to the most turbulent margins of society, where a cycle of poverty, violence, fear, and desperation had already created a booming market for aggressive dogs. Headlines about pit bull attacks on humans multiplied. Within a few short years, America's century-old love for its former mascot gave way to the presumption that pit bulls were biologically hardwired to kill.

A month or so after Sean and I adopted Nola, I met a woman named Lori Hensley. We were standing in line at a local theater making casual conversation when she mentioned that she directed a local nonprofit called the Coalition to Unchain Dogs. The goal of the organization, she said, was to invert the normal paradigm of animal welfare. Instead of removing dogs from "bad" homes and placing them with "better" families, the group worked to improve the dogs' lives in the homes they already had, thus preventing them from entering the overburdened shelter system in the first place. This was done by providing the dogs' owners with basic veterinary care, spay/neuter surgeries, doghouses, and fenced play areas for their pets, all free of charge.

"You should come see what we do," Lori offered. "We can always use more help." I smiled politely and mumbled something lukewarm in response. As noble as the group's objectives sounded, I did not consider myself an "animal person." In truth, I found certain memes of humane activism ("I love animals, but hate people!") to be alienating. I also never wanted to see another dog on a chain as long as I lived.

My family had owned nine dogs by the time I was twelve. There were two dachshunds, two Scottish terriers, a golden retriever, a Boykin spaniel, a collie, and two mutts. None lived with us longer than a few years. Most ended up roaming the neighborhood, where one was hit by a car, or chained in the backyard, where one strangled and died. All howled long into the night. Because of this, many of our neighbors viewed us with contempt: we were the family so irresponsible it couldn't take care of its own pets.

What my neighbors did not know was that the suffering of the dogs outside our home was a symptom of the suffering inside it. My parents were highly intelligent, generous, and compassionate people, but they were cursed with crippling addictions that numbed their consciences, stunted their ability to make good decisions, and rendered them unable to deal with the responsibilities of everyday life. I was six years old when my mother first told me that she had a drug problem, nine when she was arrested for cocaine possession, twelve when my father's liver failed from years of drinking, and fifteen when the failed liver finally killed him. All of us, including the dogs, were casualties of a remorseless disease. Shame and finger-pointing did not solve that problem.

"Here's what's interesting, though," Lori said before I was able to change the subject. "I don't do this for the dogs. I do it for the people." Then she mentioned, offhand, that roughly 80 percent of her clients owned pit bulls.

The first fence build I attended with Lori took place at the mobile home of a woman named Cheryl, who was pregnant with twins at the time. She and her boyfriend owned three pit bulls and a Chihuahua. Ten volunteers spent roughly two hours driving T-posts and pulling welded wire to complete the project. When Cheryl unhooked her dogs from their chains, they shook their giant heads, looked around tentatively, took a few shaky steps, then started to run, jump, and play for the first time in years, perhaps ever. Cheryl's eyes brimmed with tears, as though a crushing weight had finally been lifted.

I then began accompanying Lori on her outreach visits, during which she stopped in to see between twelve and fifteen clients each Saturday. One family sat on garbage bags of secondhand clothing because they could not afford furniture. Their sewage emptied into the yard through a plastic pipe. Another constructed a sprawling tent city out of old shipping containers and a broken-down bus. Lori sat on porches and beside hospital beds, fussed over newborns, and admired family photographs. It soon became clear that dogs were only a footnote in a much larger community project. "If the Coalition leaves any kind of legacy behind," Lori said, "I hope that it is remembered as one of the first animal groups that cared as much about the well-being of the person as it did about the pet."

Amanda Arrington, the woman who founded the Coalition in 2007, initially focused on the problem of chained dogs as an animal welfare issue. She believed it could be ameliorated with the right resources—namely free fences. When she and Lori began spending more time in the neighborhoods where chained dogs were most common, however, they found systemic social problems that went much deeper. The real enemy was not neglect or cruelty; it was poverty. Residents of lower-income communities—especially those in African-American and Latino neighborhoods—simply did not have access to the same veterinary clinics, pet supply stores, and pet care information that other animal lovers took for granted.

This was greatly exacerbated by what the women felt was a "reverse Robin Hood" mentality in the traditional animal welfare movement that disparaged those who could not spend a great deal of money on their animals. Several of the Coalition's clients reported having their companions stolen by well-meaning rescuers who did not realize how much the dogs' owners depended on them for emotional support. Other clients were afraid to ask shelter workers or animal control officers for help out of fear that their pets would be taken away. This is the gap the Coalition worked to close. Since 2007, the group has unchained 1,885 dogs in eight cities and spayed or neutered almost four thousand under Lori's direction.

The person Lori visited most often was an eighty-nine-year-old woman named Doris. She had recently lost most of her right leg to a foot infection, so she wrapped the stump in a clean cotton sock and pushed herself around in an old manual wheelchair. Doris had lived alone since her husband died in 2000, but every morning she made her bed with crisp hospital corners, scoured her kitchen from top to bottom, and set her dining table for four, though no one ever joined her for dinner. In her youth, she worked in the curing barns of a company called Central Leaf Tobacco, where she said her boss would regularly ask her to leave work early so that she could "come along and clean the house" for his wife. On Doris's right cheek was a thin scar that wound its way down to the top of her chest. She told me that many years ago, she tried to save her neighbor from being beaten by a drunken, raging husband. The man broke a whiskey bottle on the kitchen table and dragged it across Doris's throat, nearly killing her.

Other than Lori and a few volunteers from Meals on Wheels, Doris had no regular visitors until the winter of 2012, when a sickly black pit bull wandered out of the woods behind her house. The dog was old, with nubby yellow teeth, flanks pocked with scabs and sores, and leathery teats that hung so low they almost touched the ground. Fingers of charred flesh ran down the sides of a giant burn scar that spanned the length of her back, and she moved slowly, as though dragging a large cinder block behind her. Fearing that the dog would be put to sleep if she called animal control, Doris began cooking grits and eggs for her every morning and wheeling out to the front porch to keep her company. Soon the dog was sleeping in Doris's kitchen, where she answered to a new name: Pretty Girl. If Doris knew or cared that Pretty Girl was a pit bull, she never said so. There was no fretting about "breed traits." To Doris, Pretty Girl was just a dog.

Over time, I saw dozens of formerly scrawny animals like Pretty Girl transform into sleek, muscular family pets as the bond with their people intensified, but in most cases the bond had always been there; better resources just helped it thrive. "Nobody *wants* to keep their dogs on chains," a client named Marlynda told Lori. But sturdy fences were expensive ($500 minimum) and required a great deal of physical work and special equipment to install. Fence recipients were so grateful to be treated with dignity, rather than scorn, that some became volunteers themselves, spreading information about pet health around their neighborhoods. The unlikeliest of these was a single father and former backyard pit bull breeder named Rodney, who became something of an animal welfare spokesman.

At a city council meeting in Durham, North Carolina, Rodney testified that in his neighborhood pets were suffering not because their owners were cruel and depraved but because the long history of inequality in those areas poisoned everything it touched, including the lives of animals. Referring to his own African-American heritage, he drew a devastating parallel. "We used to be in chains," he told the council members. "Now our dogs are in chains."

Since it ramped up in the 1990s, the marketing of the pit bull as an icon of "urban edginess" has been incredibly effective. On some streets, every house contained a pit bull, if not two or three. No one I asked could pinpoint when the trend started or why, but it seemed to come down to two concerns: cost and safety. Few families who wanted a dog could afford to spend $500 to $1,000 on a purebred puppy, but pit bulls were always easy to acquire, if not completely free. A number of clients had tried to adopt dogs from local shelters but had been turned away, so they sought out a friend or neighbor whose dog had just given birth. Many other neighborhood pit bulls came from accidental litters or from the local flea market, where truck beds full of cheap puppies were available every weekend. The Great Recession of 2008 had left people even more desperate to make ends meet, and unlike other parts of the underground economy, dog breeding was legal and unregulated. A surprisingly large number of clients, mostly single moms and grandmothers, had taken in animals for friends or family members who lost their jobs, went to prison, or succumbed to terminal illnesses.

I met pit bulls that lived in overturned trash cans, in houses made from scraps of wood, in the beds of old trucks; pit bulls that lived with toddlers and with the bedridden elderly; pit bulls that shared close quarters with other dogs, pigs, horses, and chickens; pit bulls that loathed other animals and preferred to be with people. After several hundred of these encounters, I no longer thought of the dogs as belonging to any particular group, let alone a "breed." The only thing they shared was that their presence was a comfort and an inspiration to the people around them, to whom the death of a dog (pit bull or otherwise) was mourned as much as, if not more than, the amputation of a limb.

Occasionally, I came across the canine bruisers everyone worries about, animals whose wiring just didn't seem right, but each was off-kilter in a different way. Yes, I met two or three pit bulls with nasty dispositions, but also a German shepherd that furiously attacked its own legs when approached, a border collie I was certain would kill me if given the chance, a pug that launched herself at the necks of other dogs near food, and a chow that made even the sturdiest veterinarians cower.

As Dr. Laurel Braitman observes in her book *Animal Madness*, "Every animal with a mind has the capacity to lose hold of it from time to time. Sometimes the trigger is abuse or maltreatment, but not always." Like ours, the canine brain is far more intricate than a simple nature-nurture "debate" can clarify. It can be thrown off track by innumerable factors, including shifts in hormones and neurotransmitters, as well as injury and congenital birth defects. Also like us, many dogs that suffer from these maladies can improve with treatment, but, sadly, others cannot. Were the ill-tempered dogs I met born with their screws loose, or had a lifetime of stress mentally worn them down? There was no way of knowing. As one researcher I interviewed explained, "Nature versus nurture only exists in the media. Everyone in the sciences knows that it's both."

I was less surprised by the fraction of unstable animals than by how many dogs—from the smallest Chihuahuas to the most imposing Rottweilers—had endured years of deprivation and physical discomfort yet never lashed out at anyone. Dogs with intestinal blockages or collars embedded in their necks hardly raised a lip or bared a tooth when poked, prodded, or lifted into transport vehicles. It was a powerful testament to the affiliative nature of *Canis lupus familiaris* as a creature, something we too often look past in a culture so fixated on carnage. Even the most daunting, muscle-bound Presa Canario is, by virtue of its biology, predisposed to seek out attention from, and bond with, humans. That is yet another marvel and mystery of the dog: it is the only animal that will place our safety and survival above its own.

On one of our Saturday afternoon outings, Lori told me something that I have never forgotten. She said, "As long as there are different classes of people, there will be different classes of dogs." And the more I learned about pit bulls, the more it seemed that the dogs and their people had been pushed outside what the philosopher and ethicist Peter Singer calls "the expanding circle" of compassion. If you looked a certain way and came from a certain neighborhood, your dog was assumed to be a pit bull (or, at the very least, a "pit mix"), and your relationship to it was assumed to be motivated by greed or machismo. "Pit bull" no longer felt like a physical description to me. It felt like a social caste.

To date, "pit bulls" have been either banned or heavily regulated in more than 850 U.S. communities, including the cities of Miami and Denver, as well as in the Canadian province of Ontario and the entire United Kingdom, among other places. Many public-housing projects, apartment complexes, and homeowners' associations refuse to allow pit bulls to reside on the premises. Despite the life-saving contributions of a shelter pit bull named Howard, who served multiple tours in Afghanistan with the U.S. Army's Eighty-Second Airborne Division as a tactical explosive detection dog, pit bulls are banned from privatized housing on all major military bases. Several large insurance carriers refuse to underwrite the homes of pit bull owners, and at least one small organization has dedicated itself, with Ahab-like fervor, to eradicating pit bulls and their "close mixes" altogether, calling them "Frankenmaulers," "shit bulls," "mutants," and "undogs."

In nearly every municipality where breed-specific legislation (BSL) has been adopted, it has failed to prevent serious dog bite injuries and hospitalizations. Veterinarians, animal behaviorists, and public health experts, including those at the Centers for Disease Control and Prevention (CDC), are virtually unanimous in their denunciation of BSL on the grounds that it is both cruel and ineffective. In 2013, even the White

House released a statement against it. Only the most radical of America's humane groups, People for the Ethical Treatment of Animals (PETA), supports this approach. According to the organization's founder, Ingrid Newkirk, the dogs are kept only by "drug dealers" and "pimps."

Humane groups estimate that animal shelters around the world kill more than a million pit bulls a year. At some facilities, the bodies of dead dogs labeled "pit bulls" have been piled so high that they resembled sandbag fortresses. The level of collective fury that has been directed against this one group of animals is unprecedented, rivaled only by the annihilation of wolves on the American frontier. It calls to mind what the historian Jon T. Coleman said about nineteenth-century wolf hunters, that they seemed to treat their quarry as though it "not only deserved death but deserved to be punished for living."

After Hurricane Katrina left thousands of pit-bull-type dogs stranded in New Orleans in 2005 and the NFL superstar Michael Vick was convicted on charges related to dogfighting in 2007, America began to reconsider the dog it had once loved so dearly. An ever-growing network of pit-bull-focused animal rescues, clubs, and community events sprang up around the country to help promote the dogs as family pets, rather than pariahs. At least four reality television shows made pit bull advocacy a primary part of their narratives. Suddenly, millions of Americans were waving the banner of "pit bull pride." On Facebook, the interest page for the AKC's most popular breed, the Labrador retriever, barely tops 400,000 fans. Pit bulls? They have over 4.5 million fans—more than half the number of users interested in dogs as a general subject.

Any creature that inspires such strong emotions is destined to be both scapegoated by its enemies and overprotected by its advocates. When this happens, distortions, fables, and half-truths can run in both directions. In February 2014, for example, a four-year-old boy named Kevin Vincente wandered into a neighbor's yard in Phoenix, Arizona, where a dog named Mickey was chained. According to Vincente's babysitter, the little boy picked up Mickey's bone, and the dog bit him badly on the face, breaking Vincente's jaw and one of his eye sockets. Doctors said that the boy would need several reconstructive surgeries, but he would always be permanently disfigured and blind in one eye. Mickey's owner willingly released him to animal control officers, and everyone expected that due to the severity of the injury he caused, the dog would be humanely euthanized.

Instead, a massive social media campaign converged to save Mickey.

The Lexus Project, a New York–based legal defense fund for animals, arranged for the dog to have his own legal representation. If the child had been fully supervised, Mickey's supporters said, he probably would not have been bitten. While that is undoubtedly true, Team Mickey came across as cold and callous when its members took to social media to heap blame on Vincente's caretakers, even on Kevin himself. "Everybody is taught, from the moment they walk, you do not take a bone from a dog," the attorney representing Mickey told a reporter. The official report from the county shelter, however, didn't mention a bone. The events of that day are still unclear, but the dog's supporters vowed online that they would "save poor confused Mickey who is being punished for doing what dogs do and what children should be taught not to."

That Mickey was described as a pit bull stoked this fire considerably, with animal advocates insisting that he was being persecuted because of his appearance. According to the local sheriff's department, more than fifty thousand supporters petitioned the judge presiding over Mickey's disposition to spare the animal, even if he couldn't be placed in another home. Joe Arpaio, the notoriously bombastic local sheriff, then offered Mickey a permanent residence at the Maricopa County Jail's no-kill shelter, the Animal Safe Haven.

Today, while human inmates in Arpaio's jail are consigned to outdoor tents in the stifling Arizona heat where temperatures inside can rise to 145 degrees Fahrenheit, Mickey now lives alone in his own air-conditioned cell, complete with a $1,000 Webcam, where he will be isolated more than twenty-three hours a day for the rest of his life. Mickey's fans are ecstatic that he will remain alive, but at what cost? When meted out to humans, a lifetime of solitary confinement is considered psychological torture, and dogs, like humans, rely on regular social contact for their overall well-being.

As the "Save Mickey" saga played out in the press, the support fund for Kevin Vincente's medical bills raised less than $2,500. Numerous columnists, journalists, and bloggers around the country (including me) were disturbed by the lack of concern about the boy's recovery. Others used the battle over Mickey's fate to condemn all pit bull dogs and portray their owners as borderline sociopaths. The "Save Mickey" campaign followed a pattern typical of the pit bull wars: It devolved into demagoguery, with the protagonists reduced to caricatures and straw men. Whose suffering took priority? Whose compassion outweighed whose?

. . .

Perhaps this tempest should not surprise us. There may be no creature on earth that lends itself to as much love, hate, and mythmaking as the domestic dog. There is no animal that our culture is more invested in, no animal with whom we spend more of our lives, no animal that we see as a truer extension of ourselves. In the thirty-five thousand years since we entered the interspecies partnership that made civilization possible, the literature of dogs has mostly become a literature of longing: for home, for safety, for acceptance, and probably for some flicker of the wildness we ourselves have lost.

"The magic that gleams an instant between Argos and Odysseus is both the recognition of diversity and the need for affection across the illusions of form," noted the anthropologist and science writer Loren Eiseley, alluding to the final books of Homer's epic *The Odyssey.* After twenty years apart, the eponymous warrior's dog recognizes him when his human family cannot. That look of recognition, Eiseley says, is "nature's cry to homeless, far-wandering, insatiable man: 'Do not forget your brethren, nor the green wood from which you sprang. To do so is to invite disaster.'"

Yet our ambivalence toward the dog runs as deeply as our love for it, in both our minds and our myths. One has only to look at the sheer number of canine monsters that romp through our nightmares. For every faithful Argos, there is a knife-toothed Cerberus guarding the gates of Hades. For every Toto, there is a Baskerville hound charging across the moors, its eyes glowing with phosphorus. For every Lassie, there's a Cujo.

Dogs that remind us they are animals, rather than surrogate humans, do so at their own peril. "For whatever reason, our culture has transformed the dog into a paragon of canine virtue," writes the behaviorist Dr. James Serpell, director of the Center for the Interaction of Animals and Society at the University of Pennsylvania's School of Veterinary Medicine:

> He is our loyal and faithful servant and companion; a sort of amiable culture-hero whose friendship is proverbially better than that of our fellow humans. That such a devoted and trusted admirer should suddenly turn and savage one of us is both frightening and disturbing. For not only is this behavior disloyal—a betrayal of trust—it is also grossly insubordinate . . . As perpetual children,

> we expect [dogs] to be forever innocent, playful, and fun-loving. A murderous dog, like a murderous child, is therefore nothing less than an abomination—a disturbance in the natural order—an unacceptable threat to the perceived security and stability of the entire community.

The chaos of wildness is unnerving on its own, but more so when a creature that manifests it sleeps in our beds. In order to make sense of a confusing, chaotic world, the French anthropologist Claude Lévi-Strauss believed that we divide the landscape into strict binaries and opposing forces: good/evil, black/white, hero/monster, us/them. Knowing how to categorize things puts ground under our feet when we are frightened by uncertainty.

Dogs fit perfectly into this divided portrait. We are always looking at two dogs, in a sense: the physical dog in front of us—the one with the particular brain and heart, who likes to eat popcorn or likes to play Frisbee—and the character in a much larger story we have been telling ourselves about dogs for thousands of years. We embroider this narrative with personal anecdotes, hopes, fears, rumors, gossip, received wisdom, and the constant drone of modern media, but the story is an ever-evolving fiction; often it has little, if anything, to do with the flesh-and-blood animal looking back.

One of the most famous relics from the doomed city of Pompeii is a mosaic from the House of the Tragic Poet. It depicts a ferocious guard dog gnashing its teeth. *Cave canem,* it warns, "Beware of the dog." Just across the way, at the house of a local official named Vesonius Primus, the preserved remains of an actual guard dog that perished in the destruction were also found. Chained to its post by a bronze studded collar, the animal was unable to escape the volcano's wrath. As the dog writhed in agony, the lava and ash that overwhelmed it froze the creature in time as a symbol of pathos, not violence. Yet it is almost always the first dog, the one with teeth bared, that we remember.

Fears about "demon dogs" have cycled through Anglophone culture for hundreds of years, but none has endured as long as the controversy over pit bulls, which is now entering its fifth decade. Nor has any been fed by as much bad science, media sensationalism, political brinkmanship, moral panic, racial venom, or class prejudice.

The most inconvenient thing about obsessions is that they never announce themselves. One day you make a simple decision to adopt a shelter dog, and the next you find yourself building fences and visiting elderly pet owners in their hospital rooms. Five years later, your whole life has been consumed by the need to understand how the American pit bull became an American bogeyman. What is it about pit bulls that ignites such strong feelings? And what does that mean for us, as a society? The mascot came to be viewed as a monster not because dogs changed, but because we did. Or, rather, because we failed to.

CHAPTER 2

THE KEEP

Was there one created living thing that
had no quitting point? And was that thing the bulldog?
—RALPH G. KIRK, *White Monarch and the Gas-House Pup*

Standing in the cold shadow of the Brooklyn Bridge, the four-story brick Georgian at 273 Water Street is the third-oldest building in Manhattan. Built as a family residence and sold to a mahogany trader named Joseph Rose in 1773, it features shiny brass address plates and a smart colonial facade that springs up from the cobblestones. Modern realtors describe the Rose House as "charming," "quaint," and "untouched by time," asking more than $1 million for each of the building's four two-bedroom condos. From at least 1863 to 1870, however, the bottom floor of the Rose House served as the euphemistically named Sportsman's Hall, a saloon that fronted the country's most notorious dog pit.

The proprietor of Sportsman's Hall was a bawdy, mutton-chopped Hibernian named Christopher Keyburn, who preferred the nickname Kit Burns to the moniker given to him by the newspapers, "the Cruelest Man in New York." He arrived on the city's crowded docks as a barefoot, penniless teenager sometime around 1845, one of seven and a half million Irish immigrants who fled the starvation and disease of the potato famines during the mid-nineteenth century. Like many of his countrymen, Burns landed in the roiling cauldron of the Five Points neighborhood in lower Manhattan (roughly the same area as Chinatown today), where he maneuvered so artfully that he later became a folk hero among the city's Irish underclass. When he died in 1870 at the age of thirty-nine,

Wood engraving of Manhattan's Five Points neighborhood, 1872

a large crowd of mourners braved the December chill to walk alongside Burns's horse-drawn hearse as it made the ten-mile journey from Brooklyn to Queens.

The Five Points at mid-century was a carnival of human misery, and heroes were hard to come by. Refugees who left their home countries believing that America's streets were paved with gold spent their nights huddled together in overcrowded tenement apartments that regularly caught fire and incinerated the inhabitants. Children younger than ten years old labored in the city's sweatshops with no laws to protect them from being beaten, starved, or maimed by machinery. Sewage collected in the streets, and runoff from the stockyards poisoned the drinking water, allowing cholera and typhoid (not to mention the ever-present threat of tuberculosis) to wash across the community in noxious waves. The streets teemed with so much filth, in fact, that nurses and doctors making emergency house calls hurried along the rooftops rather than brave the sidewalks. "Lower than the Five Points," wrote the Scandinavian feminist Fredrika Bremer, "it is not possible for human nature to sink."

Native-born New Yorkers found the mannerisms and traditions of poor Irish immigrants revolting. They called them "white Negroes" and depicted them as drunken, subhuman "apes" in political cartoons, which made it even more difficult for the newcomers to succeed in legitimate professions. In 1857, one newspaper correspondent asserted that the children of Irish Catholics "are growing up in ignorance, in squalid poverty, in vicious habits, and . . . the chances are that *nine-tenths* of them will become thieves, prostitutes, depredators upon society and inmates of our jails."

The only semblance of law and order that existed in lower Manhattan was shakily maintained by the gangs that patrolled the streets. The largest of these was the Roche (or "Roach") Guard, which the press insisted on calling the Dead Rabbits. Members of the guard strong-armed local

Wood engraving of a clash between the Roche Guard (Dead Rabbits) and the Bowery Boys in the Five Points, 1857

politicians, rigged city elections, and settled disputes with iron bars and brickbats. If you flourished in the Five Points, as Kit Burns eventually did, you did so by criminal means. But to be a "rough," as they were known, was to be further dehumanized. "The facial and cranial appearance of the rough goes far to establish the truth of physiognomy and phrenology," wrote the journalist Junius Henri Browne in 1869. "All the animal is in the shape of his features and head; but the semblance of the thinking, cultivated, self-disciplined man is very nearly lost."

Gang life began for Burns not long after his arrival in America, when he gravitated toward a saloon called the Sawdust House. It was run by a hardboiled Irish pugilist named James Ambrose, more commonly known as Yankee Sullivan. Bare-knuckle boxing was illegal in New York at the time, but Sullivan's criminal allegiances allowed him to conduct his "exhibitions" with little concern about police intervention. Unlike the regulated sport of boxing today, clandestine bare-knuckle matches during the 1850s were so violent that one party might be permanently paralyzed or even beaten to death. Burns often worked as a cornerman for these bouts, but he was much more interested in Sullivan's sideshow entertainments, the rat baits (in which small terriers attempted to kill dozens of live rats in a given period) and the dogfights, most of all.

Years later, Burns would wistfully recall that his time at Sullivan's tavern instilled in him a lifelong obsession with fighting dogs, which he called his "love for animals." The bloodletting inside the pit was, he thought, much milder than the human struggle for survival in the Five Points. Outside the tavern door, you lived with your fists up, or you didn't live at all. At least the dogs were fed and kept warm at night; that was a great deal more than most of his neighbors could secure for themselves. "Dogs will fight," he liked to say, "for 'tis their nature to."

In Burns's native British Isles, a dark romance had grown up around fighting dogs and, by extension, around the men who loved them: the romance of the gladiator. This was especially true in the industrial moonscape of the English Midlands, an area so pocked with coal mines and choked with smog that it became known as the Black Country. The more downtrodden parts of Ireland and Wales had their share of fighters as well; anywhere men slaved away in foundries or coal pits, there you would find fighting dogs. Some of the dogs were so critical as family breadwinners that orphaned pups were suckled at human breasts. "The courage and fight of the Black Country man was his only asset," writes the historian Mike Homan. "The manner of meeting death was also important . . . it was only acceptable to go down fighting, this being as vital as victory."

By no means did Britons invent this idea. Many civilizations have used animal fighting as a proxy for human conflict. Both Miltiades, the Athenian general whose army routed the Persians at the Battle of Marathon in 490 B.C.E., and Themistocles, the Athenian general who again triumphed over the Persians at the Battle of Salamis ten years later, were said to have required their troops to watch cockfights before battle. According to one chronicler, Greek military strategists believed "nothing produces a greater incentive to improve one's own behavior than seeing a lowly creature reach a height of achievement beyond our own aspirations."

Roman rulers held paired combats between exotic animals imported from the hinterlands of the empire and also *venationes,* or elaborate staged hunts, during which thousands of animals might be slaughtered in a single day. But after watching these for several years, Roman citizens clamored for more "thrilling" entertainments. Roman magistrates then devised "death by wild beast" as a primary method of capital punishment. *Bestiarii,* as the condemned criminals were known, were stripped naked, bound to a "column of shame" upon which their offenses had been inscribed, then devoured by starved, agitated predators until, in the words of Martial, they no longer "had the semblance of a body."

By that yardstick, the English medieval spectacle of bullbaiting from which the bulldog earned its name was a few steps down on the ladder of sadism, but only a few. In the days before enclosures, cattle grazed freely in communal pastures, and keeping a herd together was quite challenging. Livestock that decided to go for a ramble might be stolen, destroy property, injure themselves, or, in the case of headstrong bulls, kill passersby with nothing more than a quick toss of the head. Farmers who tried to chase them down risked being gored or stomped to death, but farm dogs, specifically large guardian types called "mastives," could easily subdue the stragglers. It was not expected (and certainly not desired) that the dogs would maul, maim, or kill an animal as highly valued as a bull; they only had to grip its hide and keep it stationary until someone else arrived to yoke it. By the thirteenth century, owning at least one "catch dog" was considered essential for farmers, butchers, drovers, and anyone else who hunted or worked with cattle for a living. This necessity grew as England, the nation of "beefeaters," expanded its meat production.

Butchers soon made a bloody but cost-effective discovery. If they allowed their catch dogs to "worry" a tethered bull that was past its mating prime, the lactic acid produced as the bull exerted itself would soften its muscles, making otherwise inedible meat tender enough for sale. For a time, English law required that bulls be baited before slaughter. This practice developed into a popular type of gambling in which spectators bet on which creatures outlasted the others. Bullbaiting in England, like bullfighting in Spain, soon became a national institution, "like dancing around the May-pole or beating the parish bounds," in the words of the historian James Turner.

The Tudors transformed the size and scope of baiting spectacles by swapping out bulls for bears, most of which were imported from Russia. Queen Elizabeth, in particular, was said to have been so delighted by the sight of dogs attacking a restrained bear that she "giggled like a schoolgirl" during the events, during which she hosted many foreign dignitaries. "Bear gardens" sprouted up all over London, and it is even possible that Shakespeare's Globe Theatre was modeled on this type of baiting arena.

While there was more than enough suffering to go around, modern references to Renaissance animal baiting tend to omit that the dogs were almost always on the losing end of the equation; as many as a dozen "mastives" and their smaller replacements, the "bull dogs," could be gored, swatted, bitten, or stomped to death during a single bait. Most

Portrait of a working bulldog by Philip Reinagle, 1803

citizens refused to send their working animals to meet an almost-certain death, so a representative of the Crown traveled the countryside, conscripting villagers' dogs into service. He was so unpopular that when he arrived, commoners often hurled rotten fruit at him.

When it first stepped out of the historical record in 1609, the bulldog was a healthy, athletic, well-proportioned animal. It bore no physical resemblance to the modern English bulldog, whose shape has been contorted dramatically for bench-show novelty. (The common belief that today's bulldog has a flattened face so that the blood of a pinned bull can sluice down its wrinkles, for example, is a myth.)

It is inconceivable to most Westerners that any animals, especially dogs, could have been treated with such callous disregard, but this was the age of witch trials and the Spanish Inquisition, when public whippings and executions of criminals were widely attended as entertainment and transgressors could be drawn and quartered in the public square, disemboweled, or even boiled to death for minor infractions. Meaningful discussions about animal protection did not gain momentum during the Middle Ages, because, for the most part, the same discussions about cruelty to human beings hadn't happened, either.

It would not be until 1789, when the utilitarian philosopher Jeremy Bentham challenged the prevailing attitude that animals were soulless automatons, or what René Descartes called *bête-machines,* that a larger conversation about animal cruelty would begin. In *An Introduction to the Principles of Morals and Legislation,* Bentham asked, "The question is not, Can they reason? nor, Can they talk? but, Can they suffer?" The answer, of course, is yes. Thirty-five years later, the world's first Society for the Prevention of Cruelty to Animals (SPCA) would be formed in a London coffee shop ironically named Old Slaughter's. One of its founders was William Wilberforce, the British abolitionist who spearheaded the movement to end England's participation in the slave trade.

The early animal welfare movement was an enclave of the well connected and the wealthy during an era of rigid class divisions, and its most vocal members were deeply invested in the idea of social control. One booster for England's SPCA insisted during the 1820s that the group's ultimate goal should be to "spread amongst the lower orders of the people . . . a degree of moral feeling which would compel them to think and act like those of a superior class."

In her seminal work, *The Animal Estate: The English and Other Creatures in the Victorian Age,* the historian Harriet Ritvo points out that nineteenth-century anticruelty rhetoric often "corresponded to the line dividing the lower classes, already implicitly defined as cruel and in need of discipline, from the respectable orders of society. Sometimes this division led humanitarians to value animals more than the vulgar humans who abused them" and to "treat groups they castigated for cruelty almost as exiles from the human community."

In 1835, when Parliament outlawed animal fighting on the grounds that such violent displays degraded the nation's moral core, the working classes of the English Midlands balked. For decades, the country's coal miners, brick masons, and chain makers had literally worked themselves to death to line the pockets of England's elite. They saw the ban on baiting not as a move toward a more humane and equitable world but as a symbolic assault on what few freedoms they had left. They also knew that the wealthy patrons of blood sports, specifically the Eton boys and Oxford men who attended urban dogfights in crowded saloons, would go unpunished. Unable to set their dogs on bulls, the men of the Midlands retreated from the stockyards into the basements of taverns, where they set their dogs on each other instead. When they immigrated to America, their blood sports came with them.

Over the years, the legend of the bulldog as brave, resolute, and unrelenting took on a life of its own, and soon the line between fact and fiction was hard to discern. This had a great deal to do with the indomitability that British workers wanted to see in themselves. Dog breeds as we understand them today did not exist until the late eighteenth century, but general types of dogs were already associated with various social groups. The landed gentry favored expensive gun dogs—retrievers, setters, and spaniels—because they owned large estates suitable for hunting, while the commoners had mongrels, lurchers, and bulldogs: dogs of the farmyard, not the manor house.

As a populist emblem, the bulldog gradually took on totemic significance across the country, even among people who disapproved of animal fighting. By the late eighteenth century, it so perfectly symbolized the tenacity of the English that artists chose it as the erstwhile companion of John Bull, Britain's answer to Uncle Sam. Political cartoonists portrayed Bull's canine sidekick as a mirror image of the human character: jolly and forthright, courageous and determined, grumpy only when he had to be. "Of all dogs, it stands confess'd," wrote the poet Christopher Smart in 1755, "Your English bulldogs are the best."

But not everyone felt so warmly toward the national mascot. A number of aristocrats dismissed both bulldogs and their owners as being slow, stupid, and "savage." In 1802, a rumor was widely circulated in Parliament that bulldogs were eerily impervious to pain, like machines. "In many cases," one observer complained, "a poor man who keeps a bulldog is looked upon as a suspicious character, and a blackguard. In some such cases, no doubt, there is reason for the accusation: but in most it is the fact of his keeping a bulldog that brands him a rascal."

A good example of this assumption can be found in later editions of Charles Dickens's novel *Oliver Twist*, which was first published in 1838. Dickens never explicitly stated that Bull's-eye, the glowering canine henchman belonging to Fagin's second-in-command, Bill Sikes, came from any particular breed or type. In fact, one sentence describes Bull's-eye as "white" and "shaggy." But the animal is commonly remembered as being some type of bulldog, thanks to the illustrator Fred Barnard, who provided the art for Dickens's work during the 1870s.

Coming to the defense of this "much-maligned animal," an admirer named John Meyrick wrote,

> There is no dog about which so many foolish exaggerations are current as the Bull-dog, his former connexion with the brutal sports of the bear-garden has destroyed his reputation. Indeed, so little quarrelsome is this animal that he may be approached by strangers with far greater impunity than most other dogs, and is, as a general rule, more gentle and playful than any large dog I know of. In regard to his not being capable of education, I may observe that I have myself trained a Bull-dog as a Retriever, and found him at least as teachable and intelligent as most other Retrievers.

However a person felt about bulldogs, it was clear by the mid-nineteenth century that he or she felt it strongly. The people who chose bulldogs as companions loved them with everlasting devotion, while those who looked down on the dogs and their people could not be talked, reasoned, or wheedled out of their contempt. More often than not, however, respect and grudging admiration for the bulldog, if not love, won out.

There are no records that confirm exactly when, why, or to what extent, but around 1800 the owners of bulldogs began crossing them with now-extinct strains of terriers to produce a more agile dog for badger hunting, dogfighting, and rat killing. Terriers were considered smarter and more socially respectable than bulldogs, so this cross was thought to have a "civilizing" influence. Within a few years, English men and women from all backgrounds were keeping bull-and-terrier dogs in their homes. Working men still called them "bulldogs," as the physical shape of the animal did not change much, but gentlemen insisted that no, these were bull *terriers,* and they had nothing in common with those *other* dogs. Thus the "bull terrier" became known as "the handsomest and best of all terriers, and *the* dog, *par excellence,* for a gentleman's pet and companion, both in and out of doors."

"Such is the fancy for this dog at present," another writer noted in 1829, "that no man of the ton [*sic*], particularly in London, can be seen on a morning walk or ride without one of them either at his own heels or those of his horse." He added, "It is to be lamented that the services of this excellent dog . . . are too often misapplied." A correspondent for a New York newspaper crowed, "Of all the dogs which the Englishman possesses and breeds, give me the bull terrier for a companion, for it is in him alone that pluck, intelligence, an excellent nose, speed, and devotion to his master are all combined."

The eminent art critic John Ruskin and the Scottish novelist Sir Walter Scott added their voices to the growing crowd of bull terrier devotees. "I hate . . . seeing a bulldog ill-treated," Ruskin wrote, "for they are the gentlest and faithfullest [*sic*] of living creatures if you use them well. And the best dog I ever had was a bull-terrier, whose whole object in life was to please me, and nothing else." Scott, whose fiction features finely wrought canine characters, acquired his crop-eared "bulldog-terrier," Camp, right around the time of his marriage in 1797. Camp would become the novelist's most treasured companion for the next twelve years. "He was of great strength, and very handsome, extremely sagacious, faithful

Engraving of Sir Walter Scott and Camp by James Heath, 1810

and affectionate to the human species," Scott wrote of his friend. The dog was also "possessed of a great turn for gaiety and drollery. Although he was never taught any tricks he learned some of his own accord, and understood whatever was said to him as well as any creature I ever saw."

The entire Scott household admired Camp for being "gentle as a lamb among the children"—who, being children, were not always gentle to him. Camp was famously jealous, however, and shot daggers at Scott's greyhounds, Percy and Douglas, if ever they tried to enter the house library, where Camp liked to position himself at the writer's feet. When Camp grew too old and frail to follow Scott out riding, one of the servants would alert the dog to his master's approach. Camp would then shuffle a ways up the footpath to meet Scott on his horse, proudly walking him the final distance home.

Never did Scott's daughter Sophia see the man as gutted as when Camp died in 1809. Years later she recalled that her father chose a spot just outside his office window for Camp's grave and smoothed the turf with his own hands under the moonlight. Scott admitted to friends that the depth of his grief embarrassed him a little and that he didn't quite know what to make of it. All this, for a dog? He confided in one letter, "[Camp] has made a sort of blank which nothing will fill up for a long while."

That dog pit in Sportsman's Hall was located at the end of a dark, narrow passageway that joined it to the main bar area, where Kit Burns served homemade whiskey to a raucous mob of fellow gang members, prostitutes, pimps, sailors, dockworkers, pickpockets, opium fiends, laudanum addicts, cardsharps, and street hustlers, as well as a few brave souls from uptown. While generally considered sleazy and déclassé (a bit like how visiting topless bars is viewed today), pit sports in both London and New York always managed to draw a number of top-hatted gentlemen into their orbits. During the 1890s, opticians in the United States crafted expensive glass eyes for the fighting dogs of the wealthy, selling them for $15 a piece. In Sir Arthur Conan Doyle's *Study in Scarlet* (1887), Sherlock Holmes asks Dr. John Watson, whom he has just met for the first time, to "confess" any bad habits before the two move into 221B Baker Street together. Watson laughs and admits, "I keep a bull pup," which presumably meant that he enjoyed pit contests.

The bar's bouncer was a man named George Leese, better known as Snatchem, because it was said that he could steal anything from anybody. Burns's son-in-law, Jack "the Rat" Jennings, served as the master of ceremonies. Jennings was known for his halftime sideshow, in which he bit the heads off live rats for a dollar. If Jennings's performance failed to hold the audience's attention, Burns's daughter, Kitty, would step in to beat an unsuspecting customer senseless with a wooden club. Kitty was always a crowd favorite.

The pit itself was twelve feet square, filled with sawdust, and surrounded by zinc-lined walls three feet high. A gaslight chandelier dangled over its center, and wooden bleachers circled the periphery. According to one intrepid reporter, there was just enough space for an audience of "250 decent people or 400 indecent ones." Another writer characterized the crowd as a "brutal, villainous-looking set . . . They are more inhuman in appearance than the dogs." In addition to the usual odors—stale liquor, urine, vomit, and smoke—the entire building smelled of wet fur.

The dogs lived downstairs, in a gloomy warren of rooms that Burns called the "canine boudoir." Of the forty or so dogs that Burns owned, his favorite was Belcher, named after Lord Camelford's legendary fighting dog, a supposed 104-time winner. "Kit Burns is very proud of his dogs," a journalist reported in 1868. "They are very docile with their owner, and seem really fond of him. They are well fed and carefully tended, for they are a source of great profit." When Belcher died in the pit, Burns had the dog stuffed and gave him a permanent residence atop the saloon's bar, next to the remains of his best rat-killing black-and-tan terrier.

Some who attended fights at Sportsman's Hall were not compelled by Burns's "love for animals" and found the matches exceptionally disturbing. Writing under the pseudonym Edward Winslow Martin, the historian James Dabney McCabe described this scene in his 1868 book, *The Secrets of the Great City:*

> Two huge bull-dogs, whose keepers can hardly restrain them, are placed in the pit, and the keeper or backer of each dog crouches in his place, one on the right-hand, the other on the left, and the dogs in the middle. At a given signal, the animals are released, and the next moment the combat begins. It is simply sickening. Most of our readers have witnessed a dog fight in the streets. Let them imagine the animals surrounded by a crowd of brutal wretches whose conduct stamps them as beneath the struggling beasts, and they will have a fair idea of the scene at Kit Burns's.

The rules and culture of nineteenth-century dogfighting derived from those of bare-knuckle boxing. As in boxing, dogs in pit contests trained extensively for several weeks according to a strict diet and exercise regimen known as a "keep." The sports journalist John Henry Walsh (aka Stonehenge) wrote that "nothing was too good" for a fighting dog in keep, which was given "the best of meat—legs of mutton, even milk jellies, often enriched with a little port wine, cow heel, and boiled bullock's nose." To exercise his dogs, Burns chained them to a table with a rotating top and had them run on it like a treadmill for several hours each day.

On the day of the match, pit dogs were paired according to sex and weight, and as in boxing any dog whose weight exceeded what the handlers had agreed upon risked disqualification. Contrary to McCabe's account, rarely were the dogs "huge"; most weighed under thirty-five pounds, and many weighed under thirty, just a bit larger than modern pugs. Sometimes their ears were cropped to prevent an opponent from gaining a hold. (When it was later discovered that this was not effective, cropped ears became a purely aesthetic modification, not just for pit bulls, but for boxers, Great Danes, Dobermans, and certain mastiffs, as well.)

Once released by their handlers, the dogs advanced over a "scratch line" drawn in the middle of the pit. This was also taken from boxing, which is where the phrases "up to scratch" and "toe the line" come from. Attacks (called scratches) and retreats (turns) were scored back and forth until one dog gave up, tired out, or was too injured to continue. This could happen instantly, if one dog jumped out of the pit or simply wasn't interested, or it could drag on for hours. Until that point, both handlers acted as cornermen, shouting encouragement and sponging the competitors down during time-outs. Because handlers worked in extremely close proximity to the animals at all times, dogs that were unpredictable or aggressive toward people were generally, though not always, culled. The ideal fighting dog was trusting and affectionate to many different people (handlers, trainers, referees, sponge men), rather than loyal only to his owner. For this reason, they were often considered too friendly with strangers to be guard dogs.

Deaths in the pit were rare, but shock, blood loss, and debilitating injuries were not. Even for the winners, a great deal of physical suffering and months of recuperation usually followed each bout, with dogs sometimes being sent to live with local widows who were trained in wound care and bone setting. The myth that bulldogs could not feel pain made their plight easier to ignore. All that mattered was that they kept winning.

A bulldog that bravely faced his opponent head-on despite discomfort and fatigue was said to be "game," while one that refused to scratch was said to be a coward, or a cur. For the owner of a fighting dog, no shame was greater than having an animal "cur out" early in a match. True gameness, on the other hand, was so rare that once a game dog was found, it was guarded as closely as a holy relic.

Like the bulldog's storied courage and stoicism, gameness was central to the culture of his people. It was hard to know where the animal ended and the men began. Fighters of Burns's day looked into the eyes of their dogs and saw themselves staring back. Who was the best breeder? The best trainer? The best handler? Who was the victor and who the vanquished? Who was a gladiator, and who was a cur? The dogs never asked these questions of the men they fought for, but they paid with their lives to answer them.

At the same time that New York's roughs were crowding into Sportsman's Hall, other Americans, including some of those headed to war, were turning to pet bulldogs for comfort and companionship. In the spring of 1861, a sad-looking brown-and-white stray drifted into the station house of the Niagara Volunteer Fire Engine House in Pittsburgh, Pennsylvania. One fireman, annoyed by the dog's begging, kicked him so hard that he broke the animal's leg. His horrified colleagues splinted the fracture and

Jack, the mascot of the 102nd Pennsylvania Volunteers, 1863

nursed the dog back to health, feeling that they owed him at least room and board for the mistreatment he had suffered. They named him Jack.

Because Jack bounced back quickly and held no grudges, everyone in the station house eventually came to love him. Weeks later, when the company transformed itself into the 102nd Pennsylvania Volunteers and marched south to face the Confederates, there was no question that Jack would join them on the front lines. According to the soldiers' accounts, Jack learned all the unit's bugle calls and marched with them directly into battle, of which he saw some of the bloodiest: Fredericksburg, the Wilderness, Spotsylvania, and Cold Harbor, as well as Gettysburg and Antietam, a clash so devastating that more than twenty-three thousand soldiers either were killed, were wounded, or went missing in only twelve hours of fighting.

In July 1862, Jack was shot in the shoulder at the Battle of Malvern Hill. The men of the 102nd wept over him, prayed together, and kept vigil through the night until he recovered. The regiment then used their precious furlough days to throw a fund-raiser ball so that they could buy Jack his own silver collar, which cost them the enormous sum of $75, equivalent to $1,000 today. But several months later, Jack was captured by Confederate forces in Salem Church, Virginia. He was such a vital presence in their unit that the men knew what they had to do. They traded their most valuable bargaining chip—a captured Confederate soldier—for Jack's safe return.

A bulldog named Old Harvey joined the war effort in 1862 at the side of his owner, Daniel M. Stearns, a member of the 104th Ohio Volunteer Infantry, which would soon be nicknamed the Barking Dog Regiment. The men of the 104th wrote to their families that Harvey was impeccably behaved and conducted himself like a true "aristocrat," swaying side to side through campfire sing-alongs, keeping soldiers warm in their tents when the air chilled, and standing beside them at their sentinel posts through the night. One soldier reported that Harvey had a soft spot for small animals and once rescued an orphaned baby squirrel from the clutches of another dog. When the surviving members of the 104th Ohio Volunteers met up for their reunion two decades later, they considered Harvey so important to their legacy that they gathered around an oil painting of him in their formal reunion photograph.

The most well-known of the Civil War pit bull mascots was Sallie, a "brindle bulldog" whom the men of the Eleventh Pennsylvania named after the prettiest girl they knew back home in West Chester. Sallie was known for her punctuality (she was never late to roll call) and for her steadfastness under pressure. During battle she ran back and forth in front of the color guard, barking encouragement as though she were coaching her troops. Sallie became briefly separated from her men at Gettysburg but was found again when they returned the next day. They believed that she had stayed on the battlefield to watch over their dead. Sadly, the dog met her own death in February 1865 at the Battle of Hatcher's Run, but her companions were determined to bestow on her the dignity of a proper send-off. Despite coming under heavy enemy fire, they buried her on the field where she fell. Today the monument to the Eleventh Pennsylvania at Gettysburg bears her likeness.

The state of New York formally outlawed all forms of animal fighting in 1867, but Kit Burns had little reason to worry. There was no political will to enforce the laws, nor any legitimate mechanism by which to do so. The continued financial success of such a brazenly wicked man vexed the city's social reformers for years, and Burns enjoyed reminding them that lower Manhattan was his turf, not theirs. In 1868, he publicly announced that, thanks to the kindness of proselytizing Methodists, he had been touched by the Lord and finally wanted to "do the square thing." He would stop drinking his twenty glasses of whiskey a day, he said, and renounce dogfighting for good. As proof of his conversion, Burns then mopped up the blood in the dog pit, swept the floors, and rented out Sportsman's Hall for prayer meetings. One bystander said there was no better location for prayer, because the presence of the dog pit invited congregants to "wrastle with the lord."

Well, not quite. When the time came for the pious to gather, the smell of old blood, which stained the floorboards, and the rotting corpses of dead rats hidden under the seats drove the congregants away en masse—but not before Burns had collected fistfuls of their money. "No money, no prayers, gentlemen!" he told them. Burns and his crew hurled dead rats at the evangelists as they fled. The papers could not fathom that anyone, even the most naive do-gooders, would be so gullible as to believe that *Kit Burns* had been saved by Jesus. Forever after they referred to the debacle as the Church of the Holy Dog Pit.

By 1870, the defiant Burns seemed largely unaware that his gothic empire was on the verge of collapse. Four years earlier, a hollow-faced New York aristocrat named Henry Bergh had established the country's first animal welfare agency, the American Society for the Prevention of Cruelty to Animals, with support from the tycoon philanthropists John Jacob Astor, Horace Greeley, and Ezra Cornell. The ASPCA's charter gave it police power to enforce existing animal cruelty statutes, among other things. Armed with a cane and a lantern, Bergh devoted his evenings to walking the city's streets in a straight-brimmed silk hat and frock coat, looking for acts of animal cruelty to interrupt and perpetrators to apprehend. Critics would come to know Bergh as "the Great Meddler," a title of which he was reportedly quite proud. Within a year, a delighted Bergh announced that "unlimited sympathy and substantial aid are being showered on this humane and civilizing charity by every degree of respectable Society."

As the son of a wealthy shipbuilder, Henry Bergh was everything that Kit Burns was not—highborn, educated, socially refined, sober—but like his colleagues in Britain, the former Lincoln-appointed diplomat could not escape the prevailing attitudes of his time. Contemporary accounts describe him as a humorless dilettante who, after casting about unsuccessfully for years, settled on animal protection as a way of making his mark on the world. His deep compassion for animals, while unquestionably genuine, did not extend to a wide swath of humanity, and cartoons in local newspapers depicted a man who wept over livestock while stepping over the bodies of the human poor. Bergh actively campaigned to bring back the public whipping post so that criminals (including "drunkards") might be both beaten *and* humiliated. "The history of the world has shown that licking is a good thing for the suppression of offenses," he asserted in *The New York Times,* "but with our modern so-called civilization, . . . our exaggerated ideas on the equality of men, and such rubbish, the ruffian has risen up so high that he demands rights that do not belong to him." He then detailed, with much satisfaction, the public torture of a man he had witnessed on a trip to Cairo, describing the effects of the punishment as "magical." Bergh considered beatings "the only argument that appeals directly to [the ruffian's] understanding" but warned that they should never be applied to "the noble horse, and other animals."

Bergh set his sights on New York's dog pits almost as soon as he started the ASPCA, and specifically on Kit Burns. The city's leadership agreed that shuttering sites of such abysmal vice was necessary, but according to the historian James Turner, Bergh also "glared down upon the Irish in particular as an inferior race, whose abuse of animals evidenced a natural inclination to crime." Burns, for his part, found Bergh tiresome and sanctimonious, an opinion he broadcast in open letters to Bergh that were published in the local newspapers. "I can't bear to see people a meddlin' in what they don't understand," Burns said. But he also felt occasional tinges of respect, even admiration, for his foe.

After Bergh ignored one of Burns's many invitations to come and talk at Sportsman's Hall, man-to-man, Burns waved off the snub, saying, "I should have liked to have had Mr. Bergh to have called on and talked to me. I like him. I believe he's a good man an' in earnest, an' I'm only sorry that he didn't come along. I'd have burst the pit if he'd a said the word."

On Monday, November 21, 1870, Burns hosted what he hoped would be another well-attended evening of "sport," this time at a new dog pit on Water Street called the Band Box, not far from Sportsman's Hall. Two fighting dogs were brought out and weighed in on cotton scales: The odds-on favorite was a white dog named Slasher, the other a brindle named Old Rocks. Their match weight was only twenty-seven pounds. When the dogs entered the grimy pit and each handler took them to their respective corners, cheers and whistles pierced the thick veil of smoke. Then the fight began. Before long, the floor and sides of the pit were slicked with blood.

The battle engrossed the spectators to such a degree that they didn't notice a battalion of policemen filing through the doorway until one dropped down through a skylight. "Douse the glim!" someone shouted, and the building went black as patrons scrambled for the exits. But the gaunt man who led the police had come prepared and produced a lantern from under his topcoat. With Bergh directing them like an orchestra conductor, the officers arrested almost forty people and hustled them off to a jail known as the Tombs. One of them was Kit Burns.

Unfortunately for Bergh, Burns would not live long enough to be formally sentenced. He succumbed to either diphtheria or pneumonia just a few weeks shy of his fortieth birthday. According to the writer Chris Pomorski, Burns's only regret at the time of his death was that he would miss his wife and his dogs, one of whom, Snoozer, kept vigil by his bed. When Snoozer began to bark, Burns weakly called out to him, "Lay still, Snoozer. I'm going on a long journey."

Police seized the other dogs at Sportsman's Hall, put them into an iron cage, and drowned them in the East River.

CHAPTER 3

IN THE BLOOD

He has no right to give names to objects which he cannot define.

—CHARLES DARWIN, *The Descent of Man*

Before we adopted Nola, I never gave much thought to the concept of breed or how it originated. I assumed that the majority of American dogs could be neatly sorted into discrete kennel-club categories and that each of these implied a set of predictable, immutable behavioral traits. As certain as Kit Burns had been that "dogs will fight, for 'tis their nature to," I knew that setters set, pointers pointed, and retrievers retrieved. That is, after all, what they were "bred for."

Once the breed label on my dog's license had the potential to change the material circumstances of my life, including where I could and could not live, I began to notice how breed-focused modern American culture is. "When a cocker spaniel bites," the journalist Tom Junod once wrote in *Esquire,* "it does so as a member of its species; it is never anything but a dog. When a pit bull bites, it does so as a member of its breed. A pit bull is never anything but a pit bull."

I puzzled over this at length, not because I doubted the genetic realities of selective breeding—those are scientifically well established—but because most American dogs have been selectively bred as pets, not workers, for a very long time. More than that, however, the rigid lens of breed seems to present the dog as a series of switches and levers waiting to be manipulated for our own use, rather than as a sentient creature with an emotional life and the ability to make choices.

All domestic dogs belong to the same subspecies, *Canis lupus familiaris,* which shares more than 99.9 percent of its DNA with its closest relative, the gray wolf (*Canis lupus*). That 0.1 percent contains an ocean of biological and behavioral differences. One of these is the large amount of variation in the dog's physical shape, or morphology. There is greater morphological variance in the domestic dog than in any other terrestrial mammal on the planet, with the difference in size between a Chihuahua and a Great Dane being approximately the same as between a human man and an African elephant. But much of this physical (and behavioral) diversity is extremely recent; dogs have been formally bred by humans only for the past two hundred years.

For the other 34,800 years of our shared history, dogs lived among us primarily as vagrants or working animals, not pets. As such, their value lay in their ability to pull carts, guard livestock, or herd sheep, and they were grouped according to function, not according to physical considerations like head shape or coat length. The first catalog of English dogs, compiled by a former royal physician named Johannes Caius in 1570, lists only sixteen broad types, all of which refer back to the animals' working roles. "Terrare" (terrier) signified any dog that went to ground to hunt vermin, "bludhunde" (bloodhound) any dog that tracked quarry by scent, and so on. The "mastive" that produced the early bulldog was thus designated because it was large enough to catch cattle and discourage intruders, "masty" being an Old English word for "fat."

That shifted during the eighteenth century, when an English agriculturalist named Robert Bakewell devised a rigorous system of pairing sheep in order to produce a more robust, meatier line of stock. Before Bakewell, domesticated species (including dogs) were mated according to the vague standard of "best to best," but they also intermingled on their own a great deal. News of Bakewell's success with selective breeding for specific physical traits spread quickly across Europe and revolutionized the field of agronomy, making him a household name by the time Gregor Mendel, a Moravian friar, began his hybridization experiments with pea plants in the mid-nineteenth century. As a prime example of "artificial selection," Bakewell's breeding methods also played a significant role in Charles Darwin's understanding of natural selection, the bedrock of his theory of evolution. It was only a matter of time before Hugo Meynell, one of Bakewell's neighbors and the master of the Quorn foxhunting pack, employed this scientific approach to the breeding of foxhounds, which ushered in the dawn of the formal canine pedigree.

In 1859, right around the time that men like Kit Burns were doing battle with the early humane movement, two major events occurred that would change the relationship between humans and dogs forever. Charles Darwin published *On the Origin of Species by Means of Natural Selection*, and the first official dog show was hosted in Newcastle upon Tyne by a British gunsmith named W. R. Pape. In *The Origin of Species*, Darwin laid out the precise mechanisms by which selection, both natural and artificial, altered the physical structures of living creatures over time. He expounded upon this further in *The Variation of Animals and Plants Under Domestication* (1868). In particular, Darwin marveled at the incredible malleability of the dog's shape and how adeptly some of his countrymen sculpted it to their own tastes. "Breeders habitually speak of an animal's organism as something quite plastic," he wrote, "which they can model almost as they please." Neither Darwin nor his readers had any idea how plastic it was until Pape's conformation show in Newcastle, which had been held a few months prior, set off a major craze.

Pape never intended to invert the way people thought about dogs; he only hoped that an exhibition of pointers and setters would increase the sale of his shotguns, which it did. By judging the animals on form rather than function, Pape also made the competition much easier to win. Conformation shows for horses, cattle, and poultry had typically been the province of wealthy Victorians who owned large estates, but breeding dogs for conformation required a fraction of the time and space of livestock breeding and almost none of the manual labor. This made it much more appealing to middle- and upper-class city dwellers, especially ladies, who were now flush with industrial cash and looking for ways to spend it. Dog conformation shows sparked an obsession with pedigreed animals that quickly crossed the Atlantic. The Kennel Club, Britain's first organization dedicated to show dogs, was formed in 1873, and the American Kennel Club followed in 1884.

Once the kennel clubs' stud books were closed, dog breeding took on the characteristics of a secular religion, and a very hierarchical one at that. Practical knowledge, in training as well as breeding, was handed down from one generation to the next in an almost priestly way, while skepticism about established practices was generally discouraged. Animal breeding served an important psychological purpose as well: it allowed Victorian elites to reassert that they did, in fact, control nature, not the other way around. *The Origin of Species* shattered the long-held belief that *Homo sapiens* controlled the animal kingdom by divine right. The thought that man was himself a part of that kingdom, equally subject to its rules and whims, was too much for some to bear. To create one's own breed of dog was both a reassuring grab for immortality and, as Darwin had pointed out, an opportunity to play God with living sculpture.

Showing little regard for the animals' historical working roles, health, temperament, or well-being, Victorian dog breeders raced to mold dramatic new shapes for their pets. Producing dogs in the same way one might produce widgets on an assembly line held tremendous appeal in an age of massive industrialization. Once the now-discouraged practice of inbreeding (breeding parent to child and sibling to sibling) made it possible to "design" animals that looked much more alike, it was increasingly expected that dogs of the same breed would act alike, too. In the space of a few decades, the physical architecture of the domestic dog went from being an afterthought to being all-important, as did the perceived purity of its bloodline.

Rather than sharing one basic utilitarian shape with a few variations here and there, many pedigree dogs now had prominent foreheads, large and expressive eyes, flat faces, and chubby legs. In this way, they more closely resembled human infants than wolves. Like most predators (including bears, sharks, lions, and tigers), wolves have very small eyes in proportion to their heads, which is exceptionally disconcerting for any human trying to "read" the animal's intentions. Konrad Lorenz, one of the founders of ethology, or the study of animal behavior, called our innate preference for animals with large eyes and a "cuddly" appearance the *Kindchenschema* and believed that it triggered nurturing, positive feelings in human caretakers. (As Stephen Jay Gould explained at length in *The Panda's Thumb* [1980], Disney animators have steadily increased the size of Mickey Mouse's head and eyes over the years according to the *Kindchenschema*.) It's probably no accident that today the breeds of dog that many people find most physically intimidating, including pit bulls, are the ones whose eyes remained small. Dogs with larger eyes are also more likely to be adopted from animal shelters, though eye size has nothing to do with temperament or behavior. This is both our evolutionary wiring and our social conditioning at work.

The demand for new shapes of dog opened a new world of possibilities for commercial dog dealers, the predecessors of today's puppy mills. The most well-known of these was an Englishman named Bill George, who tinkered with the bulldog's shape until it was shorter and fatter, then re-branded it as a high-society pet rather than a common farm dog.

George kept more than four hundred bulldogs at his "Canine Castle" in London's Kensal Town, and they were so popular that he sold them in three sizes. "They are ferocious in aspect, these mighty bulldogs," Charles Dickens once wrote of his visit to the Canine Castle, "but by no means truculent in disposition, being, in fact, a living proof that it is not always safe to judge by appearances."

Another English dog dealer, James Hinks from Birmingham, attempted the same re-branding with the bull-and-terrier dog in order to socially elevate it. Hinks crossed the dogs with English white terriers, creating an animal with a longer, equine face. Elites referred to this "reformed" bull terrier as the "white cavalier." In Boston, a man named Robert Hooper bred his bull-and-terrier dogs down to parlor size and sold them as ladies' pets, thus giving rise to the "Boston bulldog," or modern Boston terrier. At the turn of the twentieth century, Boston bulldogs often weighed as much as forty pounds.

"Blood" was both a biological mechanism and an intoxicating narrative. Along with the desire to classify and control the natural world went the eagerness of Victorian high society to classify and control its own kind, with the social groups considered most undesirable being demoted to the level of "beasts." Dehumanization of this type made chattel slavery possible, but it continued long after slavery was abolished. In 1906, a Congolese pygmy named Ota Benga was exhibited at the Bronx Zoo alongside an orangutan that had been taught to do tricks. *The New York Times* reported that Benga belonged to "a race that scientists do not rate high on the human scale." The writer then added, "It is probably a good thing that Benga doesn't think very deeply." Unable to return home to the Congo, Benga committed suicide ten years later.

The stratification of mankind into "higher" and "lower" categories underscored the pronounced anxiety Darwin's work stirred up. Darwin's half cousin Sir Francis Galton championed the most notorious outgrowth of this angst, the pseudoscience of eugenics, which aimed to cleanse society of its "feeble" elements by applying the foundational principles of animal breeding to human beings. "As it is easy . . . to obtain by careful selection a permanent breed of dogs or horses gifted with peculiar powers of running, or of doing anything else," Galton wrote in *Hereditary Genius* (1869), "so it would be quite practicable to produce a highly-gifted race of men by judicious marriages during several consecutive generations." Galton went on to theorize, "It may prove that the Negroes, one and all, will fail as completely under the new conditions as they have failed under the old ones, to submit to the needs of a superior civilization to their own; in this case their races, numerous and prolific as they are, will in [the] course of time be supplanted and replaced by their betters." Madison Grant, one of the founders of the Bronx Zoo, was one of the most outspoken proponents of eugenics in the United States.

The history of eugenics and the history of animal breeding were tightly interwoven from the start. Though drastically different in scale and moral consequences, both fields were prone to the same cognitive errors—namely, a fanatical devotion to blood purity. Their language and philosophies also cross-pollinated one another heavily. The executive secretary for the American Eugenics Society (AES) during the 1920s was a prominent veterinarian and bloodhound breeder named Leon Fradley Whitney, who at one time penned a pet column for *Good Housekeeping* and wrote a number of dog-breeding manuals for the general public. During his tenure at the AES, Whitney heartily endorsed "Fitter Families for Future Firesides," in which humans were physically evaluated like livestock in formal "shows" held at Midwest county fairs. His most famous work, *The Case for Sterilization* (1934), applauded Hitler's efforts to establish a "master race" in Germany and was eventually used to justify mandatory sterilization laws in the United States. More than sixty-two thousand poor, sick, or mentally ill Americans were forcibly sterilized under such laws, and most of them were women from racial or ethnic minorities.

Hitler, as well, made no secret of his fascination with dog breeding, specifically as it related to the German shepherd, a breed originally developed in 1899 as a symbol of Teutonic superiority. The creator of the German shepherd, the Prussian cavalry captain Max von Stephanitz, wrote at length about the dangers of miscegenation and the "inferiority" of mongrels, both human and canine.

Which does not mean, by any stretch, that all dog breeders were or are eugenicists. Most of today's breeders are completely unaware of the historical link between the two disciplines. Nor does it mean that "breedism" and human racism are morally equivalent or that breeding is "bad." Without purebred dogs, our medical understanding of genetic inheritance and genetic disease would be light-years behind its current position. The physically disabled would not have the benefit of service animals, our men and women in combat zones would be even more imperiled by undetected explosive devices, and who knows how many lost souls might never have been located during search-and-rescue operations that relied on the keen noses of hand-selected, purpose-bred scent hounds. The dedication of dog breeders has both saved and improved the lives of countless millions of people.

But it does mean that how we think about breed and how we think about race inform each other, even though we may not always realize it. The very word "race" comes from the world of dogs, in fact. It was first coined in medieval France, where hunters and falconers classed their animals according to function, like the English, but also according to "nobility," in a quasi–caste system. The hounds belonging to French royalty were placed in the "highest" race, and the common guard dog belonged to the "lowest." For several hundred years thereafter, writers across Europe referred to races, rather than breeds, of dog. This was transposed onto humans sometime during the Enlightenment as naturalists, most notably Buffon and Linnaeus, expanded their taxonomies.

The idea of a social caste system for dogs flourished again during the Victorian and Edwardian periods, when eugenicists' widespread abhorrence of human racial mixing spilled over into dog fanciers' distaste for canine "mongrels." Breeds of dog were ranked according to the respectability of their people, with the mongrels of the poor being marked less intelligent, more prone to disease, and more dangerous than expensive pedigreed animals. "I somehow never feel the same respect for a man who allows himself to be accompanied by a badly-bred cur," the famous English breeder William Gordon Stables wrote in 1877, "for dog and master are so often of one type." When several rabies outbreaks caused waves of hysteria in Britain and the United States during the late nineteenth century, the dogs of the poor were usually the first to be blamed.

Thorstein Veblen considered the dog fancier's snobbery about "fashionable" pets to be yet another mark of frivolous Gilded Age excess. "The commercial value of canine monstrosities, such as the prevailing styles of dogs for men's and women's use, rests on their high cost of production," he wrote in his 1899 treatise, *The Theory of the Leisure Class,* "and their value to their owners lies chiefly in their utility as items of conspicuous consumption."

Veblen was not wrong. Roughly 75 percent of the more than four hundred dog breeds we recognize today are whimsical confections whipped up in the late nineteenth century. In some cases, the Victorians' obsession with physical novelty muted or extinguished their dogs' traditional working abilities, so grandiose creation myths were concocted to justify the animals' existence. A literally antediluvian fable that circulated until the 1960s was that the Afghan hound, a close relative of the hunting saluki favored by Bedouins in the Middle East, had been rescued by Noah and placed aboard his ark. Another claimed that the beagle of the 1880s was descended from the "mitten beagles" that Queen Elizabeth I was said to have taken riding, but no one knows what Elizabeth's beagles looked like. Breeders either guessed at standards or made up their own. A reporter covering an 1888 dog show remarked that "there are so many sizes and types of beagles that it must always be a very unsatisfactory task judging them when only one class is provided."

By transforming the world of pets into one of fashion, dog fanciers turned breeds into brands, into status symbols. Ordinary dog owners were largely unaffected by this until after World War II, when a booming postwar economy and the rapid expansion of the American suburb created new interest in lifestyle brands among the middle class. Now that many families had money to care for them and yards for them to run around in, large pedigreed dogs became essential components of the all-American 1950s family, almost as if they were new cars or brand-name washing machines. Between 1945 and 1972, the "golden age of advertising," AKC registrations of purebred puppies per capita increased by a factor of twenty.

According to Dr. Hal Herzog, a professor of psychology at Western Carolina University who specializes in the study of human-animal relationships, most of us still think of dogs according to a brand-conscious framework. After analyzing more than fifty million AKC registrations of purebred puppies from 1946 to 2003, Herzog and his colleagues found that several popular kennel-club dog breeds have followed "boom-and-bust" cycles that mimic the trend curves of clothing styles and baby names. This, Herzog believes, is part of a phenomenon known as social contagion, another way of saying that people tend to imitate those around them. One teenager starts wearing his hat backward, and soon everyone he knows is wearing his hat backward, too. It's not a question of utility, in other words, but one of identity and conformity. Marketing research has confirmed that a pet owner's loyalty to certain dog breeds is largely influenced by his role models, including his parents and other pet owners he knows. While media culture plays a part, Herzog found that breed popularity can also drift randomly, spiking and falling for no identifiable reason. Rottweiler registrations soared during the late 1990s, for

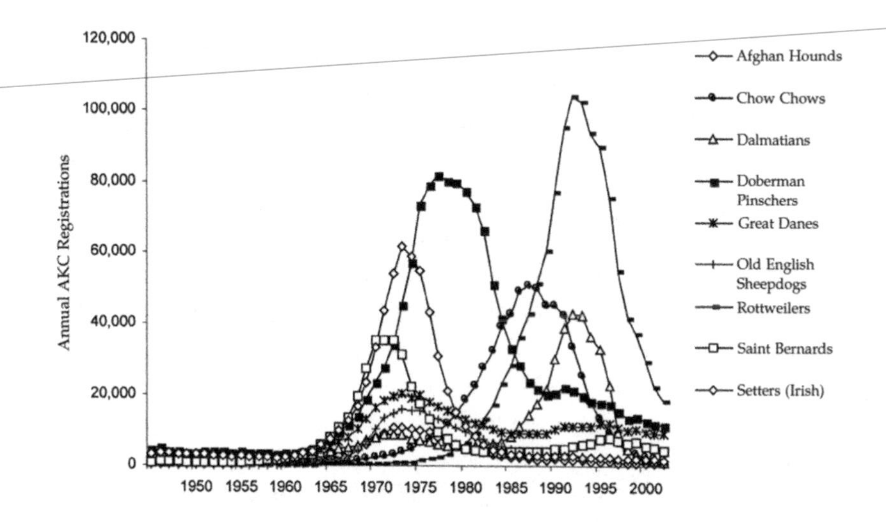

Annual AKC registrations for nine fad dog breeds (1946–2003)

example, even though Rottweilers are quite large, require a great deal of food and physical space, and have experienced several bouts of terrible press. On some level, all dogs (even mutts) can be seen as status symbols; they just denote different statuses.

The same pattern can be observed in any country where labels are valued and expensive pedigreed dogs are de rigueur. From 2011 to 2014, the demand for giant Tibetan mastiffs (which can weigh well over two hundred pounds) among China's upper crust was so great that at least two puppies sold for more than $1.5 million each, according to Chinese media outlets. By the spring of 2015, these sought-after guardians were being surrendered to shelters or sold for a pittance to processing plants, where they were butchered for meat or the manufacture of gloves. "Ten years ago, it was German shepherds, then golden retrievers, then Dalmatians and then huskies," one Chinese animal advocate told *The New York Times*. "But given the crazy prices we were seeing a few years ago, I never thought I'd see a Tibetan mastiff on the back of a meat truck."

When I spoke with Dr. Kristopher Irizarry, an associate professor of bioinformatics, genetics, and genomics at Western University of Health Sciences in Pomona, California, he made something very clear: the genetic definition of "breed" and the public's use of that term are completely different. "A breed is not so much the presence of things as it is the absence of things," he said. "The *absence* of genetic variation is what creates a breed."

To be a genetic member of a breed, Irizarry said, a dog must come from a closed gene pool of purebred animals whose lineage has been documented. My dog Oscar is a purebred pug because both of his parents were purebred pugs, all four of his grandparents were purebred pugs, and so on, all the way back to the pug's original acceptance into the American Kennel Club stud book in 1885. If one of Oscar's grandparents or great-grandparents had mated with a Boston terrier, then a geneticist would no longer consider Oscar a member of the pug breed, even if he looked exactly the same as he does today.

When the gates of a closed gene pool have been opened by outcrossing, or "mixing," the biological deck of cards that selective breeding has stacked so carefully over many generations can reshuffle itself in countless ways, only an infinitesimal fraction of which will manifest in the dog's physical appearance. This is why kennel clubs place such importance on documented pedigree (or "papers") and why the president of the AKC once testified in court that "based on looks alone he could not identify Lassie as a collie."

This is a key point. The country's leading humane organizations estimate that more than half of America's seventy-seven million dogs are not purebred. All U.S. kennel clubs combined only register about one million new dogs a year, a number that has steadily decreased since the 1990s. If we optimistically assume that each of those dogs will live to old age (say, thirteen years), that means only about thirteen million registered purebred dogs are living in the country at any given time, or a little over 15 percent of the total. This is why using AKC statistics to calculate the "relative risk" of dog bites by breed is misguided. Looking at AKC-registered dogs tells us only which trends are taking place *among the kennel-club set,* not what is happening within the larger dog population. Moreover, the AKC does not register either the American pit bull terrier or the American bully, so their statistics are useless for estimating the overall number of pit bulls.

The most common method of labeling mixed-breed dogs is to describe the pedigree breed or breeds we think they most resemble. A dog with a fluffy coat and pricked ears is assumed to be a "German shepherd mix" or a "chow mix," while a dog with floppy ears is usually thought to be some type of hound. Irizarry, who has co-authored several papers on the visual identification of mixed-breed dogs, stressed that this is not at all how the canine genome works. The majority of mixed-breed dogs in America are not crosses of two purebred parents, he explained, but multigenerational mutts, or mutts mixed with other mutts mixed with other mutts. Because the number of genes that determine the dog's shape is extremely small, and so many variations within those genes are possible, looking at a dog's physical chassis and making a guess as to its probable heritage will inexorably lead to error.

The canine genome is made up of approximately nineteen thousand mapped genes (in addition to the mitochondrial genome), which are divided among thirty-nine pairs of chromosomes. Each gene serves a specific function and "lives" at the same genetic "address," regardless of the dog's breed. The gene responsible for a dog's body size, insulin-like growth factor-1 (IGF1), is located on chromosome 15 in all dogs, for example. Genes also contain variations called alleles, one of which is donated from each parent. Some IGF1 alleles are functional and produce dogs of an average medium size. A different version, one that damages or mutates the growth factor, produces much smaller dogs. So it goes with the other genes and other alleles.

As Gregor Mendel discovered with his pea plants, certain alleles will be dominant, meaning that if they are present, a given trait will be expressed, and some will be recessive, meaning that the trait will only be expressed if two recessive alleles match up at the same location. If the dog receives the same allele for a gene from both parents, that trait is homozygous; expressed traits (which may be dominant or recessive) are called phenotypes. The specific arrangements of alleles for body size, head shape, coat color, and other physical traits give each dog its distinct appearance. A trait is considered "fixed" in a breed (requiring less rigorous selection) only when *all* members of the breed are homozygous for that trait.

"If you take an Irish setter, for example," Irizarry said, "you think of a slick coat, long reddish fur, a certain shape of ears, right? It has those attributes because all the other alleles for different head shapes, hair, and ear properties have been *excluded*. If you cross that Irish setter with another dog, you may still see remnants in the next generation, but before long you will have a dog that just looks like a general mutt. Every time there is a reproductive event, you lose 50 percent of the genetic material from the breed you started with."

When an organism grows and produces new cells, the old cells divide and replicate their DNA. Sometimes small "typos" occur during the transcription process, and these typos, or single-nucleotide polymorphisms (SNPs), can be tracked and mapped as genetic markers. In 2009, researchers at Stanford University mapped roughly sixty-one thousand canine SNPs (pronounced "snips") and discovered that only fifty-one regions of the vast genome determine the entirety of the dog's physical architecture. Slight allelic variations in only six genes control the shape of its head. The other 99.99-plus percent of the dog's genetic material controls thousands of attributes invisible to the human eye, including digestion, oxygen metabolism, sensitivity to light, blood pressure, brain chemistry, weight loss/gain, fertility, and aging. Almost sixty thousand phenotypes have been mapped in the mouse genome alone, and only a handful of them are external.

Characteristic patterns of SNPs are what genetic "paternity tests" for mixed-breed dogs, such as the Mars Wisdom Panel from Mars Veterinary, detect. These tests, which have grown increasingly popular since a company called MetaMorphix introduced them to the commercial market in 2007, match the SNPs contained in a dog's cheek swab against those of known breed profiles. So far, the tests have proven to be accurate about 90 percent of the time in first-generation (F1) crosses (purebred A × purebred B), dipping slightly with each succeeding generation. Because of the genetic nuances involved, however, even accurate results can be difficult for the average consumer to accept. The owner of a very small shaggy dog might purchase one of these tests, send in the swab, see "Great Dane" in the report, and think that the test must be wrong.

We humans are strongly oriented toward what we can see—a third of our cerebral cortex is devoted to visual processing—so if the dog does not *look* like a Great Dane, most of us (myself included) would assume that it couldn't possibly have Great Dane in its DNA. But Irizarry explained that it's entirely possible that somewhere in the dog's ancestry a number of Great Dane markers were shuffled into the deck; they just weren't the ones that correspond to the dog's exterior shape. The mixed-breed dog in question might have cells that absorb oxygen like a Great Dane's or intestinal enzymes that break down food like a Great Dane's, but a mutation of the IGF1 gene that limits the animal's body size was also shuffled in from a different donor, rendering the animal small in stature. At another point, the allele for a shaggy coat was shuffled in as well.

At the same time, it's also important to remember that these tests track statistical probabilities, not absolute certainties. They cannot guarantee that the specific breeds in the test results were part of a certain dog's ancestry, only that the dog's genetic markers most closely resemble those considered characteristic in the listed breeds. Angela Hughes, one of the lead genetics researchers at Mars Veterinary, has pointed out that several extremely divergent breeds, like the Chihuahua and certain types of mastiffs, have remarkably similar genetic signatures apart from that single IGF1 variation.

"Imagine if the situation were reversed," says the dog historian and blogger Scottie Westfall, "and dogs selectively bred us according to how we smelled. We couldn't tell the difference between odors, but they could, as long as we stayed within our separate categories. If we started to breed outside the strains the dogs had defined for us, things would get much more confused for them. That's what it's like."

In 1965, Drs. John Paul Scott and John L. Fuller published the most thorough study on animal behavior ever conducted, *Genetics and the Social Behavior of the Dog.* Over the course of twenty years of research at the Jackson Laboratory in Bar Harbor, Maine, Scott and Fuller meticulously cataloged the social development of puppies from five different breeds, as well as puppies from litters they intentionally crossed. The results of those crosses were enlightening. When they mated this pair,

Scott and Fuller's mated purebred basenji and cocker spaniel pair

they ended up with the F1 pups below. And when they mated the F1 pups,

Scott and Fuller's F1 puppies from the basenji/cocker spaniel pair

they got this F2 litter: six dogs that don't even look related to one another, let alone to their parents or grandparents. Yet there are only two breeds present in their DNA.

Scott and Fuller's F2 cross puppies

Purebred dogs can sometimes be difficult for the layman to tell apart, too. One of the dogs below is a golden retriever, and the other is a Hovawart. One was historically bred to be a biddable gun dog, and the other is a livestock guardian.

Which is which?*

* The Hovawart is on the left; the golden retriever is on the right.

Bruiser, a dog offered for adoption by the Bedford County Animal Shelter in Bedford, Virginia, is described online as a "pit bull mix."

With mixed-breed dogs, things get trickier, especially with regard to pit bulls, because the alleles for big heads and smooth coats are liberally sprinkled throughout the dog population. The Mars Wisdom Panel is now able to match the DNA of more than 250 breeds, but the American pit bull terrier is not one of them. "Due to the genetic diversity of this group," the company says, "Mars Veterinary cannot build a DNA profile to genetically identify every dog that may be visually classified as a Pit-bull [*sic*]." Some APBT bloodlines have been tightly bred for many years and constitute legitimately closed gene pools, but others have been (intentionally or unintentionally) outcrossed with other breeds. Kennel clubs do everything possible to prevent their members from falsifying pedigrees (known as "hanging papers"), but it still happens. The resulting group of dogs contains so many mutts that scientists can't isolate one signal. Only the AKC breeds, the American Staffordshire terrier and the Staffordshire bull terrier, can be genetically mapped.

What does this mean for Bruiser, a dog I found on the largest pet adoption Web site in the United States, Petfinder.com? Like most shelter dogs, his parentage is unknown.

I thought Bruiser looked a bit like a beagle crossed with something larger, like a Labrador. Or possibly a Walker hound? The red coat and white-tipped tail seemed to be dead giveaways, and the urge to guess was so ingrained that I couldn't help myself. The staff at Bruiser's shelter did not agree with me, however. To them, Bruiser looked like a "Labrador/ pit bull mix." I thought they were wrong. But the latest science indicates that I was most likely wrong, too.

In 2009, Irizarry and his colleague the veterinary behaviorist Dr. Victoria Voith set out to determine how often the breed labels given to shelter dogs match their statistical genetic profile. After collecting cheek swabs from twenty mixed-breed dogs at four California shelters, Voith, Irizarry, and their co-authors asked a number of shelter workers to look at each dog and guess its breed(s). The shelter workers' visual guesses—that is, the breeds they would have written on the dogs' kennel cards and medical paperwork—did not match the animals' DNA results 87.5 percent of the time. Additionally, the people who labeled the dogs could not agree with one another on which breeds were likely present in which dog. There was no interobserver reliability.

According to her DNA test results, Fannie is a golden retriever, chow, and boxer mix. Despite her coloring, there is no detectable trace of Rottweiler in her genetic background.

In a follow-up study, Voith and colleagues provided short video clips of twenty mixed-breed dogs to nine hundred subjects who worked in dog-related jobs, including veterinarians, trainers, animal control officers, shelter workers, and groomers, and again asked them to identify which breeds were present in which dog. For only seven of the twenty dogs did more than half of the respondents agree on the most prominent breed. Interestingly, the predominant breed they chose did not show up at all in the DNA of three of those seven animals. Subsequent research confirmed this pattern. After looking at photographs of 120 mixed-breed dogs, shelter workers mislabeled 55 as being "pit bull mixes," while missing 5 that actually were.

Voith told me that because the technology of genetic testing is still being developed, she found the lack of interobserver reliability in these studies to be more compelling than the DNA results. If trained animal professionals with years of dog-handling experience aren't good at visually identifying breeds, then what does that say about the rest of us?

The issue of breed identification is relevant not only to pit bulls but to all dogs. Numerous apartment complexes and insurance companies place restrictions on German shepherds, huskies, malamutes, Rottweilers, Akitas, and other large breeds. One woman I know has been turned away from multiple housing complexes because one of her dogs, Fannie, was once labeled a "Rottweiler/spaniel mix" by a shelter in Washington, D.C. In fact, the strongest signals in Fannie's DNA correspond to the boxer, the chow, and the golden retriever.

In light of this research, some veterinarians and behaviorists, most notably Dr. Amy Marder at the Center for Shelter Dogs in Boston, have recommended that dogs from unknown backgrounds not be labeled as belonging to specific breeds when those guesses are most likely incorrect. They say that once a breed label is affixed to a dog, it not only influences what kind of life the dog's family can have but also sets up expectations that the animal will behave a certain way, which it may or may not. What happens when the "Labrador" hates water and doesn't want to play fetch? Or when the "golden retriever" doesn't like strangers? What if that "husky" is really a "malamute"? Shelters that have abandoned using breed labels for dogs from unknown backgrounds, including Orange County Animal Services in Florida and Fairfax County Animal Shelter in Virginia, have seen the number of dog adoptions at their facilities rise significantly.

I soon realized that, despite the paperwork she came with, Nola was most likely not an American pit bull terrier. She probably wasn't an American Staffordshire terrier, either. All signs pointed to her being a pit-bull-shaped mixed-breed dog. "If we don't know for sure which dogs are pit bulls now," Voith told me, "then we probably never did."

CHAPTER 4

AMERICA'S DOG

There was in his world no such thing as the impossible.

—JAMES THURBER, "Snapshot of a Dog"

By all accounts, the nattily dressed John P. Colby of Newburyport, Massachusetts, was a hardworking husband and father who never drank, smoked, or abused his children. Nor did he keep a firearm or drive a car, preferring until the end of his life to ride through town in a buggy drawn by standardbred horses. Though his upbringing had been modest (his grandparents were Welsh immigrants, and his father worked as a ship's carpenter), Colby took great pains to cultivate an image of refinement, never leaving home without first donning a crisp tie and a bowler hat. Were it not for his love of dogfights, cockfights, and pigeon races, he might have passed for the upstanding member of society he always wanted to be.

Outside the dusty barns where Colby spent his free time, America's tolerance for blood sport was waning, and few were sorry to see the number of pit contests decline. Bare-knuckle prizefights were almost gone. Cockfighting was on its way out of the Northeast, with dogfighting not far behind it. Soon both would be relegated to remote parts of the rural South and Southwest, where large plots of land and a frontier live-and-let-live mentality kept the authorities at bay. With eyes fixed on the future, rather than on the past, the rest of the country wanted companion animals, not gladiators. In 1889, Colby began selling his small brindle-and-white fighting bulldogs as pets. He was by no means the first to do this, only the first to do it on a large scale.

Colby's plan—to keep one foot in the world of fighting and one in the world of the parlor pet—worked out well for him financially. Over the next fifty years, he bred an estimated five thousand of his pit bulldogs and sold them to a roster of clients that included the Wild West showman William "Buffalo Bill" Cody; Jack Johnson, the world's first African-American heavyweight boxing champion; and William Whiting, the secretary of commerce under Calvin Coolidge. Breeders outside Massachusetts then followed in Colby's footsteps, producing their own bloodlines of similar dogs, and within a few years pet owners who had previously been wary of pit-type bulldogs were writing rapturous letters about what cheerful, loving, and intelligent animals they were.

The newly formed American Kennel Club, however, wanted nothing to do with the human riffraff associated with pit bulls, regardless of the dogs' growing popularity outside that demographic. Ordinary families did not concern themselves with dog politics or kennel clubs, so the lack of an AKC imprimatur did not deter many people from owning pit bulls anyway. But a few fanciers—namely a Michigan man named Chauncey Bennett—found the absence of formal recognition intolerable. In 1898, Bennett established his own registry, the United Kennel Club, to exalt the virtues of salt-of-the-earth working dogs, rather than fluffy show breeds. Bennett is credited with knighting Colby's dogs "American pit bull terriers." Historians of the breed have pointed out that the title he chose was a bit of a misnomer, because the dogs Colby and his colleagues raised were neither truly American (they were derived from English and Irish stock) nor were they true terriers (they never went to ground to hunt vermin), but in the 1880s the only thing more fashionable than a terrier was a patriotic terrier. The UKC would go on to become the second-largest dog registry in the United States. (It also sanctioned dogfighting until the early twentieth century.) The first dog registered by the UKC was Bennett's own APBT, Ring.

Back in Massachusetts, Colby was criticized by people on all sides. The old guard of fighters, who called themselves "dog men," considered him a "puppy peddler" whose animals were bound to lose their gameness. Because "Colby's Famous Fighting Dogs" were openly advertised in newspapers and magazines all over the country, the Massachusetts Society for the Prevention of Cruelty to Animals (MSPCA) also put him permanently in its crosshairs. But Colby and his family would pay a much heavier price for his refusal to choose between pit fighting and pet dogs. In 1909, Colby's visiting two-year-old nephew, Burt Leadbetter, toddled out alone to Colby's dog yard, where twenty-five fighting dogs were chained, and was bitten so badly by one of the largest that he died. It was the first-ever report of a fighting dog causing a human death in the United States (since 1900, there have been fewer than ten). *The Boston Daily Globe* reported that Colby was "very secretive about the affair and declined to give out any details," but the story made headlines up and down the East Coast. Colby shot and killed the dog the next day. Charges against him were never filed.

Elsewhere, however, the idea that the homely pit bull deserved more than a life of chains and treadmills and broken bones was growing among pet owners. In 1911, a separate class for "pet pit bulls" was created at New York's Madison Square Kennel Show. As part of the exhibition, two dogs in dress harnesses, Cid and Zab, pulled a tiny wagon driven by a little girl in a frilly dress. "[The pit bull] is an ideal watch and guard dog," reported *The Evening World*, "in addition to being tractable and especially suited to the pet realm."

Unfortunately, breed labels were every bit as loosely applied in the early twentieth century as they are now. From roughly 1890 to 1945, photographs of dogs that would be widely considered pit bulls today were alternately called bulldogs, pit dogs, pit bulldogs, white bulldogs, brindle bulldogs, American bulldogs, Boston bulldogs, Boston bull terriers, bull terriers, brindle bull terriers, colored bull terriers, American bull terriers, pit bull terriers, and Yankee terriers, among other names.

In her 1903 memoir, *The Story of My Life*, the humanitarian Helen Keller wrote lovingly of her "bull terrier," Phiz: "Whenever it is possible, my dog accompanies me on a walk or ride or sail. I have had many dog friends—huge mastiffs, soft-eyed spaniels, wood-wise setters and honest, homely bull terriers. At present the lord of my affections is one of these bull terriers. He has a long pedigree, a crooked tail and the drollest 'phiz' in dogdom. My dog friends seem to understand my limitations, and always keep close beside me when I am alone. I love their affectionate ways and the eloquent wag of their tails."

Phiz is often described as a "Boston bulldog" or "Boston terrier," but in photographs he has cropped ears and a long tail, and he is much larger than modern AKC Boston terriers. (Likewise, Colby's pit bulls were sometimes referred to as Boston terriers.)

The list of possible names got even longer in the 1930s. Breeders in the U.K. and the United States wanted to maintain the shape of the old-fashioned fighting dog but not its social reputation. They severed all ties to the dubious "sporting fraternity" and began producing dogs only for conformation shows and pet homes, rather than for "work." The U.K. breed, which was significantly smaller than the American, became known as the Staffordshire bull terrier, after the mining district of the English Midlands where the dogs are thought to have originated. The "Stafford," as many called it, was formally recognized by England's Kennel Club in 1935. Its American cousin, originally named the Staffordshire terrier, was recognized by the AKC the following year. Colby's Primo was one of several dogs used to create its breed standard, meaning that the original Staffordshire terriers were indistinguishable from APBTs. Later, the AKC changed the breed's name to American Staffordshire terrier to avoid confusion.

Helen Keller with her teacher, Anne Sullivan, and one of her many dogs, ca. 1900

Bulldogs? Bull terriers? Pit bulls? Staffordshire terriers? How should we think about them? Without the benefit of modern DNA tests, we can only trace the history of dogs that *looked* like pit bulls, but it's clear from letters, photographs, diaries, and news articles from the late nineteenth century that whatever name they went by, these dogs were incredibly popular as family pets.

"There is no limit to the merits of really good bulldogs," declared *Country Life Illustrated* in 1899. "They are the best of all playmates for children, always considering their weakness and never resenting even the most humiliating indignities at their hands." *The American Book of the Dog* noted in 1891 that "the Bulldog, like all other noble animals, is fond of children; in fact, I can safely say that no more affectionate dog lives than the one under discussion."

In the summer of 1917, a four-year-old girl named Ellen Grimes wandered away from her home in Boulder, Colorado, as her mother was dressing Ellen's siblings for Sunday school. When a search party

Gentlemen with pit bull, date and location unknown

and a pack of bloodhounds finally located Ellen two days later, she had walked an incredible twelve miles. Forty-eight hours in the Colorado wilderness without supplies could prove devastating, even fatal, for the most experienced adult outdoorsman, but Ellen survived because of her family's ten-year-old "brindle bulldog," Bob, who followed her. After returning home, Ellen told reporters that she climbed under a large rock overhang when the temperature dropped and Bob kept her warm.

Laura Ingalls Wilder, the author of the Little House on the Prairie series, showcased the dogs as both playmates and protectors. In one scene from her 1932 novel, *Little House in the Big Woods,* Laura, the young protagonist, lies in bed listening to the wolves outside her family's Wisconsin homestead. "Her father's gun hung over the door with good old Jack, the brindle bulldog, on guard before it," Ingalls wrote. "Her father would say: 'Go to sleep, Laura. Jack won't let the wolves in.' "

As heartwarming as these "bulldog and baby" stories were—and there were many of them—they reflected the limited scientific knowledge of their time. A hundred years ago, leaving a child alone to be looked after by a dog was commonplace, especially in rural or working-class families. An animal that tolerated being poked, shoved, kicked, yelled at, or ridden by small children was thought to be not just patient but morally noble. This was not at all limited to pit bulls (there are just as many items about valiant Newfoundlands and Saint Bernards saving children at the turn of the century), but stories of bull-and-terriers being children's companions inspired the myth that the pit bull was once called a "nanny dog" or a "nursemaid dog." In truth, these phrases don't appear in relation to pit bulls until the 1970s. Animal behavior experts have long insisted that these nostalgic tales, while charming, place undue pressure on the relationship between children and dogs and that leaving a child alone with a dog of *any* type can be exceptionally dangerous for both parties.

left Family with Staffordshire bull terrier, ca. 1910; *right* young boy with pit bull, date and location unknown

The pit bull's escape from the darkness that produced him greatly added to his legend. It also made him ripe for projection and anthropomorphism when the public's interest in fiction that featured animal protagonists grew at the end of the nineteenth century. Several novels, including Margaret Marshall Saunders's *Beautiful Joe* (1893), Richard Harding Davis's *Bar Sinister* (1916), and Clarence Hawkes's *Pep: The Story of a Brave Dog* (1922), spotlighted bull-and-terrier dogs as main characters. Much like the bootstrapping dreamers in Horatio Alger's novels, the pit bull protagonist was said to come from humble beginnings, but he was street-smart and resilient, and, most important, he knew the value of hard work. He didn't need anyone's pity, and if threatened, he could take care of himself. His storied courage and grit became something to celebrate, rather than exploit. Later, during the 1960s, John Steinbeck would advise Land Rover that it should adopt a bull terrier for its logo, because, unlike the competition's choice of a "sneaky animal, which is easily treed" (the jaguar), a bull terrier "is forthright and headlong in both love and war. No well bred bull terrier ever gives up."

Nor was the gallantry of the "American bull terrier" lost on the humorist James Thurber. As a child, Thurber felt a profound kinship to a dog named Rex who lived in his neighborhood. "There was a nobility about him," Thurber wrote in a 1935 essay for *The New Yorker.* "He was big and muscular and beautifully made. He never lost his dignity even when trying to accomplish the extravagant tasks my brother and myself used to set for him." What defined the dog, according to Thurber, was his deep sense of loyalty and his refusal to be cowed, not unlike the character of Argos in *The Odyssey.* "There was in his world no such thing as the impossible," Thurber wrote. But, as one might expect from a classical hero, Rex's fate was ultimately tragic. Courage was his fatal flaw, as Thurber learned when Rex was mortally wounded by another dog. "When badly injured," he wrote,

> the pit bull licked at our hands and, staggering, fell, but got up again. We could see that he was looking for someone. One of his three masters was not home. He did not get home for an hour. During that hour the bull terrier fought death as he had fought against the cold, strong current of Alum Creek, as he had fought to climb twelve-foot walls. When the person he was waiting for did come through the gate, whistling, ceasing to whistle, Rex walked a few wobbly paces toward him, touched his hand with his muzzle, and fell down again. This time he didn't get up.

Of this essay, *The New Yorker*'s Adam Gopnik observed, "When Thurber was writing about dogs, he was writing about men. The virtues that seemed inherent in dogs—peacefulness, courage, and stoical indifference to circumstance—were ones that he felt had been lost by their owners . . . The dog was man set free from family obligations, Monastic Man."

This philosophy of bravery and stoicism, distilled memorably by Theodore Roosevelt in his "Strenuous Life" speech (1899), defined American culture when Thurber was coming of age. As an increasing number of women pushed for equal treatment and men spurned manual labor in favor of managerial desk jobs, fears that the country would become "soft" and "feminized" rippled through society. The traditionally masculine virtues of physical fortitude and derring-do were then elevated as national ideals.

"I wish to preach," Roosevelt said, "that highest form of success which comes, not to the man who desires mere easy peace, but to the man who does not shrink from danger, from hardship or from bitter toil, and who out of these wins the splendid triumph."

Bull-and-terrier dogs were perfect emblems of this can-do attitude, and many college fraternities rallied around them for inspiration during the early twentieth century. (In 1910, fifteen of the University of Michigan's twenty mascot dogs were pit bulls.) But admiration for the dogs and what they represented was by no means limited to men. Virginia Watrous, daughter of the prominent women's rights activist Antoinette Funk, owned a brindle "bull terrier" named Votes who accompanied her on the campaign trail as she and her colleagues stumped for suffrage in 1915.

Roosevelt's own "bull terrier," Pete, did not live up to the lofty expectations the president set forth in his speeches. In fact, Pete might have been the naughtiest pet ever to live in the White House. *The New York Times* described him as having "little sense of humor and not a bump of reverence." In 1906, Pete chased the visiting French ambassador up a tree when all the poor dignitary wanted to do was play a game of tennis. Roosevelt then promptly banished the dog to the farm of Rear Admiral Presley Rixey, surgeon general of the navy. Almost as soon as Pete returned from his yearlong exile, however, the dog bit a naval clerk, then suffered the humiliation of being badly injured by a free-roaming "plebeian pup." Reporters delighted in every detail of Pete's misadventures, because the story allowed them to rib the president indirectly. They called the dog's last dustup "Pete's Waterloo."

Roosevelt's experiences with Pete might have informed the highly public war of words he started with the author Jack London in 1907. In London's novel *White Fang* (1905), the eponymous main character, a wolf dog, struggles to survive the harsh, violent world of Canada's Yukon Territory before being sold to a dogfighter named Beauty Smith. In one of the book's most agonizing scenes, Smith forces White Fang to fight a bulldog named Cherokee, whom London sketches as merciless, unrelenting, and much stronger than White Fang, who is no wilting flower, either: he once tore the heart out of a horse. Adding a bit of extra drama, London refers to Cherokee as "the clinging death." This was a much different picture than, say, Mark Twain had drawn years earlier in "The Celebrated Jumping Frog of Calaveras County" (1865), when he created a comical, hapless fighting "bull pup" named Andrew Jackson.

The rough-riding president took issue with London's portrayal. London, he said, was obviously a "nature faker," a person who spins tall tales about wilderness adventures without ever having any. Other writers (including the pit bull enthusiasts Richard Harding Davis and Clar-

Mark Twain with dog in Bermuda, 1909

ence Hawkes) had exchanged fire over nature fakery in America's literary magazines for years, but this was the first time that Roosevelt felt impelled to jump into the fray. "[Nature fakers'] most striking stories are not merely distortions of facts, but pure inventions," Roosevelt wrote, "and not only are they inventions, but they are inventions by men who know so little of the subject concerning which they write, and to ignorance add such utter recklessness, that they are not even able to distinguish between what is possible, however wildly improbable, and mechanical improbabilities . . . A wolf that could bite into the heart of a horse would swallow a bulldog . . . like a pill." To suggest otherwise, wrote Roosevelt, was "the very sublimity of absurdity."

London upped the ante. "I would like to match a bulldog against a wolf and bet with [Roosevelt] on the fight," he wrote from his home in Honolulu. Fortunately, that never happened.

One of the ways Americans tried to counteract the much-feared "damnable feminization" of the country, as Henry James called it, was through exercise and athletics, which replaced the literal brutality of battle with symbolic approximations of it. Most of the "fighting bulldog" mascots now common on university campuses were originally pet pit bulls, not actual fighting dogs, and certainly not flat-faced English bulldogs. At one point or other, captains of sports teams at Harvard, Yale, Columbia, Cornell, Wesleyan, and Georgetown all kept some version of the bulldog-terrier in the clubhouse. The University of Georgia's football players were nicknamed the Crackers before 1894, when a white bull terrier named Trilby, who darted in and out of players' legs during practice as though she were executing plays, inspired them to become the Bulldogs.

Bevo, the first of the famous University of Texas longhorn steers, was preceded by Pig Bellmont, a tan-and-white pit bull with cropped ears who belonged to UT's first athletic director, L. Theo Bellmont. Pig, who was named after the football team's center, Gus "Pig" Dittmar,

served as the team's varsity mascot for nine years, beginning in 1914. According to one of the university's historians, Jim Nicar, Pig preferred to sleep under the steps of the university co-op but woke early in the morning to greet students and faculty on his "daily rounds," which included classroom visits and stops at the library. When not occupied with his duties as the football mascot, Pig snarled at "the slightest mention" of Texas A&M. He also checked in on the cadets at the School of Military Aeronautics during World War I, showing up for morning inspection and joining the cadets on training hikes. When Pig died after being hit by a car on New Year's Day in 1923, the university mourned his death by honoring him with a black casket and a formal parade, complete with a marching band that played taps. The dean who delivered Pig's eulogy compared him to the Romantic poet George Gordon, Lord Byron.

Men's football team (1906) and women's basketball team (1907–8) with pit bull mascots; locations unknown

. . .

Pit bull dog with footballer, ca. 1900; location unknown

The U.S. military did not officially establish a working canine corps until World War II, but adopted animals, usually dogs, served as morale-boosting mascots in all of the nation's armed conflicts, just as they had in the Civil War. During the short-lived Spanish-American War in 1898, by far the most common animal mascot depicted alongside U.S. sailors and infantrymen in the photo postcards they sent home was the pit bull, but the dog didn't reach its true height as a symbol of American patriotism and martial valor until World War I. "People who haven't been at the front don't know what a little companionship means to a man on patrol duty, or in a dugout, or what a frisky pup means to a whole company," a British lieutenant named Ralph Kynoch recalled after war's end. "The pups know when a barrage is on where they can find safety, and they go there, unless the man they look to as a master is going somewhere else. Trust the dog to stick hard by no matter whether it is in the danger zone or not."

In 1915, the artist Wallace Robinson chose to drape the "American Bull Terrier" in the national flag for a recruitment poster that placed it at the center of a lineup featuring the English bulldog, German dachshund, French bulldog, and Russian wolfhound. The caption reads, "I'm Neutral, BUT—Not Afraid of any of them." Another of Robinson's posters shows the same pit bull guarding four kittens that are nestled in the flag. "SAFE," it reads, "under the right protection."

As America readied its troops for war during the summer of 1917, a recruit in the 102nd Infantry named Robert Conroy came across a small stray brindle-and-white dog "of indeterminate lineage" wandering the streets near Yale University, where his unit, the Twenty-Sixth Yankee Division, was training. Some called him a Boston terrier, others called him a bulldog, but the truth is that we have no idea what his genetic heritage was; he was simply a smallish dog with a relatively large head and

Members of the Seaforth Highlanders, a Scottish division of the British infantry, in the trenches with their canine mascot during World War I

a bobbed tail, which inspired the men of the Twenty-Sixth to call him Stubby. During the weeks of Conroy's military training, the gregarious Stubby befriended all the camp's recruits and personnel, especially the mess hall employees, who spoiled him. Conroy also taught Stubby to "salute" by sitting back on his haunches and raising a paw to the side of his face, which would become the dog's signature trick. Not wanting to leave Stubby behind once his unit deployed, Conroy enlisted the help of a seaman aboard the USS *Minnesota,* who hid the dog in a coal bin on the France-bound transport ship until the vessel was far enough out to sea that there was no risk of Stubby's being ordered to disembark.

With Conroy's help (and that of a customized mask the soldiers made for him), Stubby learned to anticipate gas attacks and once even rousted a sleeping sergeant who had failed to hear the alarm siren. He was credited with "taking" his own prisoner of war when he alerted the men to the presence of a German spy, which earned the dog the spy's Iron Cross. After surviving a shrapnel wound in the shoulder he sustained after wandering into no-man's-land, Stubby became something of a celebrity. A group of women from Château-Thierry sewed him his own chamois "uniform," upon which Conroy pinned a series of medals (most likely his own). All in all, Stubby accompanied his men in seventeen battles.

Antique postcards of Wallace Robinson's World War I prints, 1915

After nineteen months at the front, General John Joseph "Black Jack" Pershing awarded Stubby his own special medal, and no fewer than three presidents (Wilson, Harding, and Coolidge, who invited Stubby to the White House) thanked him in person for his service. Stubby went on to lead many large parades, appear in vaudeville shows, and serve as official mascot for the Georgetown Hoyas. Today, his preserved remains are enshrined at the Smithsonian.

A pit bull named "Sergeant" Helen Kaiser also traveled to the European front with her owner, Private James White of Washington, D.C., around the same time. White served in the 372nd Infantry Regiment, part of the African-American Ninety-Third Infantry that served with the French army's "Bloody Hand" Division. Helen Kaiser was the first American dog to enter German territory. Like Stubby, she survived multiple shrapnel wounds and mustard gas attacks. Recognizing her

Stubby in uniform for an animal parade, 1921

incredible fortitude in the trenches, the French military honored her with the Croix de Guerre for bravery—twice.

During World War II, the most famous member of the bull terrier family to grace military ranks (unofficially) was Willie, the last of several white English bull terriers owned by General George S. Patton. Unlike Stubby and Helen Kaiser, Willie had a reputation for being as cowardly as Patton was brave. He dove under the furniture at the sound of thunderstorms or exploding ordnance. Much to Patton's embarrassment, Willie was also fond of shoving his nose up ladies' skirts and humping their legs whenever possible. But Patton loved Willie dearly, writing in

Bull terriers used by the U.S. Army military police lined up with their handlers in Normandy, France, 1944

his journal, "My bull pup . . . took to me like a duck to water." Patton's fondness for bull terriers must have trickled down through the ranks, because the American military police used the dogs to locate and immobilize snipers at Normandy. As if to highlight yet again the subjectivity of those words, some of the dogs looked like long-faced English bull terriers, while others were black and looked like modern pit bulls. Then as now, what bound them together was the story we told.

CHAPTER 5

DOGS OF CHARACTER

Civilize him as you please, make him whatever color you like, and Man will still worship the born fighter.

—JOHN TAINTOR FOOTE, "Allegheny"

Set against the iron clouds and evergreen spires of Olympia, Washington, Diane Jessup's carport looked like a hastily abandoned military training camp. A wooden treadmill with broken slats leaned against one wall. Empty metal crates were stacked up against another, most with their doors fallen open, as though something had escaped. Growling hellhounds on rusted metal signs warned trespassers in multiple languages—BEWARE!, *¡CUIDADO!*, *ACHTUNG!*, and finally, WARNING: MY PIT BULL WILL FUCKING KILL YOU—above cardboard boxes that overflowed with tools and duct tape and old lengths of chain. In a far corner, just beyond a frayed rope hanging ominously from the ceiling, a shelf of trophies gathered dust.

And then there was the meat—bloody hunks of beef and bone lay scattered across the concrete, turning to goo.

"Ribs," Jessup said as I got out of my rental car. "Fresh from the butcher. Backs, too. The dogs love 'em."

By the time I finally met Jessup, I had read so many of her fire-breathing epistles on the "dumbing down" of the American pit bull that I pictured her in a horned Viking helmet and armored breastplate, carrying a spear. But when she shuffled outside to greet me, she wore knee-length khaki cargo shorts and an extra-large black T-shirt that read: MAN'S BEST FRIEND, HOG'S WORST ENEMY. The text appeared below a picture of an ember-eyed pit bull, frozen mid-pounce—a reference to the hunters who use Kevlar-clad APBTs to catch feral pigs on the plains of Texas. The drizzling rain had glued her feathered brown hair to her forehead and fogged up her eyeglasses, which slid down her nose. Fifty-three years old, she took slow, labored steps, hunched over pale legs stitched with scars that bowed out sharply at the knees.

"Christ," she muttered, "I hate being a fat, old cripple." Releasing a huge sigh and swiping at her bangs, she smiled, opened her arms, and bear-hugged me as though we were old comrades newly returned from war. "Call me Diane," she said, sweeping an arm across the driveway. "You're the first person to visit me in four years."

If there were a pantheon of take-no-prisoners pit bull diehards, Diane Jessup would occupy a prominent place in it. (One of the many surprising contradictions about the world of pit bulls, which is thought to be so full of machismo, is that a significant number of its most outspoken characters are women.) Since her first kennel job at age fourteen, she has worked with and trained protection, police, and scent-detection dogs of almost every large breed. For the past thirty years, she has focused on breeding and training high-drive American pit bull terriers, almost a dozen of which have gone on to successful careers in law enforcement. To date, her personal dogs have won more than seventy titles in Schutzhund (a competitive sport that originated in Germany to evaluate protection dogs), French Ring Sport (similar to Schutzhund), obedience, weight pull, tracking, and herding while also appearing in several Hollywood films, television commercials, and print advertising campaigns. Her three books about working APBTs—two histories and one novel—are considered essential reading for anyone interested in pit bulls.

Like the dogs she loves so passionately, Jessup is a controversial and polarizing figure. Though she worked as an animal control officer for twenty years and finds dogfighting utterly repugnant, she believes that purebred American pit bull terriers from old fighting bloodlines (commonly known as "game" or "game-bred" dogs) have the most unflagging spirits of any working animals. "They aren't the best at everything," she said, "but they will try their damnedest at whatever you ask them to do." She maintains that the courage and drive that is thought to have made the dogs of yesteryear successful in the pit can now be channeled effectively toward positive goals, like police work and competitive sports such as agility, dock diving, and flyball. If the game dogs die out, she fears that all that working potential will die out with them. A good dog, she insists, is "90 percent genetics." The subset of APBT breeders who idolize Jessup agree.

Whether people love or hate her, Jessup does not care one whit. "I know what I know" is her personal credo, and she prefers practical knowledge to pointy-headed science. Case in point: her defense of inbreeding. Geneticists have found that tight inbreeding puts dogs at higher risk for heritable diseases and degrades their immune systems over time, but Jessup has no problem with the practice if it gets her what she wants, which is a dog possessed of enough strength and stamina to perform at the highest level of dog sport. Inbreeding, she says, is the fastest and most efficient way to set "type," and for her, type is everything. If a puppy she breeds is not physically healthy or shows any aggression toward humans, she will have the dog euthanized. Her loyalty is not to the individual dog in front of her but to the breed as a whole. "I'm not someone out to save every little doggie in the world," she wrote on her Web site. "I love the breed enough to believe in culling. And yes, culling may mean euthanasia . . . I deal in common sense."

These days, the main targets of Jessup's wrath are members of the animal protection movement, whom she refers to as "humaniacs" or "old ladies in tennis shoes." "We need bold, robust, exciting things in our lives," she told me. "But we are now living in the age of the insipid, spineless person. Today's pit bull is most threatened by two groups of people: those who hate them, and those who love them." The ones who hate them are trying to have them banned, and the ones who love them, she says, are "watering down" the dogs by holding up "scatterbred," "generic" shelter dogs as examples of what pit bulls can and should be.

Diane does not consider shelter dogs to be "real" pit bulls. In fact, she does not even consider show-bred APBTs to be "real" pit bulls. Both can be wonderful pets, she says, but "real" pit bulls have an internal fire that comes only from game blood. This has put her at odds with many people, including some of her oldest friends, who feel that her chest-thumping statements about pit bulls being faster, stronger, and *better* than other dogs are frightening to the public, who look to her as an example. The nail that sticks out, they remind her, is the first to be hammered down. Diane is not bothered by that; in fact, on some level she *hopes* to frighten the public. If more people were intimidated by the thought of owning pit bulls, she says, then maybe the dogs would not be so carelessly bred in such great numbers.

In defiance of those who believe that a softened image will save the dogs' lives, Diane assigns her pets tongue-in-cheek names like Freakshow, Grim, Maulie, Thing, and Hellboy, even though none of her APBTs has ever bitten anyone and she would never keep a dog around that had. Jessup has committed her life to preserving the American pit bull terrier as she believes it existed two hundred years ago and as she believes it should exist two hundred years from now: as a friendly, brave, athletic companion.

"Just *wait* until we put you in the bite sleeve," she had said darkly over the phone, referring to the protective gear worn by human decoys in bite training. "You need to see what these dogs are capable of." According to her, the APBT was a solid muscle of awe-inspiring, relentless power.

As we stood in the carport that first morning, a tan-colored puppy with a black snout bolted out from the doorway and zigzagged through my legs. She spun around like a dervish, pogo'ed up and down, and pawed at me until I finally knelt down to pet her, whereupon she licked my face furiously. I got the distinct impression that the dog was trying to enlist me in some sort of scheme. "Oh, Nell," Diane said with a sigh, leaning her head back and squeezing her eyes shut. She resembled a long-suffering Sunday-school teacher. "Nell-Nell. Come on. Please." The puppy continued to bounce around the carport as if made of compressed rubber. "Her full name is Boldog's Mad Nell," Diane said. "And you can see why." (Boldog Kennel is the name of Jessup's business.)

Diane was due at the veterinarian's office in Tacoma, where she would find out if one of her seven pit bulls had finally gotten pregnant after several expensive rounds of artificial insemination. Before leaving the house, she made sure the dogs that remained behind were adequately separated. Some liked one another just fine, but others might get into gruesome, possibly fatal, fights if she didn't rotate them into and out of the house in "teams." Normally, half of the dogs lounged with her in the living room, while the other half played in the one-acre yard, where they had rope swings, toys, and bones with which to occupy themselves. (All those ominous-looking implements in the carport, it turned out, were used for making and repairing toys and training tools.) She switched the teams every few hours, and none of them seemed to mind it. Most days, Jessup took at least one or two of her dogs swimming in cold mountain rivers or played long games of fetch with them in the shadow of Mount Rainier. They romped around like a tangle of high-school football players, tossing their toys in the air, shoving and head butting each other, and collapsed in heaps of exhausted brawn at day's end. Watching Diane attend to them was tiring on its own, and I was silently thankful that my dogs did not require that level of upkeep.

On the back of Diane's van were two bumper stickers:

CALM? SUBMISSIVE? WHY WOULD I WANT A DOG LIKE *THAT*?

And:

I THINK, THEREFORE I'M SINGLE.

On the front, a large, jagged hole had been torn from the right front fender. "Thing seems to think that my van is a bull, and she likes to grab on to it," Diane said, chuckling. "But that's nothing compared to what Damien did to my drywall!" Indeed, Damien had clawed and chewed a giant hole through one of her walls in an attempt to find a rat that had gotten stuck between the studs.

Diane pulled an eight-milligram tab of the opioid painkiller Dilaudid out of her purse and washed it down with a draw from a large Coke. Then she put the van in gear and backed out of the driveway, humming "Mother's Little Helper" by the Rolling Stones.

Chronic, debilitating pain has plagued Diane for most of her life. A doctor once predicted that she would be in a wheelchair by the time she was thirty, and that was before the back surgery, the thyroid cancer, and the double-knee replacements that for a time left her unable to function without the aid of a walker. Some of her doctors call this fibromyalgia, while others wonder if it is rheumatoid arthritis. Pain pills and a strict anti-inflammatory diet helped her get rid of the walker, but she can still do only a fraction of the things she used to enjoy. Then, in 2008, a house fire destroyed most of her belongings. All her work and travel came to an abrupt halt, forcing her to retire early, which threw her into a catastrophic depression. Ever since, she has focused every minute of her day and every cent of her income on her dogs. Without them, she says, she wouldn't be here.

Stacks of bills, magazines, and greeting cards littered the van's dashboard, along with rawhides, a broken telephone, the detached rearview mirror, some old cassette tapes, a T-shirt, fast-food wrappers, soda bottles, and a DVD of *Dark Shadows*. "I'd be a hoarder if I had any money," Diane deadpanned. A handicapped parking tag and a desiccated chicken foot dangled from the driver's-side visor. Under the console, a nervous little black-brindle APBT was curled up and trembling with anticipation. She looked up pleadingly and whimpered.

"Okay, Freakshow, hang on." Diane slowed the van to a crawl and craned her neck out the window. "My neighbors fucking hate me for this." When the coast was sufficiently clear, she opened the driver's side door and Freakshow leaped out, exploding into a sprint. Diane hit the gas. "Watch her go!" The dog flew along the ground with her mouth open wide, shoulders and flanks shimmering in oily waves until all her limbs blurred together. Diane's van trailed behind her like a support vehicle in the Tour de France. The speedometer crept upward, and soon we were clocking Freakshow at thirty-four miles an hour. "If someone else ran their dog off leash like this, I'd call them an asshole," Diane said, mouthing the straw of her Coke. "But Freaky is completely trustworthy. I don't claim to be consistent."

"There's a lot about my obsession with animals that I still don't understand," Diane told me in the vet's office. Those closest to her didn't understand, either, and that has always been a source of tension between her and them. A life devoted to dogs was not what her upper-middle-class parents had in mind. For the first seven years of Diane's life, the Jessups lived on a seven-acre farm on Vashon Island, in the middle of Puget Sound. Diane's father worked as a pilot for United Airlines, and her mother, a housewife, busied herself providing a Cleaver-esque home life for the couple's three children. When the family moved across the water to West Seattle, the change from country living to city dwelling completely unmoored Diane. None of the material comforts of suburbia came close to what she felt among the sheep, goats, and geese back on the farm. "I was a little pagan heathen, even back then," she said.

When Diane was fourteen, her parents bought her a yellow Labrador retriever named Arrow in an effort to console her. Arrow hated the water and hated other people, but Diane felt a unique sense of connection to her. A fog of loneliness lifted from her life. "With me and my dogs," she said, "there was this closeness of *friendship*. Humans and dogs are both predators, so our minds work the same, in a way. All my friends wanted horses, but ugh, why? A horse is an animal, but a dog is a *friend*. The dog is our species' soul mate. I truly believe that."

By the time she reached high school, Diane had grown into a lanky six-foot-tall tomboy in bell-bottoms who loved reading—especially British history—but hated being cooped up inside without her dog. Most afternoons she cut school to go hiking with Arrow in the densely forested mountains of the Pacific Northwest. Soon Diane and Arrow were spending all their time in the woods or roaming Seattle with other neighborhood dogs, which clustered around Diane in a big pack on walks down to the beach. The animal rights movement, which gained momentum after the publication of Peter Singer's book *Animal Liberation* in 1975, seemed to be the most natural extension of her interests, if not her conservative politics. The thought of animals being used in medical research particularly horrified her. She became a lifetime member of the American Anti-Vivisection Society and attended anticruelty vigils hosted by People for the Ethical Treatment of Animals.

It wasn't enough for Diane to be around dogs, however. She wanted to *work* with them, teach them things. At fourteen, she worked part-time cleaning kennels at a facility that trained guard dogs, one of a growing number of guard-dog businesses that sprang up as rates of violent crime across the country increased during the 1970s. Then she dropped out of high school to work in a vet's office. After graduating by correspondence course and halfheartedly attending a few classes at the University of Washington, Diane opened one of her many dog magazines and saw an ad for an official guard-dog-training school in Bakersfield, California. It occurred to her that she didn't need to spend her time in classrooms if what she really wanted to do was train dogs. With tuition money from her parents and a new Doberman named Otto, she signed up to become a professional guard-dog trainer.

It was at training school that Diane met her first American pit bull terrier, in the form of a little white female that one of Diane's roommates' girlfriends brought up to visit. Diane was immediately captivated by the dog, but she also felt a tremendous sense of . . . relief. This APBT was cheerful and easygoing, elegantly athletic, and willing to work hard without needing to be cajoled, begged, or prodded. The other dogs Diane worked with—German shepherds, Dobermans, and Rottweilers, mostly—seemed mean unless they had to be friendly, which is exactly what their training brought out in them. Otto, her own Doberman, was so nervous that he always seemed a hairbreadth from going off on somebody. This white dog appeared to be just the opposite. She was friendly with everyone unless forced to be otherwise.

The primary training protocol of that era was devised by William Koehler, a trainer for the U.S. military during World War II who went on to become the head animal trainer at Walt Disney Studios. For many years, Koehler was revered by his students for his efficiency in training and his genuine love for dogs, but Koehler believed that a dog needed to be "dominated" by its human and that the animal would obey commands only if it feared punishment. He developed training techniques so punitive that most modern trainers consider them abusive: hitting, choking, "striking" dogs with fire hoses, shocking them with electric collars, even holding their heads underwater. These methods grew especially merciless in the world of guard dogs during the 1970s. Thousands of years of evolution have inclined the domestic dog *not* to go after humans. In order to train police dogs in 1979, Diane was taught to make the dogs distrust people by punishing them whenever they relaxed.

"It was awful," she told me. "Absolutely awful. I *never* treated my own dogs that way." But Koehler's dominance-based protocol was considered the only way to get results if you wanted to be a professional dog trainer, especially a guard-dog trainer. Diane soon quit the training business and became an animal control officer instead. Years later, when she learned that treats and praise got her better results during training, she began enjoying the process much more herself. She never looked back. At present, her training group, the Washington Sport Dog Club, is the only Schutzhund club in the country that relies entirely on positive training methods.

In 1985, Diane noticed a classified ad in the local paper that had been posted by a man selling a Schutzhund-trained male APBT. She still enjoyed protection-dog sports as a hobby, so she called him up. The man with the pit bull turned out to be, as she recalls, "a nasty scumbag with a chop-shop," and the dog wasn't Schutzhund-trained at all. He offered her the chance to look at some puppies he was selling in the back instead. One handsome tiger-striped male was known as Old Carpetback because his littermates picked on him. Something about Old Carpetback resonated with Diane. She feared that if she didn't remove him from that situation, some "creep" might subject him to a terrible fate. So she scooped up the puppy and took him home. In a nod to medieval England, she named the pup Dread.

Dread was a bit aloof at first, and, as Diane puts it, "he had opinions," but he loved to train more than any dog Diane had ever worked with. "Training became fun for me again," she recalled. "I got away from all the compulsion and fear and anger I saw in the guard-dog-training world. All you had to do with Dread was offer him a toy. He was freakishly smart, freakishly obedient. And he had an uncanny way of knowing when things were *important*. No matter what you asked him to do, he showed up for it."

When Dread was a little over a year old, Diane signed him up for the novice division of a local kennel club's all-breed obedience trials. Out of 121 dogs, he was the only APBT in the competition. After completing a round of tasks, Dread finished with a score of 198 out of a possible 200, higher than any other competitor. Dread's second obedience trials were a repeat of the first. In his third, he and another dog tied for top marks. Small crowds began forming around Dread wherever he went.

Dread proceeded to rack up multiple titles in obedience, weight pull, sheepherding, duck herding, and tracking, but it was in protection dog sports, Schutzhund, and IPO (Internationale Prüfungs-Ordnung, a German police dog exam) that he truly excelled. Schutzhund, especially, required him to display strength, stamina, self-control, and a great deal of mental focus. He was so good at it that before long Diane barely had to give Dread verbal commands. "I know it sounds crazy," she said, "but we knew what the other was thinking all the time."

It would take science two more decades to discover in the lab what many owners of working dogs understand in their bones: that dog behavior may start with genes, but it can't meaningfully be divorced from its human context. From birth to death, the lives of dogs are entirely circumscribed by us. Millennia of domestication have refined them into creatures uniquely attuned to everything we are and do. Dogs not only read (and respond to) our facial expressions but also follow human pointing gestures, though their next of kin (wolves) and ours (chimpanzees) do not. When presented with a puzzle to solve, dogs immediately look to their human handlers for cues on how to proceed, a process behaviorists call social referencing. They are watching us even when we are oblivious to it, right down to the movements of our eyes. A dog and its human, then, are not separate entities that occasionally intersect; they are two parts of one dynamic whole.

Dread went on to make star turns in many television shows and Hollywood films, including *The Good Son*, which required him to bare his teeth at Macaulay Culkin, a transgression that Culkin's demented character repays with a crossbow bolt. Other trainers enjoyed having Dread on set because he tolerated the long hours and chaos of film production better than most of the other animals they worked with. Soon Dread was requested for so many engagements that Diane felt more like the dog's agent or valet than his owner.

Dread was only one in a long line of Hollywood pit bulls. Unlike Dread in the 1980s, however, the pit bulls of Old Hollywood were never portrayed as frightening. Because of their athleticism and wide, "smiling" faces, they were almost always cast as "trick dogs" in comic sidekick roles. The trend started in 1901 with a dog named Mannie, who starred

Roscoe "Fatty" Arbuckle and Luke

in *Laura Comstock's Bag-Punching Dog,* a five-minute vaudeville short from Thomas Edison's production company. Mannie went on to appear in several Buster Brown shorts, which were based on a popular comic strip of the day, before a new pit bull stole the spotlight thanks to Wilfred Lucas, an assistant director of the Keystone Kops comedies. While working on *Love, Speed, and Thrills,* with the actress Minta Durfee, Lucas promised Durfee that if she held on to the edge of a cliff long enough for him to film a key action sequence, he would give her a six-week-old puppy, which he did. Durfee and her husband, the silent film star Roscoe "Fatty" Arbuckle, named the tan-and-white pit bull Luke, after the director.

Luke grew up to become "the famous bull terrier comedian," one of Hollywood's first animal celebrities, performing tricks and comic stunts alongside Arbuckle, Buster Keaton, Harold Lloyd, and Al St. John. Mack Sennett, the Academy Award–winning director who pioneered "slapstick" comedy, considered Luke his "most dependable performer." One of Sennett's employees recalled that Luke "never had to be told more than once what the scene required. That dog could jump off the high-diving board, chase after Al St. John from flat roof to flat roof. Mr. Sennett used to have that dog driven to the studio in [various luxury cars] . . . and Luke never asked for a raise, which made him happy." In 1918, Arbuckle signed a contract that stipulated Luke would earn $50 a week—an exorbitant sum—because every time he appeared on camera with Arbuckle, the actor's mail was "flooded daily with requests for the dog's return."

After Luke, there was Pal, a small brindle-and-white pit bull with cropped ears who would eventually appear in more films than any other actor (on two feet or four) of his day. His owner and trainer, Harry Lucenay, told reporters that he found Pal right after World War I in a barn outside Bordeaux, where the dog was curled up beside the body of his dead mother, who had served as a dispatch carrier. But Lucenay was a seasoned showman (he even enjoyed a brief career as a heavyweight

Original caption, 1928: "Accompanied by their private teacher, Miss Fern Carter, the famous 'Our Gang' kids arrived in New York City for a theatrical engagement. The gang, including Farina, Joe Cobb, Mary Ann Jackson, Jean Darling, Wheezer, Harry Spear and Pete the dog, are shown receiving their first lessons in New York from Miss Carter."

wrestler), and he might well have made that up. Some say Pal was sired by a famous Oklahoma fighting dog named Black Jack. If this is true, Pal never showed it; by all accounts, he enjoyed the company of people and other dogs equally. He was insured for $10,000 and earned the princely sum of $35 for one day's work, and by the time he died in 1929, he had appeared in more than 220 films. In one of them, he actually *did* play the role of a "nursemaid."

One of Pal's pups, Pete, would become the most famous Hollywood pit bull of all time. Lucenay groomed Pete for show business when he was still quite young, securing the role of Tige for him in several Buster Brown films and training him for a comic turn in Harold Lloyd's *The Freshman* in 1925. To make his mostly white face stand out on camera, Lucenay dyed a black ring around Pete's eye, a "monocle" that would later become his trademark. Two years later, Pete auditioned for the director Hal Roach, who was casting a new comedy series called *Our Gang* and needed a dog companion for the mischievous band of child actors that would later be called "The Little Rascals."

"The dogs we were using had no personality, and couldn't do enough tricks," Hal Roach said later. "We looked at probably fifty dogs, and of all the fifty dogs we looked at, the best trained dog was the bulldog with the ring around his eye. I said, 'Great, that's the one we want; all you gotta do now is take the ring off the eye.'" But Lucenay couldn't do that; the dye was permanent. Roach said, "What the hell, leave it on," and "signed" Pete to a three-year contract at $125 a week, which later increased to $225. Children across America, including the young Fred Rogers, who once posed with Pete for a portrait, adored the dog. Between his film salaries and his public appearances, Pete made Lucenay more than $21,000 a year and traveled with his own valet, who gave Pete a morning bath, plucked his eyebrows, and trimmed his nails every day. Lucenay died in 1944, after being shot during a poker game at the American Legion's Hollywood outpost.

"Doubtless," Robert Louis Stevenson once wrote, "when man shares with his dog the toils of a profession and the pleasures of an art, as with the shepherd or the poacher, the affection warms and strengthens until it fills the soul." This shared toil was the essence of Diane's bond with Dread. But when "killer pit bull" headlines saturated the media in 1987, Schutzhund clubs informed Diane that Dread could no longer compete. The clubs said they were too afraid of potential negative press. Finding this unjust and cowardly, Diane began offering public education seminars about dog bites and dog safety. She interspersed the lessons with flourishes of Dread's spectacular training, which showed the public that there was much more to pit bulls than scary headlines. The presentations were so popular that she and Dread crisscrossed the United States multiple times and traveled from Bermuda to Great Britain.

When Diane looks back over their years together, that period on the road is what tugs most at her heart. She never felt safer with, closer to, or more accepted by anyone in her life than she felt during those trips with Dread. She never had any doubt that he would have laid down his life for her. "With Dread, I could have walked through the gates of hell unafraid," she told me. "We were a match for anything." When Dread died in 2000 at the age of fourteen, she knew that the best part of her life had ended. Diane buried Dread in the backyard, underneath her window, the same way that Walter Scott had done for Camp.

"People always say, 'You must be deficient in some way because you want a strong dog,'" Diane said. "Or, 'You must be compensating for something.' And I say, 'What's wrong with *you* that you *don't* want one? Why are you so threatened by a dog that is stronger than you are?' " As she told me this, she stared into an empty middle distance. "It's a shame we don't have better ways to talk about the friendship between dogs and people. Not love, or affection, or cutesy shit. *Friendship.* Real friendship."

On the far wall of Diane's living room, above a terrarium that housed Diane's corn snake, was a giant collection of pit bull photographs and pit bull posters from Dread's heyday, as well as professional shots of Diane's other dogs from years past, all splashing through ponds and over obstacle-course jumps. They glistened in the light like action figures.

Giant bins filled with old books, newspapers, magazines, letters, and photographs that dated back to the turn of the twentieth century crowded the floor of the small kitchen. Diane called this her "archive." She sat down and began digging through it, pulling out items for me to read. "Promise me you'll read this one, okay?" She handed me a copy of *Dumb-Bell of Brookfield,* a collection of short stories by John Taintor Foote first published in 1917. Foote was known for writing tales in which the heroism of pit bulls figured prominently. He believed that people were drawn to the dogs for a spiritual reason, the same reason that films like *Rocky* and *Braveheart* endure: because courage and valor are revered by all human cultures. Diane tapped the cover.

"Did you ever have one of those days where you walked out the door and felt like you could lick the whole world?" she said. "That's the way I feel around these dogs. I believe that what and who you're around matters. It influences your character, man. There's a reason that the Indians had totems. Brave animals *inspire* people. They just do. They always have. Courage either really speaks to you, or it really scares you." She flipped through some files and pulled out a newspaper from the 1990s. "Oh, this is one of my favorites. Here you go."

It was a copy of the English tabloid *News of the World.* On its cover was a terrifying, widely circulated stock photograph of a barking pit bull that seemed to accompany every sensational dog attack story. The picture had been doubled with an image-editing program, and the headline read "Two-Headed Guard Dogs Bred by Cops!"

Diane snorted. "This stuff is great, isn't it?" She then picked up a stack of magazines and began thumbing through them. "When I'm around pit bulls, I see everything about England that was *good,*" she said. "I see the miners that were rock-fucking-hard, solid, respectful. Take their hats off for royalty. These are not proud, arrogant dogs with people. These are not the dogs of dukes and earls. You could kick Grimbo right now"—she jerked her head toward Grim, one of her males, who was curled up in a recliner—"and he'd ask your pardon. But with their own kind, they don't step out of the way for anybody. Courage used to mean something in this country. And now look at us. Nothing but whining and padded corners. We are all so . . . insipid."

"Insipid" was Diane's favorite word, and it described her least favorite quality—in humans, in dogs, in life. But an insipid dog struck her as being particularly tragic. She believed that turning a remarkable animal that had stood beside us since the Pleistocene epoch into a spoiled milquetoast that wears outfits was the pinnacle of disrespect. A real dog, a "dog of character," on the other hand, was an ancient line in the sand, drawn against a feckless modern world that, at its core, feels contempt for the natural instincts of creatures it claims to love.

"A dog is an exciting predatory animal," she once wrote in a magazine column. "They often have strong drives, strong reactions, and wild natural passions. They must be trained to live in our world, but we must always, to be fair, let them live in their own world, too . . . How often do I meet people who tell me their pit bull (or other breed) is a total couch potato and doesn't want to do X? . . . That dog has been sleepwalking through life. Give that dog to me for ten minutes, and the dog wakes up and discovers it is a dog, not a sofa cushion. Is this cruel? Maybe, if the dog has to go back to an insipid life. Or is it the insipid life that is cruel?"

Late on a drizzly Washington afternoon, we headed out to the house owned by Diane's friend Linda, which sat on a large acreage that called to mind a tony, old-money Virginia horse farm. In Diane's estimation, Linda is one of the top obedience trainers in the country. Inside the large barn that Linda had converted into a private training facility was a big ring filled with agility equipment and competition jumps surrounded by a plastic lattice of fencing, behind which Linda's Belgian shepherds reclined in separate portable pens like black sphinxes, all facing the same direction. Linda took each one out separately and put it through a short obedience course. As expected, they performed flawlessly. Then they went back to sitting perfectly upright in their pens in a way that faintly reminded me of Grim Reapers.

Diane's three female APBTs were messier. Nell cannonballed into the plastic fencing, which tipped over drunkenly as Linda's jaw tightened. But when the time came to work, Freakshow, Guppy, and Nell snapped to attention. Diane took a few steps in her pained arthritic gait. Guppy took the same three steps, watching her. Diane backed up, and Guppy backed up, as if she were standing on Diane's feet. Diane motioned toward an obstacle or a jump, and Guppy bounded over it. The two quickened and slowed, heeled and stayed. It was the same routine that Linda's dogs had just completed, but it seemed much different. There was a noticeable sense of levity and pride freely circulating in the air. Then I realized what had changed: it was joy, entirely unself-conscious joy.

Every time one of her dogs made the correct choice, Diane whooped and clapped her hands like a mother at her child's first birthday party, showering treats and praise. "You have to be willing to look like a fool," Diane called over to me. "The dogs love it!" In return, the animals pushed themselves to exceed her expectations. "Look at that," Linda whispered to me, shaking her head. "Diane's dogs," she said, "they're just so . . . *easy,* aren't they?"

No longer was Diane the "crippled ex-dogcatcher" or the author of angry Internet screeds. Nor was she preoccupied with foreclosures or money troubles or surgeries or pain pills. In that space, she was a charismatic partner in a dance she mastered long ago. She and her dogs positively floated, as if some glowing invisible cord held them all together. The late author and animal trainer Vicki Hearne, who knew Diane through the training circuit, once said that watching Jessup work with animals was a true "joy to behold." When I reminded Diane of this, she shrugged it off, returning to her usual mode of self-deprecation. "Vicki was full of shit."

During the late 1980s and early 1990s, Hearne was one of the only public voices defending pit bulls in the press, writing odes to the American pit bull terrier in *Harper's* and *The New York Times,* as well as devoting an entire book, *Bandit: Dossier of a Dangerous Dog,* to the philosophical questions raised by the public's fear of pit bulls. When I first came across it, I fell in love with Hearne's strong voice, her easy command of history and philosophy (she was once a Yale professor), and her beautifully crafted sentences. The more I read, however, the more uncomfortable I grew with her conclusions. Like that of many game-dog enthusiasts, Hearne's work veered dangerously close to dogfighting apologia in a way that echoed the moonlight-and-magnolias nostalgia of the American South. She heaped praise upon Richard Stratton, a breed historian with strong ties to the world of dogfighting, calling him "an authentic and intelligent admirer" of the pit bull. "No one wants [dogfights] especially," Hearne wrote, "and when they occur, they are framed by attempts to stop them . . . but at the center there is nonetheless awe and admiration in the presence of a beautiful and nearly pure cynosure: when Bull Terriers fight, what we see approaches a Platonic form." Unlike Diane, Hearne remained a Koehler acolyte throughout her life, insisting that Koehler's methods had been misunderstood and misrepresented by "humaniacs," a term Koehler himself had coined.

Most of the modern game-dog crowd has no contact with what is left of the fighting circuit. For many, including Diane, the appeal of a game-bred pit bull that will never be fought is a lot like the appeal of a muscle car that will never be raced; the frame is sleek and powerful looking, but its "engine" (what the game-dog folks would call the dog's "heart") is its most impressive attribute. But a vocal constituency of game-dog owners, of which Hearne was one, are more openly hostile to the humane groups that shut down dogfighters than they are to the fighters who put their animals in such a terrible position in the first place. Isn't it unfair, I thought, to define modern pit bulls by the men who abused their ancestors? Aren't we clasping the chains that bind them?

During my last day with Diane, I joined a training session with her Schutzhund club. The dogs started by following tracks through grassy fields and locating objects by scent, then moved on to protection work, which involves locating, barking at ("alerting"), then finally grabbing on to and immobilizing human decoys in padded suits (the "bad guys"), in order to demonstrate their willingness to obey commands under stress. Most Schutzhund trainers work with German or Belgian guarding breeds (German shepherds, Dobermans, Rottweilers, or Belgian Malinois), but Diane's club was notable both for its reliance on positive reinforcement and because several members worked with pit bulls.

In a world so primed to fear these dogs anyway, I wasn't sure if training high-drive, working APBTs to go after people, even in a competitive athletic sport, was a good idea. America does not need to see any more images of pit bulls "attacking" humans. For every experienced Schutzhund trainer like Diane, there are a hundred tough-guy wannabes who have no idea what they're doing and put both people and animals at serious risk. I observed one Chicago "training class" that involved so much yelling and choking of dogs wearing prong collars that I left in tears and did not return. But if I wanted to see what a highly trained pit bull athlete "was capable of," this was my only chance. These dogs were, without question, the best of the best.

It was a brutally cold, rainy morning. Out on the baseball field where the club trained, Diane introduced me to a woman named Shade and to the only male member of the club, Gabe. Both seemed too focused on skill development to engage in much conversation. While they took turns working with their dogs, I sat in the dugout with a woman named Jenny. Both of us inched toward a space heater that someone had found to take the edge off the cold. She told me that she owned a male pit bull named Blue.

"There are two things that everyone loves to fight over," Jenny said, "pit bulls and abortion. I'd rather avoid those topics with most people as much as possible." Fair enough. We watched Shade's German shepherd repeatedly launch himself at Gabe, who was wearing a fully padded decoy suit. He lightly rapped on the dog's flank with a plastic stick. The shepherd tugged and shook and gave him hell. "We expect so much from dogs, when you think about it," Jenny continued. "They're supposed to protect us when we want them to, but also to put up with strangers when we want them to. We want them to do scary-looking Schutzhund stuff when we want them to, but not to mind when kids crawl all over them or shout in their ears or pull their tails. They should only want to play when *we* feel like it, not when they do. And they can't ever bark, chew, or dig. All the things that make dogs dogs. That's an awful lot to ask, isn't it?"

I nodded. Then I saw Diane beckoning me. It was time. "You ready?" she said.

I stood up, dusted off my jeans, and pulled up the hood of my parka. It was my turn to experience the high-octane thrill that Diane swore would change my life.

The bite sleeve was a stiff cone made out of industrial canvas and wrapped with heavy rope. It looked a bit like an oversized hockey glove without fingers, and it covered my entire arm, up past the shoulder. At the very bottom was a bar for the wearer to hold on to. The whole apparatus weighed between five and eight pounds. I slid my arm into it and walked out to center field.

The dog Diane had selected for my decoy experience was Guppy, a buckskin-colored female APBT who weighed a little more than fifty pounds. "You're going to stand right there," Diane said, pointing, "hold the sleeve in front of you, and hit it a couple times to show Guppy the target, okay?" I nodded. "When I give the command, Guppy is going to run and grab the sleeve, and you're going to try to run away. She's going to try to keep you from doing that, obviously. Whenever you've had enough, just let go of the sleeve and let her have it. It's like a toy for her. Got it?"

"Got it." Guppy's tail began to wag.

Diane led Guppy back about forty feet and turned around. "Ready?"

"Ready." I lowered my center of gravity like a wrestler and thumped the sleeve. It sounded hollow and drumlike. I prepared myself for Jack London's "clinging death."

"All right, Gup-Gup." Diane unsnapped the dog's leash. *"Go get her!"*

Guppy trotted over, grabbed the sleeve casually, let it go, and trotted back to Diane, whose face had fallen. "Aw, Gup. Come on."

We tried again, with the same result. I kept hearing Diane's words when we first spoke on the phone: "You need to see what these dogs are capable of." Now, with brows knitted together and eyes narrowed, she said, "Huh." After thinking for a second, she led Guppy back to the van. "Well," she said, "it looks like I'll just have to get Damien out." Damien was Diane's most impassive pit bull and, at seventy-two pounds, her largest. He had recently returned from a television shoot in California, where he had played the role of "demon dog" on NBC's supernatural drama series *Grimm*. Damien jumped out of the van, and Diane led him onto the field. Unlike Guppy, Damien saw the bite sleeve and immediately began to strain at his collar so hard that he chuffed and coughed, his feet wheeling underneath him and kicking up mud splatter. He thundered when he exhaled.

I swallowed hard, braced my legs, and drummed the bite sleeve again. "Ready!"

"Okay, Damie." The metallic click of the leash. *"Get her!"*

He closed the distance in what felt like both milliseconds and years. The crack that rang out when he hit me sounded like a motorcycle collision, which is a bit what it felt like. For a moment, I thought Damien had knocked the air out of my lungs, but in fact I had simply forgotten to breathe.

Nothing can prepare you for what 72 pounds of trained muscle feels like when it is hanging on to your arm. How people decoyed for 100-pound German shepherds and 130-pound Rottweilers, I had no idea. I staggered backward a few steps, dragging Damien with me. Diane

Damien, one of Diane Jessup's American pit bull terriers

called out, "Can you kind of, you know, twist around and make it fun for him?" As best I could, I feinted one direction, then staggered another. I pulled my arm across my chest, pitched to the side, then turned around and pulled it the other way. I changed my levels from low to high, high to low, and swiveled back around. My heartbeat whooshed inside my ears. Damien held fast, occasionally shaking the sleeve lightly but mostly just holding on to it.

I must give him credit: The dog was a consummate professional. He did not seem angry or frightened or even wild-eyed. There was no growling. He did not corncob up and down my arm, as if he wanted to sever it, or lunge at my throat, or anything of that sort. In fact, he seemed to be taking it easy on me. He had done this hundreds, if not thousands, of times. I never felt threatened, but I did feel physically exhausted. Fending off an untrained, legitimately dangerous dog of that weight would have been impossible, at least for me. After maybe twenty or thirty seconds, moons of sweat grew under my arms. Sucking wind, I let go of the sleeve, and it fell to the ground, where Damien chomped on it happily. I took a few Jell-O-kneed steps over to Diane, who was laughing at me.

"Nice job!" she said. "Now do you see what I mean?" She grabbed the sleeve from Damien and said, "*Aus.*" He let it go, wagged his tail, and sat down.

Back at the house, Damien put his paws on my lap and leaned his head into me, as if to tell me that there were no hard feelings. Of all Diane's dogs, he was the only one who had never seemed particularly friendly, but now it appeared that I had been admitted into his inner circle. I scratched his ears, and he closed his eyes contentedly.

Diane leaned back in her chair and crossed her arms. "So, *now* do you think pit bulls are different?"

I did not know quite how to answer this. "Depends on what you mean by 'pit bull,'" I said, "and what you mean by 'different.'" From what I could tell, they all seemed different from each other. The dozens of other pit bulls I'd met did not have the tenacity of Diane's elite sport dogs; mine didn't even enjoy long runs anymore. Nola lived what I am sure Diane would call an "insipid" life. Why not accept that the dogs of today may be mellower, instead of expecting them to be the same as they might have been a hundred years ago?

For the first time, an angry squall passed over Diane's face. "Why do you care so much about *generic dogs*?"

"Why don't you?"

She paused and stared at me as if I had just emerged from a spaceship. "Because our ancestors busted their asses to produce capable dogs, and they produced one of the most people-friendly, tolerant, trustworthy breeds *on earth,* okay? I don't think scatter-bred curs should ruin that. I think that we should be caretakers of dog breeds."

"Why?" I asked, shrugging. "Most of us don't need dogs to pin cattle anymore."

"But a hundred years from now, who knows?" She leaned across the table. "There may be another need for that dog. Look at things from a historical viewpoint. We aren't that far from needing to hunt and fight to survive. We're only a few generations away from *Braveheart.* Just because our culture right now is insipid, I don't think that we should say it's always going to be insipid. We're all just sitting around watching TV, playing our computer games and stuff, but the way the world is going with overpopulation, there's something coming. There's a reckoning coming. I don't think it's going to be pretty. I'm not a doomsday prepper, but I'm not blind. All it takes is somebody to cut the power grid, and you're going to *wish* you had a capable dog. You're going to wish to God that you didn't have a Labradoodle." She jabbed the air to make sure I didn't miss her point. "*That's* the difference between me and other people. I refuse to face an uncertain future with a fucking *Labradoodle.*"

When I arrived to say good-bye the next morning, all had been forgiven, and there was a stack of vintage pit bull magazines from the 1920s and several books on the kitchen table that Diane wanted me to take home and read.

"It's hard for me to talk about the emotional stuff sometimes," she said. "I'm not good at that shit. But when I thought about it, I wanted you to know something." She sat down next to me. "I look to these dogs for how I want to live *my* life. That's what it comes down to. You can't be around pit bulls and not be inspired by how brave they are. How much heart they have. How cheerful and loving they are, even when they are sick or hurt and feel terrible . . ." Her voice trailed off for a second, and she looked down at the floor, spinning the silver Celtic band she wore on her ring finger. "I know these guys run my life. My parents can't stand the fact that I don't have a husband and kids, that I grew up to be a broke, messy dog person. Everyone thinks I must be so sad and lonely. But I'm not. I have had a better life than anyone I know. The dogs are Peter Pans that keep me in touch with things that are exciting, with the woods, with nature."

Diane Jessup and two of her American pit bull terriers

Diane's eyes started to mist up, and her voice trembled slightly. Nell scrambled up into her lap, as if to comfort her. "All the times I've had health problems and surgeries, I have had these dogs waiting in the wings saying, Get better. Let's go. Get better. Let's go out and do bite work. Let's do decoy." She cradled Nell like a giant baby, and the dog licked her face. "If I didn't have them, I would have given up a long time ago. I owe them everything, and I want to be brave like they are." She turned her head and looked out the window. "I don't want to be a cur."

I told her I understood, because I did. She needed something different from her dogs than I needed from mine, and who was I (or anyone else) to begrudge her that? As Diane often pointed out, she and her APBTs weren't hurting anybody. For almost thirty years, she had accepted her dogs as they were, and in turn they accepted her. It was not the history of dogfighting she loved, but the spirit of the fighter. Whether her dogs really possessed it—or whether it was enough to believe they did—did not matter. They gave her strength all the same.

CHAPTER 6

TOOTH AND CLAW

Truth has rough flavours if we bite it through.

—GEORGE ELIOT, "Armgart"

As animals go, *Homo sapiens* is a relatively small, weak, and vulnerable physical specimen. Other creatures have been blessed with an infinite array of defense mechanisms to protect their vital organs—not just sharp teeth and claws, but also fur, scales, plates, horns, exoskeletons, even poisons and electric charges—but we have only our skin, which is soft and easily torn. In the scheme of things, we are bipedal oysters walking the planet without shells, so it makes sense that a heightened awareness of other animals' teeth would lodge itself so deep in our brains that thousands of years after leaving the wilderness, we still carry it around. If, as Carl Jung and Joseph Campbell believed, our myths reveal something profound about who we are, then take a look at our monsters. Almost every monster in the history of global mythology comes equipped with a gaping maw. "Among the earliest forms of human self-awareness," the journalist David Quammen writes, "was the awareness of being meat."

For those of us who are far removed from the savannas where our brains first developed that self-awareness, cats and dogs may be the only animals we ever experience up close. Cat lovers can brush off the occasional nip or scratch from their pets because cats are allowed, even expected, to retain a little of their wildness. (When a cat kills a mouse, for example, few are alarmed.) But a dog bite feels deeply personal, like a sucker punch from a spouse. To the dog, however, it's simply business as usual.

The dozen or so animal behaviorists I interviewed echoed what Jenny had said in Washington, that our expectations of dogs have become unrealistic. Biting may be scary to us, they told me, but it serves a vital purpose. Without the benefit of hands to push, shove, strike, or otherwise defend itself, the dog has only one tool at its disposal if it needs to tell us no. It has to use its mouth. It may resort to growling, snarling, or snapping, but biting usually happens only after all other options have been exhausted. In America, a country of nearly 320 million humans and roughly 77 million dogs, these types of disagreements probably happen several thousand times every day. As animals assume larger roles in our emotional lives and we push them into more stressful and unnatural situations, they will inevitably push back. Dogs have evolved to understand us better over the millennia, but in modern pet culture we appear doomed to understand them less.

Fortunately for us, dog bites almost never cause serious injury. In fact, the overwhelming majority of bites don't even break the skin, let alone require something as rudimentary as a Band-Aid. Because most bites are such nonevents, epidemiologists believe many go unreported. A study conducted by Dr. Jeffrey Sacks and his colleagues at the Centers for Disease Control and Prevention in 1994 estimated that 4.7 million Americans are bitten by dogs every year, but only about 800,000 of them (17 percent) receive medical attention of any sort. These figures may actually be high, because they come from a telephone survey of general health information.

Only about 316,200 bites (roughly 7 percent) are treated in hospital emergency departments,* and in 2008 only about 9,500 (0.2 percent) required hospitalization. The number of Americans killed by dogs is so low (about thirty-five a year) that it is statistically invisible (0.0007 percent of estimated bite injuries). While these deaths are undeniably tragic, they must be put into proper perspective. The risk of dying from a dog bite injury in the United States in any given year is approximately *one in ten million,* on par with the risk of dying from a lightning strike or from necrotizing fasciitis, the so-called flesh-eating bacteria. Compare dog bite deaths with those from other unintentional injuries, such as drug overdoses (35,563 in 2013), motor vehicle accidents (35,369), falls (30,208), and homicides (16,121), and they virtually disappear.

* Just because a bite injury is treated in an emergency department doesn't necessarily mean that it is serious. Many people seek rabies prophylaxis after any dog bite that breaks the skin, especially if the dog is a stray.

As with any potential hazard, even a rare one like a lightning strike, precautions are still essential when handling dogs. But even when put next to other animal-related injuries, dog bites are nowhere near the greatest cause for concern. Every year an estimated thirty million Americans choose to ride twelve-hundred-pound animals for fun, even though the CDC reports that equestrian sports result in approximately twelve thousand cases of traumatic brain injury and have the highest mortality rate of any athletic pursuit, including car racing, motorcycle riding, and football. Research has shown that a quarter of injured horseback riders will suffer, at minimum, broken bones, and 11 percent will need to be hospitalized. According to one Canadian study, 7 percent of patients admitted to the hospital trauma center with horseback-riding injuries died, as opposed to the 0.0007 percent for dog bites. Working with livestock also carries with it a much higher risk of death than does owning a dog. Deer cause roughly two hundred fatal motor vehicle crashes every year by running into roads and highways, making them America's most "dangerous" animal in that regard.

Though there is no alarm about cat bites in the national news, a 2014 study conducted by the Mayo Clinic found that 30 percent of people bitten by a cat on the hand required hospitalization and intravenous antibiotics. Sixty-seven percent of those hospitalized required surgery. Cats' oral flora and small, needle-like teeth increase the risk of serious infection.

The good news is that we know a great deal about how to prevent dog bite injuries, and we have had this information for more than sixty years. As Americans began adding more dogs to their homes during the consumer boom of the 1950s, reports of dog bites increased. By 1957, an estimated 611,500 people in the United States had been bitten by dogs, some quite seriously; in 1955, at least ten people were killed in dog-bite-related incidents. This increase in bite reports, and the growing concern over the possibility of rabies transmission, prompted Dr. Henry Parrish, a professor of preventative medicine at the University of Vermont's medical school, to conduct the first comprehensive epidemiological analysis of who was getting bitten by dogs, where, and why. Parrish and his colleagues examined all reports of dog bites that occurred in Pittsburgh during the summer of 1958, which they hoped would give them a basic framework for understanding what was happening elsewhere.

While looking over the data from 947 incidents, Parrish and his co-authors noticed a few significant trends:

• Children were by far the most common victims of dog bites, *especially* children between the ages of five and nine, who were more curious and mobile than infants or toddlers. Young people, noted the authors, "are intimately associated with dogs as pets, they are often abusive to pets, and, in many instances, they do not know how to care for pets properly." Additionally, children "are more likely to be engaged in activities which excite dogs, such as playing ball, running, riding bicycles, and delivering newspapers." Boys in this age group were more than twice as likely to be bitten than girls.
• Over three-quarters of bites were inflicted on the victims' extremities. When bites occurred on the head, neck, or face, the victim was usually a small child whose height was similar to the dog's.
• Bites occurred most often during the afternoons and evenings of warmer months, when children were out of school and humans and dogs were more likely to be outside together.
• The majority of victims (83 percent) were bitten on or near the property where the dog lived, which suggested to Parrish and his colleagues that the dog might feel more defensive on its own territory.
• One-third of the incidents seemed to have been completely unprovoked. The other two-thirds occurred when the victim was physically interacting with the dog, petting it, or deliberately "goading" it in some way.
• Male dogs younger than five years old were most represented in biting incidents.

Parrish and his colleagues noted that "it was not possible to single out an individual breed as being particularly vicious." As far as they could tell from the numbers, there was also "no relationship between the dogs' behavior toward other dogs and their behavior toward people." Much more significant, they found, were the human factors surrounding the events, and they believed those could be easily controlled. Rather than placing undue emphasis on the characteristics and behavior of the biting dogs, they recommended careful supervision of child-dog interactions at all times. In addition, they urged parents of children who wanted to play with dogs to instruct them thoroughly in proper pet handling so as to reduce the possibility of a bite.

Most dog bite studies conducted around the world over the next half century would find nearly identical patterns. Boys aged five through nine who encountered a dog on its home turf were by far the most at risk for serious dog bite injuries, especially injuries to the face and neck. Large dogs were physically capable of causing more damage than small ones, and sexually intact male dogs were more likely to bite than neutered males. A handful of researchers also noticed an uptick in bites from spayed females. Scientists stress that the exact relationship between sexual hormones and dog bites remains unclear.

In certain parts of the United States, Canada, and the U.K., an inverse relationship between household income and dog bites suggests that much like access to healthy foods and medical care, access to resources and information about pet health, behavior, and husbandry, as well as timely responses from animal control officers, may be another form of health disparity in economically challenged neighborhoods. In addition, children who have been exposed to domestic violence are also more likely to have been bitten by dogs. It's possible that the cycle of human violence ensnares animals, too, but there is a great deal about that we still don't know.

What nearly all researchers have agreed on, however, are the measures most likely to increase community safety: close supervision of kids and dogs at all times, strictly enforced leash and animal-containment laws, compulsory licensing and vaccines for all pets, and thorough bite prevention education for both children and parents.

If anything, the existing literature on dog bites tells us that dogs show a remarkable amount of restraint, even when sending us a message. The canine mouth is a marvelous work of engineering, capable of crushing bone, yet this almost never happens. Most dogs learn from their mothers and littermates how to calibrate their bites for different tasks (playing versus eating versus hunting, for example). If a puppy bites a playmate too hard with his sharp puppy teeth, he soon learns that playtime stops. Likewise, his mother might give him a growl or snap if he nips at her with too much enthusiasm. Over time, well-socialized young dogs learn to scale back the pressure they use when biting, a skill known as acquired bite inhibition. This is why taking puppies out of their litters too early can be a bad, even dangerous idea. A small- to medium-sized dog with no bite inhibition is much more likely to cause a serious injury than a large dog with an inhibited bite.

One of the veterinary behaviorists who has studied and written about bite inhibition most extensively is Dr. Ian Dunbar. Dunbar is an expert on canine aggression, the founder of the Association of Professional Dog Trainers, and the originator of puppy socialization classes in the United States. He was also one of the first people in the veterinary community to tamp down the flames of the pit bull panic when they started to rise in the early 1980s. Growing up in the United Kingdom during the 1950s, Dunbar told me that he came across Staffordshire bull terriers* quite often and never thought of them as being frightening or aggressive. He always considered them "great little dogs."

"The stereotype back then was, 'Oh, here's the pub dog!'" Dunbar recalled when we spoke in the courtyard of his home in Berkeley, California. Lounging next to us under a lemon tree that afternoon were Dunbar's dogs: Hugo, a French bulldog; Zouzou, a Beauceron; and Dune, a lumbering sand-colored American bulldog that looked a lot like Damien, Diane Jessup's brooding, athletic APBT. "You'd go in a pub, and there would always be a Labrador and a lurcher, a little Jack Russell, sitting on the bar stool lapping beer from a saucer or a mug, and then a little Staffie bull," he said. "He was associated with the working-class guy wearing what we called a 'flat hat'—a guy smoking a roll-up, you know. He's down at the pub having a pint and he's got his little pit bull and"—Dunbar winked and gave me a good-natured nudge—"he was probably very close to whatever side of the law, we're not quite sure."

Dunbar estimated that he had either examined, handled, or trained roughly five hundred dogs from pit bull breeds by the mid-1980s, whereas his colleagues who were wary of pit bulls had barely handled any—and that's assuming everyone agreed on what counted as a pit bull. In his experience, pit bulls fell along the same behavioral spectrum as any other type of dog. Some were "tricky" around other dogs . . . but so were dogs from many other breeds. He actually commended the dogs for their friendliness. "They're some of the easiest dogs to socialize," he said. "On par with Labradors and goldens, pits are self-socializing breeds that are so outgoing that as puppies they run up to everyone, as if to say, 'Oh, aren't you pleased to see me?' And so they get petted and kissed and they're more likely to do it the next day, as opposed to the dogs that have words like 'aloof,' 'standoffish,' and 'independent' in their breed standards."

---

* The American pit bull terrier did not make its way to the U.K. until the 1970s. The predominant pit-bull-type breed in Britain is the Staffordshire bull terrier. Interestingly, however, Britons do not consider Staffords to be pit bulls.

Dunbar leaned down to give his American bulldog a scratch on the head under the table, and as Dune rolled over in hopes of getting a belly rub, I noticed that he wasn't neutered, something extremely rare in the world of animal welfare. "It was a conscious decision," Dunbar said. "Because Dune looks like a giant pit bull, I didn't want people to think that he could *only* be a great family dog if he were castrated. I didn't want the credit to be taken away from him, in a way. I wanted to show people that dogs like Dune can just be good dogs, period. With *all* dogs, what it essentially comes down to is, are they socialized—yes or no? If they are, I don't care what the breed is, it's going to be a brilliant dog."

Whether the dog is socialized or not, inhibited or not, dog bites do happen, and they are proof positive that one party is sending out signals that the other is not picking up. When they occur, we mistakenly toss all bites into one giant bucket that we call "aggression," and we brand dogs with the dreaded label "aggressive." But animal scientists are much more judicious about painting with such broad brushes.

Like human violence, canine aggression is not one behavior but a vast spectrum of behaviors that range from curling a lip to a fatally injurious bite. Each behavior can be examined only in context. If, for example, someone were to keep a file on how often I "hit" another person over the course of my life, that information would be meaningless unless the surrounding circumstances of each incident had also been recorded. A jovial punch in the arm after a bad joke is radically different from self-defense in a dark parking lot or a punch to the face delivered in anger. What if I were consulting a map and accidentally bumped into someone on the sidewalk? What if I gave someone a high five? What if I pushed someone out of the way of a moving car? All of those scenarios would be included on my "hit record," but only one could be considered an intentionally violent act.

During the early nineteenth century, a Belgian statistician named Adolphe Quetelet theorized that the actual rates of crime in any given city were likely much higher than what social statistics could capture. How many crimes went unnoticed? How many went unreported? Which acts could be properly considered criminal? Along with the known quantities, Quetelet believed, went many variables that the former U.S. secretary of defense Donald Rumsfeld would have called "known unknowns" (the things we know we don't know) and "unknown unknowns" (the things we don't know we don't know). Known statistics stand in the shadow of their unknowns, which Quetelet called the "dark figure" of crime. Statistics, he said, told only a small part of the story.

Dog bite statistics are dwarfed by their own "dark figures" as well. They are intended to track the possibility for rabies transmission, not to diagnose patterns of behavior. The overbroad approach described in my "hit record" is normally how public health departments and animal control divisions, which are primarily concerned about possible disease vectors, collect information about dog bites. In some jurisdictions, a dog doesn't even have to put its mouth on a human to be labeled a "biter." In the Chicago suburb of Palatine, Illinois, for example, it is unlawful "to permit any dog, cat, or other animal to bite, scratch, or in any other manner break the skin of any other domestic animal." In that case, a scratch to another pet would be recorded as a "bite."

The medical records of human bite victims can also be highly inaccurate. The attending physician is entirely reliant on the patient's visual breed identification of the biting dog and his or her version of the events that led up to the injury. Like the famous Rashomon effect of eyewitness testimony in human criminal cases, the emotional and psychological trauma of a dog bite injury can alter the patient's memory of relevant details. In addition, a person who might have teased, mistreated, or chronically neglected a dog is unlikely to admit that to authorities. The same goes for a caregiver who might have left a toddler unattended. For these and many other reasons, the American Veterinary Medical Association (AVMA) warned in 2001, "Dog bite statistics are not really statistics, and they do not give an accurate picture of dogs that bite."

Most of us aren't equipped with the scientific training or the practical experience to properly interpret the behavior of our own species, let alone another's, so dog aggression can be especially challenging for us to understand. Animal behaviorists divide aggression into at least eight types (which have countless subtypes depending on the circumstances): play; territorial/possessive; maternal; intraspecific (dog-dog); irritable (the dog is sick/injured/tired/hungry); predatory; dominance; and fear. Each of these can crop up in different environments, at different times, during different phases of development, for different reasons, and can be directed toward different targets (human strangers or owners, for example, or female dogs but not male ones). Many behavior experts actually prefer the term "reactivity," because "aggression" implies an intention on the part of a dog that may or may not be present. If a dog barks incessantly on a leash and bares his teeth when he sees another dog, that's an observable reaction, but it could arise for a number of different reasons. One animal might be frightened and acting defensively, another might be frustrated by the leash, while a third may actively want to injure the other dog, and so on. Along these same lines, behaviorists generally prefer the word "bite" to emotionally charged verbs like "attack."

According to both Dunbar and Jean Donaldson, the founder of the Academy for Dog Trainers (not to be confused with the Association of Professional Dog Trainers, which Dunbar established), most dogs bite out of fear—not malice or vengefulness or dominance—when a human pushes the animal beyond its stress threshold or forces it into a situation it feels it can't escape. Bite victims often mistakenly believe that the bite "came out of nowhere," when in fact the dog was sending subtle signals about its level of discomfort for quite some time. Lip licking, yawning, looking away, and shaking themselves off are just a few of the ways that dogs communicate anxiety.

Much of Donaldson's writing focuses on mediating the "culture clash" between dogs and humans. One of the first things she stresses to her clients is that any pet owner who expects a companion animal to put up with everything, all the time, fails to appreciate how bizarre, scary, unpredictable, and irritating the human world can be from the dog's point of view, or what the cognitive scientist Dr. Alexandra Horowitz calls the *Umwelt* of the dog. Donaldson writes, "For every anecdotal report of an otherwise (allegedly) perfectly friendly dog who nailed the burglar, there are scores, hundreds, thousands of dogs that, for *identical reasons,* nailed the neighbor, the delivery guy or a child in the park . . . The mental hurdle people seem to have is accepting that *the dog decides what is spooky or threatening.*"

Donaldson told me that our "quasi-Madonna-whore-complex" with regard to dog aggression is misguided. "Just a generation ago if you went near a dog when he was eating and the dog growled," she explained, "somebody would say, 'Don't go near the dog when he's eating! What are you, crazy?' Now the dog gets euthanized. Back then, dogs were allowed to say no. Dogs are not allowed to say no anymore . . . They can't get freaked out, they can't be afraid, they can never signal, 'I'd rather not.' We don't have any kind of nuance with regard to dogs expressing that they are uncomfortable, afraid, angry, or in pain, worried, or upset. If the dog is anything other than completely sunny and goofy every second, he goes from a nice dog to an 'aggressive' dog."

Infants and small children can't be expected to decode the nuances of animal behavior, and they should never be personally blamed for "provoking" a dog bite. But children can appear very strange, even frightening, to dogs because their movement, vocalizations, and behavior are so different from those of the adults that most dogs are used to. According to the veterinary behaviorist Dr. Karen Overall, "Dogs may view children as small and erratic variations of adult humans. Children's normal expressions of affection can be loud, shrill, and quite physical, and their movements are often rapid and chaotic . . . A behaviorally abnormal dog may deliver a substantial bite because it views the child as a threat, regardless of the child's behavior, whereas a behaviorally normal dog may bite because it cannot escape a child's attentions and is fearful."

Animal behaviorists stress the role of parents, babysitters, and other caregivers when children and dogs are around each other: the adults in charge are the responsible parties 100 percent of the time.

"Dogs don't have lawyers, they don't have e-mail, they don't have anything," Donaldson said. "Biting is how they resolve conflict. We should understand this. Everybody gets angry. Everybody has probably cursed at other drivers in traffic. Everybody has said things they regret to people they love. Should you be locked up for life, should you be muzzled, or should we kill you because you argued with your spouse? If I sue somebody or I write a letter to the editor that's a little bit tart, or if I'm passive-aggressive to a colleague because I had a bad day, is that equivalent to felonious assault or murder? Of course it isn't. Dogs have the physical machinery to amputate digits like *that*." She snapped her fingers. "But look at how rarely they use this capability. What does *that* tell us? They should be getting gold medals in terms of agonistic behavior. Humans will go after each other with bare hands."

Donaldson believes that we owe it to our companions to respect their boundaries and to remember that some level of aggression is essential for any animal's survival. Therefore it's our job to learn their limits, control our expectations, and, if it comes to it, admit when we are in over our heads, seeking help from trained professionals. Every dog has individual quirks and preferences. Donaldson's own dog, a chow named Buffy, tended to get into nasty fights when playing with other dogs.

"Could I train this out of her if I wanted to?" she asked. "Sure. Could I make her socialize with other dogs? Of course. But what would be the point? Buffy's very clear about not enjoying it, and there's no reason why she has to. This is just who Buffy is. It's my responsibility as her owner to keep her safe and to keep other dogs safe by not setting her up to fail. Keeping your dog on a leash and away from other dogs is actually a very easy thing to do."

Like Dunbar, Donaldson has worked with many pit bulls in almost thirty years of professional dog training. Has she noticed any frightening pattern of deviance? "Oh, please," she said, smiling. "Among trainers, pit bulls are considered cheap dates, actually. They have a reputation for being incredibly easy to train. They'll pretty much do whatever you want. Some of us want a bit more of a challenge."

The big question, perhaps the *biggest* question, that the American public seems to want answered is, are some breeds simply "dangerous"?

Janis Bradley, a former educator at Donaldson's Academy for Dog Trainers and the author of *Dogs Bite: But Balloons and Slippers Are More Dangerous,* began delving into the scientific literature on dog bites after reading a host of sensational headlines about dog "attacks" that implied serious bite injuries were on the rise. She had never been seriously bitten by a dog in the entire course of her career, nor had most of her colleagues, and it was their job to deal with problematic dogs all the time. What she found was that tiny numbers had been extrapolated into giant figures without being put into context. Bites that did not break the skin were grouped with life-threatening injuries, and this drastically distorted the overall picture. Then academics used public health data to make inferences about dog behavior that the data were never intended to describe, which made things much more confusing.

"'Where there's smoke, there's fire' is not a scientific position," Bradley told me. "To my knowledge there is no other injury vector that is studied at the level where it does not cause injury. Nobody counts how many people trip on something and don't fall down. But we measure bites—or *try* to—because we're hardwired to react with fear to predators with sharp teeth. We don't react to them in the same way that we react to a tree root, regardless of the relative danger. I don't think that association can be disconnected, but we can learn to question our responses. We can't teach ourselves not to flinch when the cobra at the zoo strikes from behind the glass, but we can learn that we don't have to run screaming outside because of the terrifying cobra. I think that's the difference."

At one point in her training career, Bradley noticed herself making judgments about a dog's behavior based more on the dog's breed stereotype ("pointers point, retrievers retrieve") than on objective evaluation. If a client's cattle dog nipped someone, she thought, "Of course it does, it's a cattle dog. No big deal." But if a Labrador growled over its food, that was unacceptable, because . . . well, Labradors were always supposed to be friendly, weren't they? These preconceived ideas even influenced how she viewed her own dogs (greyhounds). "Dog professionals are as prone to these biases as everyone else," she wrote. But when Bradley examined dog behavior more critically, she found that most dogs exhibited the same behavior patterns, just in different ways and to different degrees. "Preconception is an incredibly powerful thing," she told me. "There's a very widespread idea that racing greyhounds can't be trained to sit, which is ridiculous," she said. "It's just that at the track, they've never received any reinforcement for sitting."

The rhetoric around the so-called dangerous breeds troubled Bradley deeply because so much of it seemed to her to be similarly rooted in confirmation bias and expectation, rather than in empirical science. "It's very intuitive for us to scapegoat," she said, "so if we see something that worries us, it worries us much less if we can say, 'Yeah, but it's only *those* ones that do it. Mine is fine. My people and my dogs are completely benign. All we need to do is get rid of *those*, that group over *there*. They're the ones.' Much more threatening to say, 'You know, we're all kind of capable of this.' "

In 2000 and 2001, researchers from the Institute for Animal Welfare and Behaviour at the University of Veterinary Medicine in Hanover, Germany, put 415 dogs from the allegedly aggressive breeds that had been banned in Lower Saxony (American Staffordshire terriers, Staffordshire bull terriers, Rottweilers, and Dobermans, among others) through behavioral tests that mimicked twenty-one situations that might occur between dogs and humans (including the approach of a stranger and the approach of a "menacing" stranger) and fourteen situations that a dog might experience in its everyday environment (noises, approaching vehicles, umbrellas opening, and so forth). They found that 95 percent of the dogs behaved appropriately during the test. One hundred fifty-eight of them displayed no aggression at all, even when threatened. Another 201 only escalated their behaviors to the level of barking or backing away. Only 37 (9 percent) bit when threatened, and of those only one did not give some type of warning before biting. When these results were compared with a control group of 70 golden retrievers, researchers found "no significant difference" in the behavior of the two groups. As a result, Lower Saxony lifted its breed ban.

In the Americas, things don't appear to be much different. A 2011 Canadian study found no significant difference in the behaviors of forty pit-bull-type dogs adopted from animal shelters and forty-two dogs from other breeds. The only differences they noted were that the pit bulls were adopted by younger owners and were more likely to sleep in their owners' beds. In 2008, researchers at the University of Pennsylvania asked more than five thousand dog owners to complete the 101-item Canine Behavioral Assessment and Research Questionnaire. Of the thirty-three breeds represented in the sample, "pit bulls" (yet again classed as one "breed") scored lower than average on all scales of human-directed aggression. On owner-directed aggression, they scored even lower than Labradors. Pit bulls scored slightly higher than average on aggression directed toward other dogs, but several other breeds, including dachshunds, equaled or surpassed them on that scale. The pit bulls were well within the range of normal.

But owner surveys such as these have their own limitations. How reliable are these breed identifications? Where do mixed-breed dogs fit in? How accurate is the behavioral assessment of the average pet owner? Which pieces of context might be missing? The same goes for the oft-cited behavioral exam administered by the American Temperament Test Society, which has evaluated 1,656 dogs from pit bull breeds, 1,427 of which (86 percent) passed. That's great news for pit bull owners, but unfortunately these data do not really mean anything. The average dog owner is highly unlikely to take an unsound animal to be evaluated by such a test, which creates a considerable sampling bias.

So, what do we know about breed and behavior? Does the phrase "pointers point, setters set, retrievers retrieve" hold up? Yes and no, says Bradley. In order to catch and kill its prey, a hungry dog must proceed through a full sequence of predatory motor patterns, which the biologist Ray Coppinger defines as "orient > eye > stalk > chase > grab-bite > kill-bite > dissect > consume." The behaviors that are considered characteristic in the working lines of certain breeds, such as pointing, retrieving, and herding, are pronounced because humans have taken dogs whose behavior was interrupted at key points in that sequence and selected offspring that exaggerated one or more of those particular traits.

Pointers exaggerate the stalking phase of the sequence. Shepherds, the stalking and chasing phases. Heelers, which often work with stubborn livestock, have well-developed grab-bite instincts. Vermin hunters, such as the wirehaired terriers, follow the sequence all the way through. Bradley explained, "In most cases, breeders simply ignore the behaviors they don't care about, so they appear more or less randomly, while the targeted behavior appears more consistently. There are some exceptions, where specific parts of the sequence are actively selected against, notably in the finishing parts of the sequence. A retriever, for example, wouldn't be much use if he dissected and ate the birds. Greyhound breeders, on the other hand, are working with dogs who will never have any opportunity to do anything other than the chase part of the sequence, so the selection is 'blind' to the other behaviors, with the result that some greyhounds do them and some don't."

But breeding for behaviors is not like shopping from a catalog. Rather, it is more like pushing a car uphill. The "gravity" of evolution naturally pulls domestic dogs into one large bell curve. The job of the pedigree breeder is to exert a constant "push" in the other direction by way of controlled mating and rigorous selection. If all goes according to plan, this push will increase the incidence of the desired behavior in a tightly controlled population. Every dog will display the behavior on a spectrum, and some may not display it at all, which is why someone in search of a working dog will often buy a "proven" adult rather than a puppy. "Dog breeding is a well-established art, but a crude, unestablished science," Dr. Jasper Rine, a geneticist at the University of California at Berkeley, told *The New York Times.*

A 2015 meta-analysis of forty years of research on these characteristic traits revealed that, surprisingly, their heritability is low. This means that in similar environments, a relatively low percentage of variation in those traits among individuals can be attributed to genes. There can also be a great deal of overlap in various behaviors among breeds. Pointing, for example, has been observed in roughly fifty breeds of dog. It is likely tied to fewer than five genes that are common in the dog population.

"Even people who have done rigorous, selective breeding expect to fail most of the time," Bradley said. "They're breeding large numbers of dogs to try and get that *one* that will exhibit these very, very specific traits. Nobody ever expects to have a whole litter of winners." Guiding Eyes for the Blind in Yorktown Heights, New York, which is one of the most elite service-dog breeding programs in the country, has a "washout" rate of approximately 40 percent.

The Victorian dog-show mania of the mid-nineteenth century not only created hundreds of new breeds but also created two possible categories of bloodlines within many of them: working bloodlines, in which behaviors were most important, and conformation or show bloodlines, which prioritized appearance over behavior. The "washouts" from the conformation lines usually went on to pet homes. The dramatic increase in the number of breeders also allowed for more physical and behavioral variation within each breed, with the most popular dogs also being the most varied. Today, Labradors from American show lines are much shorter and fatter than they were even twenty years ago, while Labradors from British field lines are leaner and leggier. Dogs from these two strains may not only look different, they may also have drastically different behavioral profiles.

When breeders stop pushing, the car rolls back down the hill, and canine behavior drifts back to the middle. Exaggerated traits that are not selected for and not adaptive will mellow out and disappear over time, which is what appears to be happening in both the American and European dog populations. The overwhelming majority of modern dogs live as pets, rather than workers. Great Danes are no longer used for boar hunting. Siberian huskies do not pull sleds. Rhodesian ridgebacks do not bay lions, and most dachshunds will never see a badger, let alone kill one. Rather, these animals are physical reminders of the way the world once was. As the historian Scottie Westfall says, "Dogs are artifacts." Though it is common to attribute a dog's behavior to the task it was historically "bred for," many of us fail to consider that most of today's dogs are "bred for" the work of being companions, and have been for many generations.

In 2005, Kenth Svartberg, a zoologist from Stockholm University, collected data from more than thirteen thousand dogs from thirty-one breeds that had been subjected to a standardized behavior test and sorted them according to behavioral traits such as "playfulness," "curiosity/fearlessness," and "sociability." After analyzing the data, Svartberg and his colleagues found that there was "no relationship . . . between the breeds' typical behavior and function in the breeds' origin." He did, however, find that dogs from working lines (not breeds, but lines) retained more of their historical working traits than dogs from show lines, leading him to conclude that "basic dimensions of dog behavior can be changed when selection pressure changes, and . . . the domestication of the dog is still in progress."

Pit bull breeds are not exempt from this trend. Unlike pointing or retrieving, both of which increase a dog's ability to feed itself and its offspring by hunting, fighting isn't one behavior but a complex series of behaviors that put the animal at tremendous risk. As highly social creatures that negotiate and renegotiate their relationships over time, most dogs depend on shared resources for their survival. If removed from human society, a dog that indiscriminately attacks or kills its own kind doesn't live very long. While it's certainly possible to breed for certain types of aggression (toward humans or other animals), it's much harder to breed dogs that match the profile that fighters say they want: an animal that is indiscriminately accepting of humans, selectively reactive around other dogs in a specific environment—the pit—but tolerant of dogs outside of it, one that "doesn't signal its intentions," and also "doesn't feel fear or pain." They may as well be describing the American unicorn terrier, because these are all genetic dead ends.

No researcher has yet located an "aggression gene" or a set of aggression genes, despite years of genomic analysis. While conducting his research at Bar Harbor, John Paul Scott considered aggression "a poor scientific term [that] chiefly functions as a convenient handle to relate phenomena described in more objective terms to practical human problems." At best, today's scientists can only make educated guesses about certain components of canine reactivity, like the startle reflex (which multiple studies indicate is heritable) and individual pieces of the agonistic repertoire (freezing, fleeing, defensive postures, vocalizations, etc.). But this requires that researchers clearly define and isolate the behaviors they are observing, which is always a challenge. It's possible, for example, that what was once called "rage syndrome" in certain lines of the English springer spaniel and English cocker spaniel is not one condition but several that were mistakenly grouped into one category. A few studies in mice and dogs have shown that disruption of the 5-hydroxytryptamine (5-HT) receptors in the brain, which regulate the neurotransmitter serotonin, may be linked to specific types of impulsive aggression, but in both animals and humans, the 5-HT receptors can be damaged by stress and trauma that occur both in utero and after birth. Yet even these possible neurological links have been observed only in dogs from *tightly closed* gene pools. They are not widely passed from dog to dog in an open breeding system, like the passing of a disease.

"Let's assume that you and I are working to breed the most dangerous aggressive fighting dog in the world," Kris Irizarry, the geneticist at Western University, told me. "And we want this dog to turn and attack any human being, child, or any other animal relentlessly and never stop until it dies, 100 percent of the time. That's our goal, okay? Now, let's make the crazy assumption that we achieve that goal, and we produce, I don't know, fifty dogs, a hundred dogs, even a thousand dogs that all have the same amount of this supernatural trait. For our purposes, we'll call them 'Crazy Dogs.'"

As he previously pointed out, "The moment our dog mates with any other type of dog, half of that genetic material is lost, so now you have a litter that's only 50 percent Crazy Dog. If that litter reproduces, then their offspring are only 25 percent Crazy Dog. Then it goes down to 12.5 percent, 6.25 percent, et cetera. Within only seven generations, you're at 1 percent Crazy Dog, and that's assuming you were 100 percent successful at the beginning, which we know isn't true of any breeder or any type of dog. Especially when you're talking about complex behaviors like fighting, it just doesn't work that way. There are probably constellations of genes, maybe even hundreds or thousands of genes that are contributing to that behavior. You have to get the right neuron shape, the appropriate amount of neurotransmitters, all these things.

"So," he continued, "the idea that any dog that has an ancestor—however many generations back—that had a head shape that cast a shadow against a wall that looked like the shape of a dog that bit someone in the pants . . . the idea that this dog is now going to be biting people is absolutely ludicrous! Americans watch too many zombie movies."

A number of other studies have confirmed that dogs lash out most frequently from fear and anxiety, not "rage." Not every dog that displays these behaviors has been abused, neglected, or formally trained, but overwhelmingly, the factors most highly correlated with dog aggression, such as the dog's early development, its level of socialization with people and other dogs, how it is contained, and which training methods the owner uses, are completely within the owner's control. Research indicates that these factors are far more important than the physical shape of the dog in determining its behavior.

Our own perceptions and expectations of the animals we encounter play a role in this, as well. "Dog breeds develop reputations," writes the biologist Ray Coppinger, "and those reputations color people's interactions with them."

> The fearful responses of people to a perceived aggressive breed "teaches" the shepherds or pit bulls to be aggressive with people. As the dog walks the streets, some people, almost imperceptibly, will take a step back or away from the dog. In two weeks the dog can become aggressive toward people. If people treated a golden retriever the same way, in theory one would get the same results.
>
> Are shepherds genetically aggressive? Yes! Where are the genes for aggression? In their coat color and shape. It is a feedback system,

where each time a person steps back from the shepherd because of its coloring and shape, the dog becomes more responsive to the move, and the people react more demonstratively to its movement, and so on. Can you train the dog not to be aggressive once it has learned to be? Probably not satisfactorily.

Okay, then can you *breed* people-aggressiveness out of shepherds? Of course! I'd start by breeding shepherds to have yellow coats and floppy ears.

"Gameness," however one defines that elusive quality, has never been studied in the laboratory with other variables held constant. Nor is it defined with any consistency. That's not to say it doesn't exist—there's much anecdotal evidence that it does—but we have no way of measuring it. And, as we know, not all pit bulls come from fighting stock, anyway. The Stafford and AmStaff are show breeds, as is the American bully. Most APBTs come from conformation/pet lines as well. So, the selective pressure for "gameness" was relaxed for most pit bulls between 80 and 150 years ago. As a result, many have retained their looks but not their historical working drives.

If we want to own dogs, their teeth come along. It is up to us to learn how and when dogs use them and to keep our dogs out of situations where they feel they need to. Aside from that, we must also accept that sometimes accidents and misunderstandings, even tragedies, can happen. As much as we may want them, there are no simple answers.

CHAPTER 7

A FEAR IS BORN

> If they can get you asking the wrong questions,
> they don't have to worry about answers.
> —THOMAS PYNCHON, *Gravity's Rainbow*

The guard-dog fad that Diane Jessup found herself swept up in during the 1970s was part of a larger national trend, one that fundamentally changed the relationship between many Americans and their dogs from one of companionship to one of violence. Guard dogs have existed in some form or other since the dawn of domestication, but in America there was never a formal movement to "weaponize" the dogs of private citizens until the 1960s, when graphic media coverage of several high-profile murders (notably the "Boston Strangler" murders, which occurred from 1962 to 1964, Richard Speck's murder of eight student nurses in 1966, and the Manson murders in 1969), combined with the political assassinations of John F. Kennedy, Robert Kennedy, and Martin Luther King Jr. and the backdrop of race riots in Harlem (1964), Watts (1965), Newark (1967), and Baltimore (1968), led many Americans to believe that they were no longer safe in their homes.

By 1969, owners of guard-dog kennels in New York, Chicago, Los Angeles, and Philadelphia were reporting that they could not keep attack-trained dogs in "stock." Faced with such implacable demand, some companies rented out "attack dogs" by the hour, particularly in low-income urban communities. Celebratory articles about dogs taking "a bite out of crime" and ads for guard-dog kennels swamped newspapers and magazines. The police, who were themselves afraid of growing violence and the rise of black militancy, routinely ignored calls from inner-city neighborhoods, leaving citizens to feel that their local governments had abandoned them. Among residents of these areas, who could not afford the expensive new technology of alarm systems, nor could they hire private security guards (a growing trend among the wealthy), the guard-dog business surged.

As citizens' fears of one another increased, so did the size of their dogs. American Kennel Club registrations of the German shepherd, the Doberman pinscher, the Great Dane, and the Saint Bernard quadrupled, displacing longtime favorites like poodles and cocker spaniels, as more families bought dogs that projected an imposing image. From 1965 to 1975, the AKC registered more than 1.1 million new German shepherd puppies, averaging more than 90,000 a year, while Labrador retriever registrations hovered around 30,000 annually. While only a fraction of these dogs were professionally trained to guard or attack, the sudden swell in the popularity of dog breeds with formidable reputations marked a significant change in how many Americans viewed the dog's role in modern society. This increased steadily during the social turbulence of the "macho" 1970s.

Race-baiting in the media boosted this trend on both sides of the color line. In 1969, one writer for *The New York Times* noted that in Washington, D.C., whites' fear of "violent Negro crime" had turned daily life into a Wild West showdown. "Men and women go armed," he wrote. "Guns are kept under store counters. Fatal duels between storekeepers and holdup men are commonplace." On the other side of town, he noted disapprovingly, "poor Negroes who cannot afford them have bought dogs for protection." The *Chicago Tribune* reported that "up to 90 percent" of guard-dog sales in fifteen major cities were made to "Negroes and other residents of high-crime, inner city areas."

The situation in New York City was particularly bleak. By 1971, guard dogs had become, in the words of one writer, "standard fixtures" in Harlem and Queens. In Bushwick, Brooklyn, residents took their German shepherds and Dobermans on foot patrols of their own blocks in an attempt to provide a sense of law and order, while those in the East Bronx resorted to pulling fire alarms instead of calling police when they needed help. After being either mugged or robbed at work eleven times, one Bronx resident told a *Times* reporter that feeling so vulnerable and powerless in his neighborhood was "no way to live." Unable to secure a permit for a handgun, he bought a "vicious" German shepherd because he simply did not know what else to do.

With the guard-dog boom came consequences. A significant number of family pets in American cities had gotten larger not only in size but also in the levels of aggression their owners either expected from them, encouraged them to display, or were willing to tolerate from them, at least for a while. Poorly bred and improperly trained dogs from large guardian breeds were suddenly a noticeable public health problem. As early as 1966, reports of guard-dog businesses defrauding their customers by selling them unsound German shepherds and Doberman pinschers were causing alarm. When these animals proved to be more than the purchaser bargained for, thousands of hastily acquired dogs were abandoned to roam the streets and reproduce at will.

Add to this the proliferation of commercial dog-breeding operations ("puppy mills") that sprouted up to satisfy Americans' desire for purebred pets and the relative lack of spay/neuter services that are now routinely provided by large animal shelters, and you have an explosion of dogs in cities whose infrastructure could not bear it. By 1974, New York City's Department of Health and the ASPCA estimated that the city's canine population topped 700,000—only 375,000 of which were registered pets. *Time* magazine declared that dogs should be demoted to the rank of "man's worst friend" and the world's "most obnoxious minority group."

Then came the bites. In 1972, thirty-eight thousand dog bites were reported in New York City. (New York's 2011 total was around thirty-five hundred.) Los Angeles reported more than forty thousand. Officials in New York's health department called this an "unrecognized epidemic." Dog-bite-related deaths began to rise soon after. Their numbers were (and still are) very low, but with a growing human and dog population they were bound to increase, especially with the proliferation of dogs acquired for protection.

"There is substantial evidence that there has been a change in the character of the dog population in recent years, with a trend toward large, feisty, aggressive dogs," Dr. David Harris, the associate director of New York's Mount Sinai Hospital, reported in 1974. "The onset of the rise of reported dog bites in the mid-1960s coincides with heightened public concern about street crime and burglary, often associated in large cities with drug addiction."

Doberman "devil dogs" had been lauded for their bravery when serving the U.S. Marine Corps in the Pacific during World War II, but when two Dobermans killed their sixty-four-year-old owner in 1955 and another "prize-winning" Dobe killed its owner in 1960, those two unrelated incidents five years apart were enough to create an urban legend that the dogs were canine "traitors," a narrative immortalized in the 1972 James Garner film, *They Only Kill Their Masters.* The guard dog fad placed thousands more Dobermans and German shepherds—some trained, some not—into powder-keg social environments. Today, the visual image of the Doberman elicits greater physiological fear responses in patients with cynophobia (fear of dogs) than any other breed—even pit bulls.

After that, the fatalities of three children killed by Saint Bernards—one in 1972 and two in 1974—led to reports of police having to shoot rampaging Saint Bernard "attack dogs." A Massachusetts surgeon wrote that Saint Bernards were "more vicious and unpredictable than ever before." "The Saint goes along on an even keel until something triggers it," one veterinarian told his local newspaper, "then you've got a whole lot of mean dog." Stephen King most famously captured this stereotype in his 1981 novel, *Cujo,* about a rabid Saint Bernard with homicidal inclinations.

This was mainly an urban problem, but not exclusively. In the rural Southeast, unaltered strays that had been turned loose to fend for themselves not only threatened humans but also formed packs and ravaged livestock. In 1975, officials in Georgia reported that loose dogs had killed nearly eighteen thousand pigs and more than seven thousand cattle in a single year, causing more than $700,000 in losses to local farmers. In addition, game wardens estimated that more than three thousand deer had been taken down by stray dog packs. Georgians took to calling them the *real* "devil dogs." Kindled by harrowing personal close calls and stoked by the media, the public's fear of dogs grew, which fed into the pit bull hysteria yet to come.

The widespread belief that American dogs were "turning" on their owners was well under way by August 15, 1974, when readers of *The New York Times* opened their papers to even more disturbing news: the crime of dogfighting was spreading across the country, despite being illegal in all fifty states. "There is no question that [dogfighting] has gotten bigger," Duncan Wright, executive director of the American Dog Owners Association (ADOA), told *Times* reporter Wayne King. "In the past few years it has quadrupled. Ten years ago, you wouldn't have organized rings, now you do. It's getting bigger and nobody is noticing." Wright estimated that there were five thousand professional dogfighters scattered across the country, a thousand dogfights held in various locations each year, and between forty and fifty breeders of fighting dogs.

At that time, Duncan Wright was one of the most vocal animal advocates in the United States. A former physicist and Great Pyrenees breeder from California, Wright founded the ADOA in 1970 to investigate commercial dog-breeding operations, which were keeping hundreds of dogs at a time in appallingly cramped, filthy conditions, and to oppose what he considered overly restrictive dog-ownership laws. He was one of the first activists to bring the plight of puppy mill dogs into the national conversation and would later be appointed executive director of the national ASPCA. Wright believed that cruelty to animals and cruelty to other people were closely linked. "A society that encourages abuse of animals, as ours does, will never solve its human problems," he said in a 1977 interview. "Turn your head when a kid throws an animal off the roof of a building, just for fun, and you can damn well bet that when he's sixteen or seventeen he'll be mugging people in the subway."

When Wright and his colleagues shifted their focus from puppy mills onto dogfighting, they faced a significant problem: Only a tiny portion of the American public knew about this form of cruelty, and even fewer seemed to care. Neither did law enforcement. Faced with the rising rates of human violence and a growing drug trade, police officers and sheriff's deputies resisted spending precious man-hours and money to enforce local animal cruelty statutes that resulted only in misdemeanor charges or negligible fines. Frank McMahon, then director of field services and investigations for the Humane Society of the United States, complained that in Maryland "you can be fined $500 for littering and $50 for dogfighting." In places like Texas and Louisiana, law enforcement officials actively participated in underground fights or looked the other way when sufficiently paid to do so. This relationship between police and pit sports stretched back more than a century, to the days when major fights were prominently advertised in Richard K. Fox's *National Police Gazette.*

Humane advocates decided to push for federal legislation that would stiffen the penalties for animal fighting and place the responsibility for the enforcement of dogfighting laws under the jurisdiction of the U.S. Department of Agriculture. But to do that, they needed public support, and to secure it, they needed the help of the popular press. Duncan Wright partnered with HSUS, which had conducted a few high-profile raids on dogfighting operations in the Deep South, and the Massachusetts Society for the Prevention of Cruelty to Animals, one of the humane groups most known for its rigorous enforcement of laws to punish cruelty.

HSUS knew how effective help from the media could be when it came to mobilizing voters. During the early 1960s, its members had made several unsuccessful attempts to gain sufficient public backing for laws protecting animals used in scientific research. Because most Americans never saw or interacted with research animals in their daily lives, it was easy for many to think of them as abstractions, as someone else's problem.

Then, in 1965, a Dalmatian named Pepper was stolen from her family's yard in Pennsylvania and sold to a research hospital, where she was cut open to test pacemakers. The process killed her, and when the family learned what had happened, they enlisted the help of New York Representative John Resnick to ensure that other pet owners would never suffer the same loss. The following year, *Life* magazine published photographs from an HSUS raid on a dog-dealing operation in Maryland in which Frank McMahon and others stood among a hundred sick and starving dogs, most of which were obtained from municipal pounds and slated for sale to research labs, a legal process called "pound seizure." The layout highlighted several dogs that were not strays but, like Pepper, stolen family pets, creating a narrative about animal research that hit much closer to home: *This could happen to you.* In response to the spread, outraged readers sent more letters to the magazine than they had written about any other subject, including the civil rights movement and the Vietnam War. President Lyndon Johnson signed the Laboratory Animal Welfare Act into law within a few months of the story's publication.

Once members of the animal protection movement set their sights on federal dogfighting legislation, there was every reason to think that an all-out "media blitz" in major national newspapers, magazines, and television news programs would be the most effective strategy by which to make that happen. But these well-intentioned efforts backfired, in part because the resulting press coverage favored wild speculation over solid facts, which were gruesome enough on their own. The flurry of news coverage then triggered a wave of countercultural "gonzo" magazine stories that glamorized and glorified the machismo of dogfighting rather than forcing it into obscurity.

Numbers are an essential part of social activism, and the more dramatic they are, the stronger and more immediate the public's response is likely to be. Advocates of all types must contend with basic human curiosity about scale. *Okay, there is a problem, but how* big *is the problem? How likely is it to affect me or someone I care about?* A disease that afflicts only ten people isn't likely to inspire much concern, but one that affects ten million most probably will.

For this reason, Joel Best, professor of sociology and criminal justice at the University of Delaware, cautions against the uncritical acceptance of "official" statistics. Figures presented by advocates and interest groups, people that Best refers to as "problem promoters," are often exaggerated (many times unintentionally) in order to spur the public to action. Sometimes real data exist, but it is more likely that the numbers have been derived from tiny samples or biased populations or that they are simply guesses. "Numbers are created and repeated because they supply ammunition for political struggles," Best warns, "and this political purpose is often hidden behind assertions that numbers, simply because they are numbers, must be correct."

Following the abductions and murders of six-year-old Etan Patz in 1979 and six-year-old Adam Walsh in 1981, for example, the national child protection movement demanded action from federal lawmakers by presenting the public with wildly overblown numbers that the media accepted and disseminated without skepticism. "Stranger danger" was presented as a major national crisis, with some well-respected advocacy groups claiming that fifty thousand children were abducted by strangers each year. As far as grabbing the public's attention is concerned, Best writes, "big numbers are better than little numbers; official numbers are better than unofficial numbers; and big, official numbers are best of all."

When Best and his colleagues examined the actual data from the FBI and other law enforcement agencies around the country, however, they found that the true number of children abducted by strangers every year was fewer than a thousand nationwide. While those families no doubt endured terrible suffering, child protection advocates greatly distorted the true level of the threat.

It would be cynical and unfair to assume that social activists knowingly misrepresent their data. In many cases, advocates (few of whom have professional training in statistics) genuinely believe those numbers to be true, and their insistence on holding the public's attention comes from a sincere desire to help. But bad data beget more bad data, and bad data lead to bad policy, which can then trigger a domino effect of unintended consequences. Once you draw the public's attention to an issue, you cannot always control what happens next.

There's reason to be skeptical of Duncan Wright's 1974 statements to Wayne King—that there were five thousand dogfighters scattered across the country and that the number had "quadrupled" over the past ten years. The biggest names in underground dogfighting separately attested to a very small fellowship of enthusiasts. Ralph Greenwood, the onetime president of the American Dog Breeders Association who knew many of the country's most notorious "dog men," estimated that in 1957 there were roughly fifty fighters left across the country. Jack Kelly, the editor of *The Sporting Dog Journal,* said that there were two hundred when he first entered the dogfighting milieu in 1958. Pete Sparks, the so-called dean of dogfighting, knew of only six professional owners in all of Florida in 1966. The subscriber list to his magazine, *Your Friend and Mine,* never topped four hundred. Almost all of these were white men (and a handful of women) clustered in rural areas of the South, the West, and the Midwest.

Match reports from the early 1970s support these low numbers. In fifty years' worth of fighting publications, I struggled to find a single event that drew more than two hundred people, and roughly half of those were wives, girlfriends, and acquaintances of the actual participants. The majority of events drew no more than fifty people, and the attendance lists at each "convention" seemed to contain the same names across the country. According to the animal protectionists of the 1970s, however, this crime was *everywhere.*

On August 15, 1974, Duncan Wright told *The New York Times* that the main fighting hubs were Chicago, Texas, Mississippi, Florida, and Oklahoma. Ten days later, he said that the "six major dog-fighting centers" were "the New York City area, lower New England, Southern California, Florida, Illinois, and Pennsylvania." Southern Illinois, he said, was particularly worrisome because fights there were "still in the bush-league but beginning to expand." When a local reporter checked with police, however, he couldn't find any evidence to substantiate that claim. In a separate interview, Frank McMahon added Louisiana, Utah, Montana, and Indiana to the list. Soon Wright urged that all of New York state, not just the city, was rife with dogfighting. By October, he said that Connecticut was a "major" concern, but the local humane society said that it had never received any complaints.

In addition to speaking with Duncan Wright, Wayne King of the *Times* attended a dogfight in Chicago. His two primary contacts were the longtime fighters Pat Bodzianowski and Sonny Sykes, who told King that they tied up kittens in burlap sacks and allowed their dogs to attack them and that they punished losing dogs with ice picks to the chest. According to Jack Kelly, who knew both men, "Mr. King asked them so many foolish questions about the dogs that Pat and Sonny decided to have some fun with him. They regaled him with blood curdling stories about the dogs, stories that both Pat and Sonny knew were too ridiculous to be believed."

If Bodzianowski and Sykes intended to shock King with their callous bravado, they did an excellent job of it. In turn, they shocked millions of American newspaper readers, who didn't need much convincing that fighters were sadists who had to be stopped. Animal activists then repeated their worst fears as though they were facts. Building on a theme of frenzy, Wright told a reporter for the *Chicago Tribune* that a fighting pit bull "kills one hundred dogs and cats during training." Frank McMahon said that each dog was trained on "kittens or smaller dogs, and it's not uncommon for the animals to be splashed with blood to excite their instincts to attack and kill." Roger Caras, also from HSUS, repeated this terrifying detail on the *Today* show. Nowhere in the underground fighting literature, in which training regimes are painstakingly detailed, are these practices ever mentioned.

Despite its shaky foundations in reality, the story about "live bait" (which seems to have been confused with the live bait used for greyhounds at dog racetracks) morphed into one of the most notorious urban legends about pit bulls, the legend of the "bait dog." Chris Schindler, the current head of animal-fighting investigations for HSUS who oversees eighty-plus investigations a year, explained to me that this idea is mostly a myth. If "bait dogs" exist today, he said, it is only because media outlets have repeatedly published incorrect and outlandish stories, which then inspired the actions of disturbed individuals—who were not involved in professional dogfighting.

"[Fighters] want a dog to build confidence, and they want him to get his butt kicked a little bit so that he's going to have to fight through that," Schindler told me. "They don't want to let him hurt some dog that couldn't really properly fight back. That doesn't teach the dog anything." The dogs train *up*, the same way human boxers spar with experienced trainers, not the other way around. Terry Mills, director of the ASPCA's Blood Sports Field Investigations and Response Team, who attended more than eighty dogfights when he posed as a professional dogfighter for eighteen months of undercover work that resulted in the arrests of 103 people in eight states, also denied that "bait dogs" were ever used.

Schindler and his colleague Janette Reever said that if there was one term that they wished both activists and the media would stop using, it's "bait dog." It divides the world of pit bulls into aggressors and victims, which is both inaccurate and untrue. Most are neither. Assumptions made about an animal's history can also determine how it is housed and handled and whether or not it is given a chance at adoption. Certain shelters will not allow dogs with scars to be adopted, for example, based on the mistaken assumption that all scars come from fighting. Some dogs assumed to be "bait dogs" may also be suffering from other types of severe injuries that require much different treatment. (In Chicago, I encountered a suspected "bait dog" that had in fact been hit by a car. Had a medical team not recognized this and attended to his internal bleeding, he likely would have died.) The focus on "bait dogs" originally functioned as a well-intended means of generating sympathy for victims of cruelty, but it now perpetuates the stereotype that all pit bulls come from fighting backgrounds, when in reality only a tiny fraction of them do.

However sensationalized the news stories from the 1970s might have been, the tight-knit culture they described was, without a doubt, barbaric, and there is evidence that it was getting worse as the old guard of fighters left the "game" and a new generation that prioritized carnage over gameness took over. In 1900, the life of a fighting dog involved cuts, bruises, and broken bones, but in 1975 that life was increasingly filled with steroid injections, amphetamines, and electric cattle prods. Because fighters operated in tentacled, multistate networks, apprehending them was difficult.

Within a few weeks of Wayne King's first dogfighting story in the *Times,* Representative Thomas Foley of Washington and Senator Harrison Williams of New Jersey, both Democrats, added their support to an anti-dogfighting amendment to the Animal Welfare Act that would make transporting dogs across state lines for purposes of fighting a federal crime, enforceable by the U.S. Department of Agriculture, rather than at the discretion of under-resourced local law enforcement officials.

When the Senate hearings about the amendment convened at the end of September 1974, members of the panel watched graphic video footage of a Florida dogfight, which was followed by testimony from Duncan Wright and investigators from Boston and Chicago. Instead of placing the emphasis on the human criminals who tortured the animals, attention shifted to the dogs, which several investigators described as "bloodthirsty." No claim was too outrageous. One man testified that killing "bait dogs" gave pit bulls "a feeling of victory and a taste for blood" and that "a woman's wig soaked in blood" was a popular training tool. The state's attorney from Cook County, Illinois, asserted that prior to one match "a referee twisted the head off a live pigeon and poured its blood over the head of one of the dogs to show the animal's desire for blood and to entertain the spectators."

The panel was not only shocked by all this but completely disgusted. "Rarely have I seen a more blatant case of cruelty to animals than the one

I seek to remedy with this bill," said Senator Williams. The amendment was finally signed into law by President Gerald Ford in 1976. This should have been a great triumph, but the horror stories used to win it soon got out of control.

So dark and taboo was the criminal underworld of American dogfighting that from 1975 to 1982 major features on it cropped up in every large newspaper in America—not just *The New York Times,* but the *Los Angeles Times,* the *Chicago Tribune, The Christian Science Monitor,* the *Miami Herald*—plus popular general-readership magazines including *People, Esquire, Harper's, GEO, Maclean's, Texas Monthly,* and *Atlanta,* as well as the renegade publications *Hustler* and *High Times.* A few of these pieces, including a 1982 *High Times* article, quoted only what others had written rather than consulting original sources. In others, such as the three-page profile of Pete Sparks ("the dean of dogfighting") that appeared in *People,* infamy conferred celebrity. After his *People* interview, Sparks went on to appear as a guest on *The Mike Douglas Show.* But nowhere in any of these articles did a journalist challenge the assertions made by dogfighters about the capabilities of the dogs they abused.

While the newspaper items misrepresented the facts, the magazine stories tended toward drug-addled bits of Hunter S. Thompson–esque gonzo reportage, dispatches from the anarchic realm of sleazy outlaws. Tens of thousands of dollars, as well as drugs and guns, were said to change hands in a single night. Fighters never worried about getting caught, reporters said, because the laws against dogfighting were hardly ever enforced, and when they were, the punishments were laughably weak. There was an element of slumming to these stories, a peep-show approach to dogfighting that stopped just short of overtly endorsing it.

"Dogfighters . . . tend to be very moral individuals," wrote Ike Abbott in *High Times.* "They are strong-willed, believers in the American principles of self-reliance, hard work, fierce individualism and the private settlement of grievances." The average dogfighter was, he said, "a self-made man, born into poverty, who has by his own initiative and sweat made a small place for himself in his society."

Writing for *Hustler,* David Epstein concluded, "I can no longer say that dogfighting is an unconditionally cruel sport and that those who participate in it are cruel people. There is no doubt in my mind that the dogs I saw enjoyed getting into the pit and fighting . . . I've got no choice but to say that [three of the dogfighters] are now among my favorite people."

Articles in the same vein were also printed in magazines geared toward an upscale intellectual readership. *Harper's* published "An American Pastime," which trotted out the now-familiar legend about kitten slaughter. The glossy travel and culture magazine *GEO* (then one of *National Geographic's* main competitors) ran a long center-of-the-book story that featured graphic close-ups of dead and bloodied animals opposite an ad for the ultra-luxe Beverly Wilshire hotel. The writer, Benno Kroll, admitted doing speed with his subjects to gain their confidence. Very little attention was given to the suffering of the animals at the center of the crime, with writers instead focusing on the cowboy swagger of the men who fought them.

Later that year, *Esquire* published a nostalgic essay, "A Day at the Dogfights," by the novelist Harry Crews. "I am not defending fighting dogs," wrote Crews, who acknowledged that he had attended dogfights all his life. "I just wonder why we can't tell the truth about blood sports, which would go a long way to telling the truth about ourselves. We are a violent culture."

Writing for *Texas Monthly,* Gary Cartwright claimed,

> For two thousand years or longer, pit bulldogs have been bred for a single purpose—to fight . . . To attack anything with four legs. They do not defend, understand. They are worthless as watchdogs unless the intruder happens to be another dog, or a lion or an elephant. No, they attack. That's their only number. They were bred that way.

He added that the pit bull had "an undershot jaw capable of applying 740 pounds of pressure per square inch (compared to a German shepherd's 45 or 50)."

Gradually, the claims about pit bulls grew more hyperbolic, and soon the pit bull was no longer a domestic dog but, in the words of one newspaper, "an efficient, nearly inexhaustible fighting machine." Regardless of the obvious exaggerations (two thousand years? an elephant?), Cartwright's point about super-canine bite strength, in particular, would enjoy a long shelf life, even though there is nothing anatomically unique about the pit bull jaw, and its bite pressure/force has never been scientifically established.

The true bite force of any living animal is difficult to quantify. In order to arrive at an estimate, researchers must fit the animal with a special electrical transducer, then have it bite down on a plate or rod containing a sensor. The researcher conducting the experiment can never be sure the animal is biting at full capacity or whether, in the case of dogs, the subject is inhibiting its bite. (As Ian Dunbar had explained, dogs employ different levels of bite pressure depending on what they are biting and why.) Studies conducted on the biting force of domestic canids have traditionally used approximations of other variables (such as the width and length of the jaw as lever arm), or they have measured the amount of force required to shatter the tooth of a deceased dog, but these estimates can differ according to the size and development of the temporalis and masseter muscles in the dog's jaw. The unit of measurement for these estimates of force is the newton meter, not pounds per square inch (psi), with one newton being roughly equivalent to 0.000015 psi. According to the latest research, the main determinant of a dog's bite strength is body size, not breed.

Cartwright's "740 pounds of pressure per square inch" is what Joel Best would call a "mutant statistic," and it seems to have been carried over from the figure attributed to attack-trained German shepherd military working dogs (MWDs) at Lackland Air Force Base in 1969. In a paper for *Minnesota Medicine,* two authors noted that at Lackland MWDs had a "bite power" of "400 to 450 pounds per square inch by the end of training," supposedly twice what they started with, but no information was given about how that number was derived. Nevertheless, the following year this number grew to "750 pounds of pressure" in a *Palm Beach Post* article on guard dogs, and soon the "pounds of force" figure became a popular motif in many similar dog stories as a measure of terrifying strength.

The 1978 mondo film *Faces of Death,* which purported to be a pastiche of real suicides, murders, drownings, and animal attacks, included footage of a dogfight so gory and gratuitous that it was cut by British censors. In an interview for the Blu-ray DVD edition of the film, its director, Conan Le Cilaire (a pseudonym), explained that while the footage was shocking, it was also faked: "It looks really savage and cruel and mean in the movie. But these dogs were the most playful dogs in the world, we just smeared them with jelly, they were just playing around, they weren't doing anything wrong at all, in fact the footage itself is so laughably cute, we couldn't believe that anyone would buy this, but you add sinister music and some sound effects and cut it a certain way, and it looks like these dogs are killing each other."

Crime reporters often face a thorny paradox when it comes to informing the public. The more sensational and violent the crime, the more likely it is that media attention will inspire copycats. Teenagers who have been exposed to antidrug advertisements, for example, may actually be *more* likely to show an interest in drug use, rather than less. For many years sociologists have also mapped the clusters of suicides that tend to follow the deaths of major celebrities, and some researchers have theorized that mass shootings are "contagious" in the same way. Sensational coverage inspires curiosity, even experimentation. "It's excessive media attention that creates the copycat phenomenon," the criminologist Dr. Jack Levin has said. "We make celebrities out of monsters."

Thanks in part to all the media coverage it received during years of pronounced social ferment, the crime of dogfighting didn't go away; it got worse. Throughout the late 1970s and the 1980s it cropped up in places where it had not existed for a hundred years, mostly poor inner-city neighborhoods. For the unemployed and the disenfranchised, the prospect of making easy money without facing consequences was all too seductive.

While dogfighters can come from any position on the social ladder, the culture of dogfighting does not generally take hold in places where people are safe and happy and feel that they have options. Rather, it flourishes in pockets of society where the bottom has dropped out. The countries where it is most popular today are Russia, Afghanistan, Brazil, Mexico, and the Philippines, places with extremely high levels of poverty, low educational attainment, and existing problems with other forms of violent crime.

In depressed American neighborhoods, owning a dog for protection was already thought to be necessary for survival, and for many people it probably was. As a result of these two forces colliding—the culture of guard dogs and the media frenzy about dogfighting—pit bulls became the latest fad breed to be enlisted in a dangerous game of musical chairs. They would proliferate quickly because, unlike German shepherds or Dobermans, they were smaller, easier to hide, and cheaper to purchase. As one cruelty investigator told me, "You can't push around a hundred-pound Rottweiler the way you might be able to with a thirty-pound pit bull." Common folk beliefs held that pit bulls could not feel pain and remained loyal to their owners despite harsh treatment, which added even more to their popularity in high-crime neighborhoods.

Once the pit bull was portrayed as an "inner-city dog," however, it became a magnet for racial fears about crime and the American underclass. "Do killer breeds exist?" was the wrong question to be asking, and it was destined to produce the wrong answers. The right questions—about why people feared one another, why the inner city had been left to decay, and why a certain group of men (and sometimes women) found cruelty and power so appealing—would not be asked for years.

CHAPTER 8

THE SLEEP OF REASON

El sueño de la razón produce monstruos.
(The sleep of reason produces monsters.)

—FRANCISCO DE GOYA

Historically, Americans have annihilated the animals that most frightened them. We have done this with wolves, bears, mountain lions, alligators, and virtually every other large predator within our borders. Some, like the alligator and the wolf, have rebounded with the help of aggressive protection programs. Others, like the eastern cougar, were extirpated long ago. As I write this, a brief encounter between a shark and an Australian competitive surfer that did not involve a bite, let alone a serious injury or death, has prompted Brian Kilmeade, a host on television's *Fox & Friends,* to say, "You would think they would have a way of clearing the waters [of sharks] for a competition of this level, but I guess they don't."

This attitude is not limited to wild creatures. We have also adopted an attitude of extermination toward certain types of American dogs, and almost always the true root of our fears has stretched back to the people associated with them. Historically, it started small. In Massachusetts during the 1690s, for example, Puritan villagers put several dogs to death for acting as the "familiars" (diabolical agents) of suspected witches. One was shot and killed in 1692 after a young girl alleged that it had "bewitched" her. When the dog fell over dead (rather than doing something supernatural), the Puritan minister Cotton Mather declared it innocent and pardoned it posthumously—not that it did the dog any good.

Our European neighbors had been doing this sort of thing—seeking vengeance on animals for alleged human misdeeds—for several hundred years. During the Middle Ages, hundreds of pigs, beef cattle, horses, donkeys, sheep, dogs, cats, and even vermin were not only killed if they injured (or merely inconvenienced) their human caretakers; they were also put on formal trial in actual courtrooms. In 1386, in the northeastern French town of Falaise, a large sow was tried for "having torn the face and arms of a child and thus caused its death." Despite having the aid of a human lawyer and appearing before the judge dressed in a man's jacket and breeches, the sow was declared guilty by an ecclesiastical court, which sentenced her to death. In their minds, any animal that lashed out against a human did so for one of two possible reasons—intentional malice or demonic possession—and allowing either to go insufficiently punished was tantamount to courting the Devil himself. Defendants that could not be physically present in the courtroom, like flocks of birds or swarms of insects, were tried and sentenced to death in absentia. The courts felt that the cathartic spectacle of a trial provided the community with a proper sense of vindication, retribution, and punishment.

In 1738, a bulldog owned by a Scottish butcher named James Grieg reportedly attacked several other dogs at the Edinburgh Flesh Market. When a number of witnesses confirmed the events to the local magistrate, he ordered that "all dogs of the mastiff kind be put to death" and that if dogs of *any* type were seen in town after noon the next day, the town officers would be authorized to kill them—which they did—with great enthusiasm. According to the historian Mike Homan:

> The Town Crier's drum beats had barely stopped when all the citizens set about killing every dog in sight. Many were hung from shop signs. People were stopped with difficulty from killing the dogs that led the blind. Ladies were attacked and their lap dogs killed. A detachment of the City Guard was ordered down to the Butcher Market and all the dogs there were hacked to death . . . The next day the Magistrate of Leith ordered all dogs of the [entire] Town be put to death. The dogs were rounded up and in batches driven from the harbor walls into the sea to drown. Similar action was taken in various towns across Scotland.

"I cannot think that a whole species should have been destroyed on the account of one," wrote a despondent London journalist. "I am very

German man with spitz dog, ca. 1910

sorry to observe that our brethren and fellow subjects in Scotland should level their chief resentments against Bull dogs—what comparison can there be between the death of a thousand or so brutes than that of one."

Lacking sufficient scientific knowledge to explain the mechanisms by which the rabies virus was transmitted, a number of American physicians theorized in the 1870s that certain breeds of dog must be responsible. Linking the disease to the entire subspecies would have been socially unpalatable in a culture gone dog crazy, so they singled out one in particular, a favorite with German immigrants that had recently gained immense popularity as a ladies' pet: the fluffy white spitz. Much like the term "bulldog" or "bull terrier," the word "spitz" was only loosely defined back then, and it included what we know as the Pomeranian.

American doctors believed that the "disproportionate" number of rabid spitzes in New York City must be linked to the dog's shape, as opposed to the high number of spitz dogs in the city at that time. Physicians across the country latched onto this view, writing in various medical journals that not only was the spitz "useless," "snappish," and uniquely susceptible to rabies, but it was not even a domesticated animal; therefore it should be heavily regulated, if not wiped out. Here is a representative entry from an 1876 issue of *The Cincinnati Lancet & Observer:* "Another case of hydrophobia in New York from the bite of a Spitz dog. The fact that most of the cases of this horrible malady are caused by this kind of dog should certainly lead to increased precautions in the care of such dangerous creatures. The Spitz is said to be the progeny of the Arctic fox and the Esquimaux fox, and its snappish disposition and ungenial manners are reasons why it should be prevented from having an opportunity to exercise its hydrophobic propensities in any other than its native wilds."

The Medical and Surgical Reporter concurred, warning its readers that spitzes were "especially dangerous as pets." Some of these criticisms were barbed with disdain for the women's suffrage movement, which was just beginning to gain momentum. "Spitz dogs are the fashion," one doctor wrote in the *Atlanta Medical and Surgical Journal,* "and if women will have dogs instead of babies, they must expect to suffer the real or imaginary penalties that attach to the unnatural fancy."

Sure enough, more reports of biting spitzes began appearing in daily newspapers. At one point, some 200 out of 240 bites in New York City were blamed on the spitz, which caused the social temperature around it to rise at a much faster rate. "There are but four venomous beasts among the *fauna* of the United States," wrote the *New York Times* editorial board. "These are the rattlesnake, the copperhead, the moccasin, and the Spitz dog, and of the four, the latter is by far the most aggressive and deadly in its hostility to man." The newspaper went on to call the dog "misanthropic, malicious, and useless; and if we can truly judge a man by the dogs with whom he associates, the keeper of a Spitz is deficient in those qualities which secure the respect and love of high-toned and intelligent dogs." The author called upon the state legislature to order "the immediate slaughter of every Spitz dog in the State," pointing out that "the few misguided persons who are the accomplices of this noxious beast cannot have their perverted tastes gratified at the cost of the whole community."

Sanctioned by medical professionals and inflamed by the press, panic about rabid spitzes turned into an all-out war against them. New York dogcatchers rounded up spitzes and clubbed them to death, shot them, or drowned any real or suspected members of the breed. One of the only people to defend the spitz was Henry Bergh, the founder of the ASPCA, but even his considerable influence could not allay the public's terror. Fearing reprisals, owners of spitzes either abandoned their pets or preemptively killed them. Furious columns in the city's papers called the dogs "deadly," "wicked," "vicious mongrel curs," and "dangerous animals in whose blood there is a wildness and savagry [*sic*] not characteristic of the breeds of dogs which have been longer domesticated." "No trust or dependence can be placed upon this breed," one claimed, "for it is impossible to tell at what moment they will turn upon their keepers and betray their origin by a savage attack."

The most unsettling aspect of the 1870s spitz panic is that Americans seemed to take so much pleasure in it. The act of hating both the dogs and the people who owned them energized large swaths of the populace.

While dogs were being mercilessly attacked, certain newspaper columns sounded absolutely gleeful. "As a dog of fashion the barbarian little Spitz is no more," wrote the *New York Herald*. "He has had his day and now passes into history as having been the main instrument in introducing hydrophobia into good society."

Anti-spitz attitudes in New York spread throughout the Northeast like spilled ink across a map, first in Massachusetts, where a statewide ban was considered, and then in New Jersey, where the town of Long Branch banned the spitz in 1878; it was the first formal breed ban in America. Even after scientists discovered that any mammal could succumb to the rabies virus if bitten by an infected carrier—that there was nothing "poisonous" or "beastly" about the spitz—medical textbooks continued to describe spitz bites as "the worst kind" well into the 1890s.

Shortly thereafter, the press brought a new four-legged killer onto the national stage: the "Cuban bloodhound." The Cuban bloodhound was not a bloodhound in any way we would recognize today, with the wrinkled brow and pendulous ears of the Saint Hubert's bloodhound; rather, the now-extinct Cuban variety was an enormous Spanish mastiff that tipped the scales at 150 pounds. Sir Arthur Conan Doyle was probably envisioning a Cuban bloodhound when he described the "Hound of the Baskervilles" as a "bloodhound-mastiff cross" in 1902.

Originally imported to the United States from colonial governments in the Caribbean to subdue Florida's Seminole Indians during the 1840s, Cuban bloodhounds were purchased by plantation owners across the South to track, catch, and in some cases torture or kill slaves. Captain Henry Wirz, the commandant of the notorious prison camp at Andersonville, Georgia, during the Civil War, was said to have kept a Cuban bloodhound named Spot for torturing Union soldiers who tried to flee. It is unclear from the testimony of several witnesses at Wirz's trial, however, if the captain actually had bloodhounds or if the very idea of the bloodhound was so terrifying that it was simply assumed he did. A number of witnesses testified that Wirz kept run-of-the-mill "plantation curs," foxhounds, and possibly even a bull terrier as well.

The canine giant's imposing looks, frightening reputation, and long association with human depravity made him a perfect villain for the popular traveling stage plays of Harriet Beecher Stowe's abolitionist novel *Uncle Tom's Cabin,* known as Tom shows. In a trend that foreshadowed the "creature feature" films of the 1950s, the Tom show troupes with the scariest "bloodhounds" drew by far the largest crowds. Interest-

Playbill for a live production of *Uncle Tom's Cabin,* 1881

ingly, a number of famous stage dogs from the Tom shows weren't bloodhounds at all, but "Ulmer hounds," or Great Danes. No one seemed to know the difference. Wanting nothing to do with bloodhounds of any type, the state of Massachusetts banned the Cuban bloodhound, the Saint Hubert's bloodhound, and any dogs that even *resembled* the Cuban bloodhound—including the Great Dane—in 1886.

Reports of Cuban bloodhound depredations in the newspapers were harrowing (150 pounds is a lot of dog!), but there is no indication that the public was ever truly endangered by Cuban bloodhounds, because there weren't that many of them left after the Civil War. The real fear and contempt Americans felt wasn't toward the dogs but toward their owners, who had used animals to perpetrate horrific acts against defenseless people. Among abolitionists, the word "bloodhound" turned into a symbolic shorthand that dredged up everything they hated about slavery and carried with it all the psychological trauma of a war most people desperately wanted to leave behind.

Subsequent military conflicts stoked hostility toward any dogs associated with "the enemy." While Wallace Robinson and the postcard artist Bernhardt Wall wrapped pit bulls in the American flag during World War I, English and American artists drew dachshunds, which

Anti-German postcard from World War I featuring the "treacherous" dachshund

by then had replaced spitzes as stereotypical German dogs, in spiked *Pickelhaube* helmets, often marching dutifully behind the kaiser. One of Wall's postcards depicted Uncle Sam's large disembodied hand strangling a dachshund above the caption "I like dogs, but not this breed," and another showed a dachshund being squashed to death under a giant American thumb.

The public got the message. In 1920, a resident of Navarre, Ohio, reported that the town's mayor had shot and killed his dachshund "for being German." The dogs were "completely driven off the streets" in Cincinnati. Londoners feared walking their dachshunds in public, lest the animals be stoned to death. Reports of children beating, kicking, and "siccing" other dogs on dachshunds throughout England and the United States were common, and AKC registrations of dachshunds dropped to the low double digits, even as breeders scrambled to rename them "liberty hounds" and "liberty pups." Seventy years later, shelter directors in New York and San Francisco would attempt a similar re-branding of pit bulls, naming them "New Yorkies" and "Saint Francis" terriers. Neither effort succeeded.

American soldiers themselves showed no ill will toward German dogs after World War I and brought a great many home with them. They were particularly excited about the possibilities of a new breed, the German police dog, later named the German shepherd. Developed by a Prussian cavalry officer named Max von Stephanitz in 1899, the German "shepherd" was intended not to herd but to guard property and in some cases attack humans. Wealthy families were the dogs' main boosters, especially during and after Prohibition. "If a man or his wife has not got a police dog," wrote the humorist Ring Lardner, "they certainly are not worth the wile [*sic*] I would spend on them." In 1914, a New York newspaper proudly reported that "several of the young women of the social group that likes to call itself 'exclusive' have purchased [police dogs] to accompany them on their trips through our masher-infested streets."

Police dog training to apprehend bootleggers, 1923

By 1920, however, the shine of these trendy guardians had worn off as thousands more German shepherds were imported into the United States. More than two thousand were registered in the borough of Queens alone. Complaints about the dogs running loose and biting children quickly soured the public's goodwill. The dean of the Veterinary College at New York University expressed his displeasure that police dogs had surpassed collies in popularity, saying that the German shepherd "was not likely to be loyal and had an erratic disposition."

As the months passed, reports of bite injuries caused by German shepherds grew more severe, using words such as "mutilate" rather than "nip." On Long Island, one shepherd almost severed a boy's leg. Animal control officers began shooting the dogs in the streets. "Being the fashionable dog of the moment," wrote *The New York Times* after the then secretary of commerce Herbert Hoover's shepherd bit a small child in 1924, "the police dog has become the terror of many communities, ranging far, like the wolves he so much resembles, terrorizing all the gentler dogs he is so fond of shaking out, and making of himself an almost intolerable nuisance for everybody except his own devoted people." It added that the dog's "strength and agility make his attacks always dangerous."

"The police dog situation is one that the city will soon be unable to cope with," pleaded James Conway, the city's magistrate, in letters to the Queens Department of Health. "In the city at the present time there are thousands of these savage dogs. They are bred from wolves. Hundreds of persons have been victims of dog bites during the past year, and the majority of the biting dogs were police dogs. The dogs should be barred from the city." The *Times* supported Conway, insisting, "So strong is the animal's guarding instinct developed that he becomes a real menace to both adults and children." Fearing that the dogs would mate with dingoes and produce vicious, sheep-killing mutants, Australia banned the import of German shepherds in 1928 and didn't officially lift that ban for more than forty years.

Dogs identified as "police dogs" were then implicated in several U.S. child fatalities. In rural areas, they also decimated flocks of sheep, and in cities they attacked police attempting to serve warrants. In an eerie parallel to the media narrative about pit bulls, there were numerous accounts of officers having to shoot and kill German shepherds owned by bootleggers who used the dogs to guard their caches of illegal liquor, and one tied the dogs to Al Capone's Chicago mob of gangsters. In New Jersey in 1930, a "large, muscular" shepherd was said to have "withstood a tear-gas attack that drove gasping policemen into the street and it maintained its fighting pose while six police bullets tore into its head, three into its body, and one into its leg. It was finally killed when an eleventh bullet hit it squarely between the eyes." Hyperbole reigned, while AKC registrations of German shepherds plummeted from 21,596 to 788.

Then, almost as quickly as it had started, the police dog controversy evaporated from the national culture. Police departments that used the German shepherd responsibly to protect citizens from harm boosted the breed's standing more than the negative anecdotes damaged it. Hollywood made valiant canine superstars out of two German shepherd actors, Strongheart and Rin Tin Tin, which attached a sentimental glow to the animals that overpowered the fear simultaneously swirling around them. When a woman was attacked by a police dog in Galveston, Texas, a reporter defended the animal by noting that "the highly developed watch-dog instincts of this same dog recently saved its owner from serious injury or perhaps death, when he was attacked by a highwayman." In New York, another journalist wondered if perhaps the German shepherd that killed a seventeen-month-old boy "had been crazed by the heat."

But the factor that most insulated the German shepherd was that it was much beloved by the wealthy, to the point of being called a canine "aristocrat." As long as the shepherd was owned by elites, it would never be banned or threatened by punitive legislation. The wealthy owned the papers, and the wealthy made the rules.

After the spitz, Cuban bloodhound, dachshund, and German shepherd frenzies, a new breed took on the role of most despised dog with every changing decade. Like trends in dog breed popularity, trends in dog breed villainy crested, then crashed once another breed was targeted. For a time during the 1960s, Dobermans were thought to have brains too large for their skulls, which could "explode" without warning. The public health fallout from the guard dog boom dragged up old fears about the shepherd. Then, in the early 1970s, it was the Saint Bernard, which had grown "more vicious and unpredictable than ever before," as mentioned earlier. The Rottweiler, which had the misfortune of being cast as the devil's house pet in the 1976 film *The Omen,* was subject to its own flurry of panicked news articles in the United States and the U.K. during the 1990s.

All of these panics, with the exception of the Rottweiler's, eventually withered away as the public moved on because they occurred before the technology existed to put frightening images and gory details on a constant loop in the public's mind. When the German shepherd scare was raging in the mid-1920s, television had not yet been invented, and many Americans didn't even have telephones. The majority of people learned what they knew about the world from what their friends and neighbors told them, from personal written correspondence, and from newspapers. Gossip, rumor, and misinformation spread through the culture at a much slower pace.

The rise of twenty-four-hour news and the Internet changed everything, and the cornucopia of media outlets available to us today makes it much harder to discern which "facts" are true. The culture critic Malcolm Muggeridge once called this constant background noise "Newzak." One downside to its ubiquity is that people with fringe beliefs now have access to much larger information platforms and anyone can declare himself an authority. If you don't want to accept climate change, numerous professional-looking Web sites, blogs, podcasts, social media groups, and other dubious organizations are more than happy to filter out any evidence that contradicts you.

The pace at which information comes at us is also much faster than ever before. Numerous technologists have pointed out that a Maasai warrior with a Google-enabled smart phone now has access to more information in rural Kenya than the U.S. president had access to in the Oval Office fifteen years ago. When the public relied only on daily newspapers and evening news broadcasts rather than the Internet, journalists "had time to digest before disseminating," Michael Gartner, the former president of NBC News, has said. "Now it's regurgitation—fact, rumor, innuendo, often with no context and no distinction between fact and rumor or important fact and irrelevant fact. You have to be a lot smarter to be a news consumer these days than you did a generation ago."

Despite being able to find out more about the world than at any other point in human history, the American media have unfortunately let down their guard when it comes to scientific skepticism. Scientific consensus is now treated as a matter of partisan political opinion, rather than one of experimentation and evidence. The astrophysicist Neil deGrasse Tyson says, "The good thing about science is that it's true whether or not you believe in it," but if you don't believe in it, you have myriad resources at your disposal to make trouble for those who do.

This dramatic increase in the speed of information and the precipitous decline in critical thinking have been disastrous. Today, 33 percent of Americans reject the theory of evolution, almost a third of Americans don't believe there is solid evidence of climate change, more than a third believe extraterrestrials have visited earth, and 23 percent have not read a book in the last year. One 2015 poll revealed that four in ten Americans believe that our military located weapons of mass destruction in Iraq after invading the country in 2003, despite an untold number of articles and news broadcasts that reported unequivocally that we did not.

The moral psychologist Jonathan Haidt, who has conducted extensive research into political belief systems, maintains that when thinking about a controversial subject, most of us decide what we believe based on our emotions and intuitions, not on the facts. Once we have made an intuitive judgment, we search for the facts that will support our position, then surround ourselves with people who agree. Haidt likens our emotional brain to an elephant, and our rational brain to the elephant's rider. When there is a moral or political component at stake, we are even more deeply invested in being right, because our morals form the core of who we are. That is why the scientific method, in which hypotheses are tested and experiments reproduced to ensure the accuracy of their results, is so important—and why politicizing science can have such devastating cultural consequences. Says Haidt, "Morality binds and blinds. It binds us into ideological teams that fight each other as though the fate of the world depended on our side winning each battle. It blinds us to the fact that each team is composed of good people who have something important to say."

Once misinformation takes hold, actual facts can do very little to dislodge a false belief. In fact, a "backfire effect" can occur in which attempts to *correct* misinformation end up strengthening belief in it. As the old saying goes, you cannot reason someone out of what he was not reasoned into. Like struggling in quicksand, the more you try, the worse it gets.

This is the social and psychological vortex that pit bulls were sucked into. The definition of "pit bull" continually expanded to include more dogs, and the nonsensical claims in the media were never borne out by science, but no amount of evidence could dislodge the emotionally gripping idea that underneath there was just something unnatural about these dogs, that somehow the otherwise virtuous American dog population had been tainted by impure pit bull blood. The dogs were then linked to a much-feared social group, which would sweep them into a human drama of proxy and projection. The eradication of these animals then became, for some, a moral imperative.

CHAPTER 9

THE PHANTOM MENACE

To him who is in fear everything rustles.

—SOPHOCLES

In 1972, a South African sociologist and professor at the London School of Economics named Stanley Cohen first laid out his theory of what he called "moral panic." Moral panic, Cohen said, was a sudden swell in "fundamentally inappropriate" hysteria about a novel, obscure, or previously ignored phenomenon that causes the members of a society to fear not just for their personal safety but also for their entire way of life. Cohen believed that moral panics arise during times of increased social tension, when they serve as psychological distractions from much more frustrating and intractable issues, like poverty, unemployment, or racial unrest.

In some cases, the panic begins as legitimate concern about a real problem (say, violent crime or terrorism), then transforms into a steamroller of groupthink and scapegoating. In other cases—most famously the witch hunts that swept across medieval Europe—the actual "threat" to society's moral fabric is either minuscule or nonexistent. Both types of panics allow the "good" members of a society to unite around the common goal of defeating "those people." Anger and anxiety are highly contagious feelings, and studies have shown that they spread more quickly in social networks than any other emotions. Having one's righteous indignation validated by your peers just *feels* good.

The "stranger danger" of the early 1980s is a good example of the first type of moral panic. The murders of Etan Patz and Adam Walsh were legitimately alarming, but the fears they ignited were grossly disproportionate to the actual risk of stranger abduction. The same goes for the Red Scare of the 1950s: Americans' concern about Communist spies was understandable, given the tense political climate of the Cold War, but in no way did those concerns justify Senator Joseph McCarthy's demagogic crusade to hunt down alleged Communist "traitors" within the United States government.

Another panic of the 1980s, the widespread claims of satanic ritual child abuse (SRA) in schools and day-care centers, was a moral panic of the second type. Despite waves of international press coverage, numerous criminal trials, and dozens of public symposia devoted to combating the "scourge" of SRA, not one case of a child being abused as part of a satanic cult ritual was ever confirmed. Yet if you opened a newspaper or watched the evening news in 1985, the chances were fairly high that you would encounter any number of "experts" who proclaimed that SRA was a rapidly growing phenomenon that must be battled by any means necessary.

Moral panics cohere around the wrongdoing of what Cohen called "folk devils," or members of a social group who can be easily portrayed first as different, then as deviant. Any group can come under scrutiny, as long as its members can be framed as cultural "outsiders" whose actions pollute and threaten the lives of good, decent people. Religious, ethnic, and political minorities are almost always the easiest targets.

Like other forms of motivated reasoning, moral panic is impervious to fact. Violent crime in the United States has declined every year since 1993, for example, yet polls indicate that Americans fear crime now more than they ever have. Dozens of case-controlled medical studies have failed to show any causal link between vaccines and autism, yet dedicated anti-vaccine activists continue to insist that there is one, muddying the waters with pseudoscience until the public doesn't know what or whom it should believe.

The pit bull panic fit into the first category: it was a dramatic overreaction to the legitimate problem of serious dog bites in American cities that was never placed into proper perspective. Its first rumblings began in southern Florida in 1980, where residents' most pressing concern was crime, most of which was linked to the booming Latin American and Caribbean drug trade. Instances of violent crime in Miami that year increased 21 percent over the year before, and the murder rate jumped a terrifying 70 percent. Latino immigrants seeking asylum in the United States crowded into southern Florida, stirring racial resentment and intergroup strife as more Floridians competed for jobs and housing. By 1987, Miami would have the highest murder rate of any city in the world. In the words of the Miami Beach commissioner Alex Daoud, "An absolute war is being fought in our streets at night."

The same cycle playing out in New York, Philadelphia, and Chicago played out in southern Florida: crime, violence, and racial conflict led to a general sense of paranoia that boosted the sales of protection dogs—mainly German shepherds and Dobermans—that were now being selected, trained, and often abused by fly-by-night breeders in order to elicit from the animals maximally erratic and threatening behavior. Thanks to the recent surge in media hype about dogfighting that framed pit bulls as uniquely savage, the popularity of American pit bull terriers increased among those who wanted to protect themselves, their families, and their property. A higher tolerance of, even preference for, dangerous canine behavior led to more bites and more serious human injuries. As the citizens of southern Florida soon realized, however, the casualties of this trend weren't always the violent criminals they hoped to keep at bay. More often they were family members, neighbors, and—even worse—children.

In February 1980, a six-year-old boy named David Pride was severely bitten by a dog that attacked him as he stepped off his school bus in Sarasota. The dog, identified alternately as a German shepherd and as a "German shepherd–husky mix," latched onto Pride and began dragging him down the street before a passerby intervened. The dog punctured Pride's skull and almost completely severed his left ear, requiring him to undergo multiple hours of surgery. After an initial flurry of short articles in the local paper, Pride's story faded with the usual news cycle. The story that did *not* fade, however, was a similar incident involving a six-year-old boy that occurred a few hours away in Hollywood, Florida. That event would be covered by papers across the state throughout 1980 and for years afterward.

Frankie Scarbrough and his neighbor, Shane Milone, had been playing with dump trucks in the Milones' backyard on December 3, 1979, when the Milones' dog, reported to be an "85-pound pit bull," rushed over and bit Scarbrough in the face. Like the dog that mauled David Pride, the Milones' dog had to be beaten off, this time by Shane Milone's seventy-eight-year-old grandmother. The location of Scarbrough's injuries (the face, rather than the ear and arm) required more extensive and complicated surgery, which local townspeople banded together to help fund. In the end, they donated more than $100,000 to Frankie's family. The Miami Dolphins football team even held a special fund-raiser for his treatment.

Local humane officials used the attack on David Pride to demand stricter animal control laws, but the attack on Frankie Scarbrough became a statewide referendum on pit bull dogs, animals that were incorrectly assumed to be owned only by dogfighters. Several months later, Scarbrough was readmitted to the local hospital with a serious infection. His mother and stepfather had failed to care for his wounds or bring him in for routine checkups, despite being repeatedly urged to do so by his doctor, who publicly charged the boy's parents with neglect. His stepfather was subsequently arrested on domestic violence charges. Soon after, Scarbrough's parents abandoned him, and by 1989 he had lived in five separate foster homes.

Could there have been a longer history of abuse or neglect that contributed to what happened in the Milones' backyard? Why, for instance, were two small children left alone with an eighty-five-pound dog in the first place? The media chose not to focus on that issue.

Two weeks later in Miami, an elderly woman named Ethel Tiggs was seriously injured by her two dogs, which were also described as pit bulls. Two incidents in separate Florida cities then became a "trend." *The Miami News* ran a series of pro- and anti-pit-bull articles under a headline that sounded as though an invading army had landed: "The Attack of the Pit Bulls." An accompanying sidebar quoted Frantz Dantzler, the cruelty investigator from HSUS. Though there was no evidence that the dogs involved in these incidents were linked to fighting operations, Dantzler said, "[Pit bulls] are borderline dogs. They're right on the edge all the time."

Bad publicity only made pit bulls more popular among the people who wanted dangerous dogs. After the *Miami News* series ran, local pet shops reported phones "ringing off the hook" with pit-bull requests. One retailer reported that inquiries about the dogs doubled. "Some people don't even know what kind of dog it is," the salesclerk said. "They just say they want the kind of dog that bit the old lady . . . [Customers] think [pit bulls]'ll make good watchdogs because they're mean," he said. "But I have to explain to them that they're pretty good-natured."

Not only did the public not understand the dogs' history or temperaments, most did not even know what they looked like. Florida's newspapers labeled a number of shorthaired mutts as "pit bulls" or "pit bull terriers," including one that closely resembled a yellow Labrador. The town of Hollywood then passed a formal ordinance requiring owners of suspected "pit bulls" to carry $25,000 in liability insurance.

Gradually, these stories took on a single thematic shape: they focused on the sordid history of American dogfighting and the dark "instincts" that all pit bull dogs, no matter how far removed from that history, must possess as a result. Never mind that the American Staffordshire terrier and the Staffordshire bull terrier had been bred for dog shows since the mid-1930s, or that several other breeds that have never been subject to media scrutiny, such as the boxer, Akita, and shar-pei (as well as the modern English bull terrier and the Boston terrier), were once used for fighting. Even more telling was that reporters labeled pit bulls "killers" well before reports of pit bull–related human fatalities began appearing in the news.

Gone were the days of Luke, Pal, Pete, and Stubby. Gone were the pets of Helen Keller and Gary Cooper. Shelter workers, animal control officers, and animal rights activists all maintained that pit bulls were stronger and more aggressive than other dogs, relentless, and—most frightening of all—completely unpredictable. Even loving family pet pit bulls, they said, could "turn" at any moment. More often than not the dogs' most vocal critics had no scientific background in veterinary medicine or animal behavior. An advocate from one humane organization implied that all pit bull owners were criminals, telling a reporter that "the [people] who don't fight their dogs own them for the same reasons as the people who do." The media's caricature of the pit bull made it a perfect folk devil: it was publicly associated with violent criminals, it was said to be unpredictable and uncontrollable, it was "everywhere," and, even better, it was hard to define.

Around the same time, California cities were dealing with social turmoil similar to Florida's. The recent economic downturn, the dramatic rise in unemployment, and the near-total lack of financial investment in the inner city exacerbated the size and scope of youth gangs, the membership of which increased sharply in both New York and Los Angeles during the 1980s and the early 1990s. The underground drug economy that strangled southern Florida also plagued South Central Los Angeles,

Watts, and Long Beach, turning parts of L.A. into war zones between street gangs and police.

The way Americans absorbed their news was also changing. In June 1980, the media mogul Ted Turner debuted the Cable News Network (CNN), the first network to provide twenty-four-hour news coverage, rather than the hour or two of local news to which most people had grown accustomed. Now all news was local, in a sense. Conflict between social groups and trend stories about crime were critical pieces of CNN's narrative framework, because the producers of television news understood that fear was very good for business. Not only were there human killers hiding in the alleys, the anchors said, but they were peddling "killer drugs" and carrying "killer diseases," such as the newly discovered AIDS virus.

And, of course, they owned "killer dogs," which is how the *Los Angeles Times* began referring to pit bulls. When the paper ran a large feature to promote an episode of the television cop drama *Lou Grant* about illegal dogfighting, the reporter returned to the now-familiar "fact" about pit bull bite pressure/bite strength/bite force. Her unsourced figures were substantially higher than those in Gary Cartwright's *Texas Monthly* article a few years earlier. "A German shepherd or Doberman . . . can exert 800 to 900 pounds of pressure in his bite," she asserted, "an American pit bull terrier, 2,000 to 2,600. Once it bites, it does not let go."

Sergeant Dan Burt of the L.A. County Sheriff's Department added, "Pit bulls are something like sharks. Once they get into a frenzy, there is no stopping them. They are bred that way." There was no mention of the fact that the Los Angeles Police Department had been successfully using a pit bull named Frog for narcotics detection since 1978. Frog served the LAPD until he died of cancer in 1986, yet his work received only a footnote in the local papers.

Psychological studies conducted during the 1960s found that no word elicits a stronger fear response in humans than the word "shark"—not even "snake," "spider," "rape," or "murder." "Why do sharks, which human beings come into contact with so rarely, frighten us so profoundly?" the government security expert Gavin de Becker asks in his book *The Gift of Fear.* "The seeming randomness of their strike is part of it. So is the lack of warning, the fact that such a large creature can approach silently and separate body from soul so dispassionately. To the shark, we are without identity, we are no more than meat, and to human beings, the loss of identity is a death all by itself." Becker points out that our fear of being "prey" distracts us from a much more unsettling truth: "Man is a predator of much more spectacular ability."

Two days after the *Los Angeles Times* ran another article on "killer pit bulls" in early 1982, Wayne King (the same reporter who had written the first major exposé on dogfighting in 1974) followed with a trend story about California pit bull "attacks" in *The New York Times,* again drawing no distinction between pit bulls and fighting dogs and again claiming that pit bulls were "blooded" by eating kittens. The story drew from five incidents that occurred in California over a period of several months, none of which resulted in the victim's death, and was followed by a major ABC News television broadcast the same night.

King interviewed two young black men for his article—twenty-one-year-old Kevin Thompson and his younger brother Carl—both of whom bred pit bulls in San Francisco for extra money. Kevin told King that he loved pit bulls because they made good companions. "They're great to be with," he said. "I like taking walks with them and having people stop and admire your dog . . . I raise them for pets and sell them." Carl, on the other hand, took the opposite position. "That's all they're good for, fighting, and that's all they know how to do." King gave Carl, not Kevin, the last word in the article, and the *Times* gave Carl a large photograph in the layout. Carl, the African-American breeder of fighting dogs, was now the face of pit bull owners everywhere.

It was at that point that the animal protection movement, as a whole, started to back away. There were continued raids on fighting establishments, but stray pit bulls that ended up in shelters were almost always euthanized without any chance at adoption, regardless of temperament. There is no reason to believe that animal protectionists harbored any conscious ill will toward pit bull dogs. Years before, the Humane Society of the United States had selected a "Yankee terrier" to be an honored guest at one of its major fund-raising luncheons. It's more likely that animal welfare activists found dogfighters so repugnant that they distanced themselves from everything even remotely associated with them, including pit bulls.

"Our industry contributed to the pit bull hysteria by not thinking about the consequences of our rhetoric," Richard Avanzino, the former president of the San Francisco SPCA, told me. "Our investigators were in a sense brutalized themselves by having to see the cruelty and the abuse of dogfighting, and those images became generalized to all pit bulls everywhere. In the press, we talked mainly about the atrocities and the horror. Positive statements about pit bulls were throwaway comments, maybe 3 percent of the conversation. If we made any big mistake, it was giving the public the impression that all pit bulls were fighting dogs."

The sad irony for today's animal lovers is that, almost without exception, the most inflammatory statements made about the dogs during the early days of the panic came from members of the animal welfare movement. While the media bears responsibility for continually broadcasting pit bull horror stories, there would have been nothing to broadcast had it not been for the assertions of animal activists that pit bulls were not "real" dogs. It would take almost twenty years for a new host of forward-thinking animal welfare leaders, who armed themselves with new science and new ideas, to put the old urban legends to rest. The late 1970s and the early 1980s, however, were a hotbed of fearmongering and reactionary storytelling. It was the age of the pit bull profiteer.

From its establishment in the mid-nineteenth century until today, animal welfare has been an almost exclusively white middle- and upper-class social movement. As of 2005, African-Americans made up only 4 percent of animal welfare employees, according to one survey. Sixty-two percent of animal welfare organizations had no African-American employees at all. That had changed little by 2013, when the U.S. Bureau of Labor Statistics reported that 96.5 percent of veterinarians were white. Veterinary medicine continues to be less racially diverse than any other professional field.

Cultural stereotypes become much harder to dissolve when the landscape is so homogenized. In a 2003 paper on the lack of racial and ethnic diversity in veterinary medicine, Dr. Ronnie Elmore of Kansas State University (who is white) wrote that "blacks generally tend to see animals as valuable only to the extent that they can be used or have a purpose, while European-Americans generally tend to see animals as objects of sentiment." He went on to describe research that seemed to indicate "the health benefits realized by whites owning pets might not extend to blacks." Several African-American veterinarians understandably found such statements both untrue and deeply offensive. Drs. Kechia Davis and Rhonda Pinckney from Madison, Wisconsin, responded, "Reading this article, there is little question in our minds why people of color are underrepresented in veterinary medicine, and we seriously doubt it has anything to do with the assertion that we don't own or love enough pets."

The chasm between the communities where pit bulls became popular during the 1980s and the animal welfare establishment could not have been greater. It's unlikely that the animal advocates who freely spoke about the lives of pit bulls in "those" neighborhoods during that era had spent any meaningful time there. They would come to learn that you can't push an animal into the margins of society and then blame those on the margins for identifying with it.

As time passed and pit bulls were increasingly mentioned in stories about drugs, the more dramatic the speculation about them became—more outsized, even, than the boasts of dogfighters had been. In 1987, Dale Dunning, then the executive director of the Arizona Humane Society, maintained that the pit bull "is aggressive to anything that moves":

> In my opinion, the pit bull is more dangerous than a loaded .357 magnum. People who buy a pit bull are getting a bomb without a detonator, because you cannot predict what will trigger an "explosion." Most dogs give adequate warning of an impending attack and then bite and retreat. A pit bull, on the other hand, typically gives no such warning of an attack . . . Other breeds fight to protect territory, food, or mates. A pit bull fights because it has been bred to fight. In short, it is a four-legged fighting machine and a killer.

Dunning co-authored an "educational" pamphlet that pushed these statements into the realm of utter anatomical impossibility, alleging, "[Pit bulls] chew with their back teeth while holding with their front teeth." The pamphlet was distributed to shelters and animal control agencies nationwide. For a time, it was widely believed that the dogs had locking jaws, double jaws, or jaws that could unhinge like those of a snake.

In 1981, the Medical College of Ohio's division of laboratory medicine agreed to house thirty-two APBTs seized from an illegal fighting operation. The division's director, a veterinarian named Donald Clifford, then extrapolated the behavior of those thirty-two dogs to all pit bulls. Clifford began traveling the country giving presentations about the dangers of pit bulls to veterinary groups, humane societies, animal control agencies, and shelter employees. Newspapers quoted him without skepticism. If he made any effort to handle, examine, or even meet average family pit bulls, he never said so. Diane Jessup took him to task for this in multiple letters, offering up her dogs' long list of accomplishments as counter-examples. Clifford wouldn't hear it. "You are not going to change my opinion that most pit bulls are derived by fighting stock and they are dangerous and unpredictable," he wrote to her in 1991. He, too, voiced his concern that "pit bulls used by drug traffickers are a threat to the police."

One tale built on the next. Soon *U.S. News and World Report* declared the pit bull "the most dangerous dog in America," one that could "chomp through chain-link fences." *The Guardian* called pit bulls "the dogs of war who can bite through concrete." *Newsweek* dubbed the pit bull "the macho dog to have." UKC registrations of the APBT climbed to roughly fifty thousand a year by 1986.

Persecution and threats of violence were not far behind. "This is *a dog in name only*," one man wrote to the *Los Angeles Times*. "Actually, it is a killing machine . . . [H]e is a creature that does not think or behave like other dogs. He is an animal of instinct and reflex responses." A columnist for the *Seattle Post-Intelligencer* took that sentiment even further, writing, "Every pit bull in the country should be bashed with a baseball bat, skinned, and the skins stitched into suits for the poor."

The increased flow of coverage created a phenomenon that the legal scholars Cass Sunstein and Timur Kuran call an "availability cascade," or information cascade. The information cascade, as they defined it, is "a self-reinforcing process of collective belief formation by which an expressed perception triggers a reaction that gives the perception increasing plausibility through its rising availability in public discourse." Simply put, the more we hear about an idea, the more we believe that it's true, whether or not the belief is supported by credible evidence. Pit bulls would set off one of the biggest information cascades of the decade.

CHAPTER 10

KNOWN UNKNOWNS

The monster never dies.

—STEPHEN KING, *Cujo*

In 1986, the main voice missing from the national shouting match over pit bulls was that of science. In June of that year, Dr. Franklin Loew, dean of Tufts University's School of Veterinary Medicine, set out to change that. He invited a number of veterinary professionals, animal behaviorists, animal control officers, and humane advocates to a workshop in Boston titled "Dog Aggression and the Pit Bull Terrier." The event's organizers expected that only thirty or forty people might attend a discussion on such a narrow topic, but they had no idea how contentious the subject of pit bulls had already become. According to Andrew Rowan, who compiled the event's proceedings, the university received "a barrage of phone calls and letters from judges, attorneys, victims, and owners. The media attention increased and it was clear that we would have to hold a separate press conference since we could not have accommodated all those who had told us they were coming plus another twenty to thirty media in our small conference room."

The panel came to several conclusions, one of them being that the press's coverage of pit bulls was misleading the public. "The media maintain the state of hysteria," Rowan wrote, "by reporting any pit bull attack but ignoring incidents involving other breeds. Under these circumstances it becomes increasingly difficult to present a calm and reasoned argument." Another key finding was that "the available data did not support the claim that pit bull terrier–type dogs were overrepresented among biting animals."

In fact, most animal control records revealed that German shepherds still topped the bite lists, which was to be expected; they were (and are) one of the most popular breeds in the country. Some members of the workshop believed that the APBT's origins as a fighting dog inclined its bites to be more severe, but this was pure speculation; none cited any measurements or experiments to confirm this hypothesis. The American Staffordshire terrier and the Staffordshire bull terrier were not discussed.

Rowan also pointed out that the academy had directly contributed to the growing panic in at least one instance by not getting its facts straight before attempting to calculate a "breed-specific risk" for dog-bite-related fatalities (DBRFs). In 1982, Drs. Lee Pinckney and Leslie Kennedy, radiologists at the University of Texas Southwestern Medical School and Children's Medical Center in Dallas, used newspaper clips and medical records to compile a list of seventy-four DBRFs that occurred in the United States between March 1966 and June 1980. When they added up the total DBRFs by breed, German shepherds led the pack with sixteen DBRFs, followed by huskies (nine), Saint Bernards (eight), and "pit bulls" (six). By their own admission, the authors "made no attempt to verify the news stories" and acknowledged that their method of data collection was "incomplete and may not be entirely reliable." Nevertheless, they concluded that "in relation to its small registration the bull terrier (pit bull) was responsible for the highest number of deaths." The study was subsequently published in the peer-reviewed medical journal *Pediatrics*.

Pinckney and Kennedy had no professional knowledge of dog breeds or canine science, nor did anyone on the journal's peer-review board. Their expertise was in human medicine. In order to determine relative risk by dog breed (the number of deaths in proportion to the number of dogs), they used AKC registration data for a single year, 1976, without understanding that there were three distinct pit bull breeds at that time, not one, and the most popular of these breeds, the APBT, was registered not by the AKC but by the UKC and ADBA. Unaware of these issues, the authors used AKC data for the (English) bull terrier, only 929 of which were registered in 1976, instead. By the 1970s, the bull terrier was no longer considered a pit bull breed, and in 1976 the UKC registered 23,500 APBTs, not 929. This number was significantly higher than the number of Siberian huskies, Saint Bernards, Great Danes, and Alaskan malamutes registered by the AKC and nearly as high as the number of new AKC golden retriever registrations. Had the authors realized their mistake, they undoubtedly would have come to a different conclusion.

Pinckney and Kennedy's erroneous paper was the first to indict pit bulls as being "disproportionately" responsible for human deaths. It would be cited in eighty-two other scientific journal articles and legal opinions.

The main problem with making declarations about breed and risk, Andrew Rowan noted after the Tufts conference, is the same problem that stumps epidemiologists today: in order to know whether or not something is disproportionate, you have to know its correct proportion. Even if solid information about breed demographics *were* available, he said, "reported dog bites are not a reliable and representative index of actual dog bites" because "bites by strays were more likely to be reported than bites by owned animals," as other researchers had noted.

One of the participants in the 1986 Tufts workshop was Dr. Randall Lockwood, an animal behaviorist from the Humane Society of the United States who directed its higher-education programs. Lockwood, who is now a senior vice president of anticruelty initiatives at the ASPCA, specialized in the study of animal cruelty and cruelty prevention. He presented his own paper on DBRFs, co-authored with his assistant Kate Rindy, titled "Are 'Pit Bulls' Different? An Analysis of the Pit Bull Terrier Controversy." The paper appeared to answer that question in the affirmative. After the conference, Lockwood became the media's go-to expert on pit bulls, dog aggression, and dog bite deaths throughout the 1980s.

I first met Lockwood in Leesburg, Virginia, when I attended a seminar on dogfighting investigations that he and his ASPCA colleague Terry Mills conducted for members of local law enforcement. With neatly parted gray hair, glasses, a close-trimmed beard, and a subdued manner of speaking, Lockwood came across as fastidious and scholarly. "The insanity over pit bulls is why I got out of the dog bite business," he told me. "All the media wanted to do was publish data tables out of context. None of the real facts made any difference to anybody. It just became too crazy."

Some months later, when I visited Lockwood at his home in Falls Church, Virginia, a black-masked pug-beagle cross, or "puggle," named Tonka greeted me at the door. I reached down absentmindedly to pet him, and he answered me with a percussive snap. "He's very protective," Lockwood said apologetically, scooping the dog into his lap as Tonka scowled at me. Every time I shifted in my chair, the dog emitted a low grumble. "If a policeman were to come here when I wasn't home and knock on the front door," Lockwood said, "he would assume Tonka was a pit bull and shoot him; I can almost guarantee it."

For a while, Lockwood kept files on police shootings of dogs, but they quickly became too numerous for him to track. He receives at least three hundred reports a year and estimates that another 1,000 go unreported. A lawsuit in Milwaukee uncovered that from 2000 to 2008, the city's officers shot 434 dogs, or one every eight days. Between 2009 and 2012, law enforcement officials in southwestern Florida shot at animals 111 times. Police in Buffalo, New York, shot 90 dogs between 2011 and 2014, killing more than 70, with a single Buffalo officer responsible for 26 of those shootings. The fear of dogs among law enforcement officers is now so prevalent that all types of dog are at risk, including golden retrievers and Chihuahuas. According to the investigative journalist Radley Balko, "Dog shootings are part of the larger problem of a battlefield mentality that lets police use lethal force in response to the slightest threat."

"It's an evolving phenomenon," Norm Stamper, a former Seattle police chief, told Balko when asked about dog shootings. "It started when drug dealers began to recruit pit bulls to guard their drug supply." In 2014, a police officer shot and killed a "pit bull" he claimed "lunged at his face" from a parked van, but in fact the dog was a two-year-old black Labrador.

Switching the light on in one of his offices, Lockwood showed me a display shelf with a dozen or so animal skulls arranged in order of size. The largest, a museum cast, belonged to the *Tyrannosaurus rex,* while the fox's was no larger than a key lime. "Those are my canids," Lockwood said over his shoulder, nodding to the right. "Foxes, wolves, coyotes. And those," he said, smiling and pointing to several that were clustered together, "are my so-called super-predators."

The Rottweiler's skull was broad, heavy, and evenly proportioned, roughly the size of a cantaloupe. The Doberman's was sleek, elegant, and finely sculpted along the sides. The skull of the Presa Canario, probably the closest modern approximation we have of the extinct Cuban bloodhound, resembled that of a small horse. Next to them, the skull of the American pit bull terrier looked so . . . small. And unimpressive. Like an ossified softball. In proportion to its head, its teeth were no larger, or longer, or sharper than those of any of the others. The jaw was no thicker, and it possessed no special pulleys or clamps or locks to do whatever reporters had once insisted it could.

Holding this APBT skull in Lockwood's office, I was reminded of the apocryphal story about Abraham Lincoln's meeting with Harriet Beecher Stowe during the 1860s, the one in which the Civil War president supposedly said, "So you're the little woman who wrote the book that made this great war." Here was the little dog that started an enormous battle.

The first pit bulls Randy Lockwood ever encountered up close were a litter of six puppies seized during a Toronto dogfighting bust in 1979. They appeared basically sound to him, but they did fight with one another, so he recommended they be placed in homes without other pets. He never heard any complaints about them after that, and as far as he knows, they did just fine.

Then, in March 1984, a nine-year-old girl named Angela Hands was badly mauled by her uncle's two American pit bull terriers while walking home from school in the village of Tijeras, New Mexico. (The dogs had escaped their enclosure.) Loose dogs of all types had been killing livestock and injuring humans across New Mexico for years, but Tijeras, a tiny town of three hundred people, had no animal control officer to deal with the problem, which was so bad that the local papers warned readers about the "spring slaughter" visited on farm animals by domestic dogs every year. Two months after the attack on Hands, the local government passed a total ban on pit bull ownership.

Unlike the ordinance in Hollywood, Florida, Tijeras's pit bull ban was covered by papers around the country, including *The New York Times* and *The Wall Street Journal.* Soon other small towns began looking to breed-specific legislation to put their citizens' mounting fear of dangerous dogs to rest. The advice columnists Eppie Lederer and her twin sister, Pauline Phillips (better known to the public as "Ann Landers" and "Dear Abby"), both took firm positions against pit bulls and their owners. Even small children were taught to fear the dogs in 1986, when the *Weekly Reader,* an educational newspaper distributed in elementary school classrooms, featured the terrifying story of the Angela Hands attack, information about dogfighting, and the "facts" regarding pit bull viciousness on its front page.

As noted previously, when coverage of pit bull incidents increased, the AKC registrations of the more expensive AmStaff dropped steadily, but UKC registrations of the APBT tracked upward. One of the main reasons for this dramatic increase was the dirt-cheap price of APBT puppies, which were now being bred in American backyards by the thousands. AKC Doberman, Rottweiler, and AmStaff puppies regularly sold for $600 and up, but APBT breeders were now selling pit bull pups for as little as $50—in some cases, as low as $5. The pit bull had now been officially deputized, in the words of one California breeder, as "the poor man's guard dog."

Predictably, the sharp increase in pit bulls among those seeking guard dogs led to more pit bull bite incidents, especially when owners let their animals roam freely and wreak havoc. Contrary to the picture painted by the media, however, the number of Americans *killed* by dogs actually dropped four years in a row as the number of APBTs rose, from twenty-four DBRFs in 1983 to fifteen in 1987.

Lockwood and other academics who studied human-animal relationships wondered why they kept reading about pit bulls in the press. What if there *was* something going on here? If so, couldn't that knowledge save lives, given the dogs' increasing popularity? If not, couldn't some of the fog be cleared surrounding these animals? Lockwood began his own research into recent DBRFs to see if dog-bite deaths involving pit bulls did, in fact, stand out from deaths caused by other breeds of dog.

Immediately, he realized this study would be trickier than he thought. There was no central public health database that logged either bites or dog bite deaths. The Centers for Disease Control and Prevention had started tracking dog bite fatalities by breed in 1979, but the byzantine system of medical coding used by county coroners to fill out death certificates resulted in a number of DBRFs slipping under the government's radar. Lockwood supplemented the CDC's data with the running file of DBRFs kept by HSUS, as well as the media reports he found through a news clipping service, as Pinckney and Kennedy had done. He didn't know how many DBRFs might be missing from his overall tally, especially in rural or remote areas, but "we recognized at the time that we probably were missing a lot," Lockwood told me.

The media-centered approach to DBRF studies inverted the normal relationship between the academy and the press, wherein news outlets relayed scientific research to the public, not the other way around. Using media reports as data assumed that the reporters covering these fatalities were thorough, their breed attributions were correct, and the sources quoted in their articles were always truthful, which Lockwood suspected from the beginning was not always the case. Nevertheless, he and Rindy proceeded as best they could, analyzing news reports and weeding out the ones with obvious inconsistencies. They compiled their findings for the Tufts conference and later published them in the peer-reviewed journal *Anthrozoös*.

The authors located information on twenty-one total DBRFs from two time periods: October 1983 through December 1984, and 1986 (they skipped 1985 for reasons Lockwood doesn't recall). Their data seemed to show that 67 percent of dog bite deaths "involved" pit bulls, which meant that a pit bull was reported to have been present on the property at the time of the attack. Despite Lockwood and Rindy's cautions about the potential weaknesses in their research and their recommendations against knee-jerk policy-making, that 67 percent soon floated free from its context to take on new meaning in city halls across America, where politicians used it as evidence that breed bans were necessary. The real numbers told a much different story. What Lockwood and Rindy had actually captured was the extent of the media's preoccupation with pit bulls, not a clear picture of which dogs were killing whom and, more important, why. But that discovery was a long way off.

Right around the time that Lockwood published "Are 'Pit Bulls' Different?," a string of highly publicized pit bull attacks lit the fuse for another burst of panic. In November 1986, three pit bulls belonging to a man named Hayward Turnipseed escaped their yard through an open gate and killed a four-year-old boy in Decatur, Georgia. A judge sentenced Turnipseed to five years in prison for involuntary manslaughter. Lockwood later evaluated Turnipseed's dogs with the assistance of another behaviorist, Dr. John C. Wright of Mercer University in Georgia. Lockwood remembers the dogs as being even more disturbed than the hundreds of fighting dogs he has handled in the years since. One attempted a play bow, then lunged at Lockwood's face. He wondered if there was something neurologically wrong with them.

Lockwood felt that, almost from day one, the press misrepresented what was happening with regard to dog bite injuries and deaths. While the Turnipseed case had been horrendously tragic, dog bites in DeKalb County, Georgia, actually declined 54 percent between 1975 and 1986 because of well-enforced leash laws. In 1985, there were only 573 bite incidents, despite a countywide dog population that topped 100,000. Nationally, pit-bull-related DBRFs didn't *add* to the DBRFs involving other breeds; they *replaced* them, which led Lockwood to believe that the same person most likely to seek out a spooky Doberman in 1975 was now likely to seek out a spooky pit bull in 1986. It was the proverbial "switching seats on the *Titanic*." Turnipseed, for example, had been the subject of many neighborhood complaints for constant mismanagement of his animals; he likely would have been equally reckless with any dog he owned. If pit bulls truly represented a new threat to the American public, Lockwood said, the overall numbers of fatal dog attacks would have jumped much higher.

Over the years Lockwood estimates that he has evaluated more than seven hundred pit bulls seized in fight busts and cruelty cases, and he has never been bitten. "I still have all my fingers," he said, smiling and waving his hands in front of his face. Of those seven hundred pit bulls—highly stressed dogs all born and raised in the most hellish of circumstances—he said, "We've seen maybe three that we were a little uncomfortable with in terms of direct aggression toward people." Besides babesiosis, a parasitic blood infection often seen in fighting dogs, the most common health issue his team confronts when handling pit bulls from fight busts is "happy tail," an injury caused when an excited dog whacks its wagging tail so hard against a hard surface that the impact breaks the skin.

As Lockwood told Malcolm Gladwell in 2006, almost every DBRF is "a perfect storm of bad human-canine interactions—the wrong dog, the wrong background, the wrong history in the hands of the wrong person in the wrong environmental situation . . . These are generally cases where everyone is to blame. You've got the unsupervised three-year-old child wandering in the neighborhood killed by a starved, abused dog owned by the dogfighting boyfriend of some woman who doesn't know where her child is. It's not old Shep sleeping by the fire who suddenly goes bonkers. Usually there are all kinds of other warning signs."

In April 1987, a sixty-seven-year-old retired surgeon named Dr. William Eckman was killed by two dogs in Dayton, Ohio, while visiting a thirty-two-year-old woman named Joetta Darmstadter. Darmstadter had a long arrest record for prostitution, and parties on all sides of the subsequent court case agreed that Eckman had come to her house looking for sex. After she rebuffed him, Darmstadter told authorities, Eckman tried to enter her house forcibly, whereupon her two pit bull dogs attacked and killed him. Darmstadter and her boyfriend, Wilbur Rutledge, were put on trial for manslaughter but were eventually acquitted.

The public outcry over the Eckman case and the media attention that followed it (*U.S. News and World Report* ran a full-page feature but did not mention Darmstadter's claims) prompted Ohio legislators to pass a bill declaring all pit bulls to be "vicious" and therefore subject to a long list of extra insurance and containment requirements, though the animal control officers tasked with enforcing these laws ran into the age-old problem of not being able to identify which dogs were pit bulls and which were not. An aide to one Ohio state representative complained, "We can't get a veterinarian anywhere to define one." Ironically, Ohio's (then) laws on exotic animal ownership were notoriously weak, making it much easier to keep—even breed—a six-hundred-pound Siberian tiger (or eighteen tigers, seventeen African lions, three mountain lions, six black bears, two grizzly bears, and multiple monkeys and wolves, as a Zanesville man named Terry Thompson did before he committed suicide in 2011) than to own a certain shape of domestic dog. Ohio overturned its statewide pit bull law in 2012.

The tragedy that most tipped the scales of public opinion against pit bulls was the death of a two-year-old boy named James Soto in the San Jose, California, suburb of Morgan Hill, which the media turned into a summerlong horror circus. On June 13, Soto's mother, Yvonne Nunez, left her son alone in the backyard to play with a new Tonka truck. When she returned, the gate was open and her son was gone, having toddled into the yard of a shady neighbor named Michael Berry, who kept a chained pit bull in his yard to guard a plot of marijuana plants. Later an investigation would reveal that Berry had acquired the dog with the specific intention of its being aggressive toward people. Berry was not home at the time of Soto's death, so a jury acquitted him of second-degree murder but convicted him of involuntary manslaughter. He was sentenced to three years and eight months in prison. (Lockwood testified as an expert witness in his trial.)

News outlets in countries as far away as Japan reported on James Soto's death, which was written about in over seventy American newspapers. The *San Jose Mercury-News* noted that pit bulls "have locking jaws that some trainers estimate can exert as much as 3,500 pounds of pressure per square inch." Other dog-bite-related deaths that year were barely reported at all. A four-year-old boy named Timothy Nicolai was killed by a pack of chained sled dogs in Alaska in April 1987, but his death received only the slightest mention in the Sitka *Daily Sentinel* and an even smaller note in the Anchorage police blotter. The following year, a four-year-old girl named Jennifer Evon was also killed by one of her family's dogs in the *same* Alaskan village where Nicolai had lived, yet her tragic passing barely warranted a mention in the local news. The thirty-six other children killed by non-pit-bull-type dogs between 1983 and 1986 were also given minimal coverage in the press. The American media seemed to be interested in dog bite deaths and public safety only when pit bulls were responsible.

Early in the 1980s, reporters hit on a formula for pit bull stories that immediately drew readers into heated arguments. It presented the dogs in terms of opposing forces—good/bad, nature/nurture, villain/victim—behind an interrogatory headline. Florida, 1980: "Pit Bulls: Fine, Loyal Pets—or Dangerous Killers?" South Carolina, 1981: "Do [Pit Bulls] Deserve Halos or Horns?" Florida, 1987: "Pit Bulls: Born Killers or Trustworthy Pets?" Vancouver, British Columbia, 1987: "Pit Bull Terriers: Misunderstood Puppy, or a Shark on Paws?"

The framing of the subject implied that what might be true of one pit bull was true for all of them, a ridiculous assertion. These stories hinted at a knowable, metaphysical certainty that existed just beyond the reach of the article. In reality, however, one side of the scale was always heavily burdened with quotations and sources that made the animals look as frightening and cold-blooded as possible. The British journalist Andrew Marr warns, "A headline with a question mark at the end means, in the vast majority of cases, that the story is tendentious or over-sold. It is often a scare story, or an attempt to elevate some run-of-the-mill piece of reporting into a national controversy and, preferably, a national panic. To a busy journalist hunting for real information a question mark means 'don't bother reading this bit.'" The cognitive neuroscientist Ullrich Ecker and his colleagues at the University of Western Australia have shown that headlines—in particular, headlines about highly emotional subjects—can dramatically influence which information from an article a reader retains and which is passed over and forgotten. Once you present an audience with the possibility of a "born killer" or a "shark on paws," that will likely be all they remember.

Within days of James Soto's death, owners of pit bulls in California's Bay Area began fearing for their own safety, as well as that of their pets. "There's a kind of mob psychology out there," said Richard Avanzino, then president of the San Francisco SPCA. "I'm advising people who walk their pit bulldogs to do it in the wee hours when nobody else is out." A woman from Gilroy, California, tried to sell her pit bull's ten puppies, only to walk outside and find her adult female dog had been beaten to death with a shovel. In St. Louis, a pit bull was "firebombed" in its own yard. In Revere, Massachusetts, a pit bull's bullet-ridden body was found in a local river with a noose around its neck, and two dead pit bulls were found hanging in California.

More than three hundred pit bulls were surrendered to California shelters in the first six months of 1987 (most in Los Angeles County), while other owners had their family dogs euthanized by the local vet, sometimes voluntarily and sometimes under duress. "People come in here with their dogs with tears in their eyes, hugging and kissing them," said Warren Broderick, the director of the Humane Society of Santa Clara Valley, but the owners had their pets killed "because neighbors are threatening to shoot or poison the dog." A seventeen-year-old San Jose resident named Arthur Chacon Jr. was forced to give up his dog when his neighbor told their landlord that he no longer wanted to live next to a pit bull. "It was really just heartbreaking," Chacon said at the time. "It was just painful. I look at his dog house, and just because of what happened on the news it isn't right to have my dog taken away from me."

From that point on, American families who decided to keep pit bulls as pets became much less likely to register them, or they purposefully registered them as other breeds. In cities like Detroit, animal control professionals estimated that 80 to 90 percent of the city's dogs were not registered at all. Over time, these low registrations extinguished any hopes of ever knowing how many American pit bulls there really were.

But the press was not done with pit bulls yet. It was the summer of a nonelection year, when national news is typically at its slowest—so slow, in fact, that journalists often refer to it as "the silly season" because frivolous non-issues are magnified to fill space. The week after Soto's death, a camera crew from a local CBS TV affiliate accompanied a young Los Angeles animal control officer named Florence Crowell as she investigated reports of an aggressive dog at the home of a woman named Edlyn Hauser. The dog, a pit bull named Benjamin, had a long history of dangerous behavior and had been impounded the previous year for biting. Shortly after Crowell showed up with the news team in tow, a very erratic and out-of-sorts-seeming Hauser did not prevent Benjamin from rushing out the door and biting Crowell on her arm and chest, all of which was broadcast that night on *CBS Evening News.* There it was: the "crazy" pit bull owner in a muumuu "siccing" her vicious dog on a public servant, for all the world to see. Few asked why a television news crew followed Crowell on this call in the first place, unless they expected (or hoped) that something would happen.

In the press, this sequence of events—Eckman's and Soto's deaths, plus the televised attack on Crowell—turned the information cascade into a fire-hose blast. Within days of the Crowell broadcast, Angelenos began seeing loose "pit bulls" where none existed. "Officers from animal shelters countywide reported they have been barraged by telephone calls . . . from people reporting that pit bulls are running loose in their neighborhoods," reported one Los Angeles paper. "Animal control workers said many of the frantic callers have mistaken other kinds of dogs for pit bulls." Over the next three years, the AKC would receive complaints about dogs from twenty-seven different breeds being misidentified as pit bulls, including Chesapeake Bay retrievers and Dalmatians.

On June 30, the rhetoric took a radical turn that ensured that the pit bull panic would not spike and fade away, as the other dog panics had. United Press International reported that pit bulls were now the "weapon of choice among drug dealers," who, according to the ASPCA in New York City, were "using pit bulls to guard crack houses." Police officers in "major cities" told journalists that "ghetto youths often keep pit bulls as status symbols."

To some extent, this was true; the illegal drug trade was booming in poverty-stricken neighborhoods where owning dogs for personal protection had been a well-established tradition since the late 1960s. But reports of police tangling with frenzied dogs on drug-and-alcohol raids went all the way back to the days of Prohibition, when German shepherds guarded moonshiners' illegal liquor stills. Fifty years later, drug dealers in large cities frequently kept German shepherds, Rottweilers, and Dobermans to protect their operations, as did the Hells Angels and the Ku Klux Klan. But the press seized on the pit bull as being mysterious, "different," and uncontrollable, as though guard dogs were an entirely new and bewildering development in the criminal underworld.

The reporters implied that the pit bull was not an animal that had lived as a family pet in America for a hundred years, but a demented marionette used by the crack dealer, who in turn was described as an urban "predator" who sought out the weak and the vulnerable. By fending off police, pit bulls were portrayed as crack-dealing "accomplices" who abetted the rapid growth of the drug trade. To see just how closely the terms "crack cocaine" and "pit bull" were linked in the media, one need look no further than the Google Ngram Viewer, which charts word frequencies in published materials. From 1986 to 1990, the two terms followed a curve that is almost identical.

The human parallels are important here, because the legend of the urban pit bull would become a literal companion piece to America's failed war on drugs. When a dog scare collides with a drug scare—especially one as racialized as the crack "epidemic"—the effects are multiplicative. Had it not been shackled to the moral panic over crack cocaine, media coverage of pit bulls would never have tipped over into global hysteria.

Though crack is almost chemically identical to powder cocaine (its only added ingredient is baking soda), news coverage of crack took a completely different path from the bemused shrug it adopted toward white users of powder cocaine in the 1970s. Horror stories about crack dealers and crazed crack users, as well as "crack whores" and "crack babies," radically shifted how consumers of American news, especially white consumers, viewed life in the inner city. There is no doubt that drug distribution and drug-related violence increased with the arrival of crack, but violence had been steadily increasing in large American cities since the late 1960s. Between 1970 and 1990, a time when many large production plants were shuttered and jobs were shipped overseas, the number of Americans living in high-poverty neighborhoods doubled. Multiple studies have shown that the use of crack cocaine by a small segment of the urban poor never reached "epidemic" levels, as the media insisted. Nor was crack "instantaneously addictive," as drug warriors often claimed. Many of the worst crimes committed by crack users were committed by people who were already mentally ill before they took the drug.

According to Robert Stutman, the director of the Drug Enforcement Administration's New York office in 1985, the dramatic rise in the press coverage of the "crack epidemic" during the mid-1980s was no accident; rather, it was the result of a conscious strategy by Reagan officials to drum up public support and congressional funding for the politically crucial drug war. The more frightening the plague of crack cocaine appeared to American taxpayers, the fewer questions were asked about how the money allotted for the war on drugs was spent. As we now know, a great deal of that money funded the Contra War in Nicaragua and propped up a murderous dictatorship in El Salvador. Stutman later wrote, "In order to convince Washington, I needed to make [drugs] a national issue and quickly. I began a lobbying effort and I used the media. The media were only too willing to cooperate, because as far as the New York media was concerned, crack was the hottest combat reporting story to come along since the end of the Vietnam war."

Stutman's plan worked. The number of crack stories that appeared in *The New York Times* swelled from forty-three at the end of 1985 to more than three hundred in 1986. Taking their cue from the *Times*, television news channels and magazines began churning out crack stories as well. CBS then aired a horrifying television special, "48 Hours on Crack Street." By 1988, the *New York Times* editorial board realized that the issue had been overblown and walked back its sensational coverage. "America discovered crack and overdosed on oratory," it said.

The sociologist Katherine Beckett has observed, "Much of the drug-related news coverage during this period emphasized the spread of crack-related violence to white communities, the threat of random (drug-induced) violence to which this 'epidemic' gave rise, and the need for enhanced surveillance and policing in order to establish control . . . Because these stories highlighted the threat of random violence, they appear to have contributed to growing support for a quick and dramatic response to the drug problem." She could have just as easily been describing the media narrative about pit bulls.

On July 2, 1987, *Rolling Stone* (with a readership then estimated at 7.3 million) ran a feature by Mike Sager titled "A Boy and His Dog in Hell," which focused on the culture of inner-city dogfighting that, like all other crimes, now appeared to be flourishing in North Philadelphia. Sager later recalled that his editor, Jann Wenner, sent him on the assignment after reading the *U.S. News and World Report* article that presented the pit bull as "the most dangerous dog in America." Wenner found the "ghetto" angle especially intriguing.

Sager's pit bull narrative confirmed white suburbia's worst fears about where the country was headed. The story's two drug-addled teenage protagonists, Tito and Chic, lived in a predominantly Puerto Rican section of North Philadelphia where the nights were "sharp with the promise of dogs and drugs, blood and adventure." They spent their money on "pork rinds or Newports or drugs or dog food" but viewed the animals only as props in the macho pursuit of street status, not companions. Tito and Chic stole pit bulls from other people's yards, fought them, and callously dumped their bodies in back alleys filled with "tin cans and bedsprings and car parts and pieces of foam and Pampers that cover the ground like mulch."

The analogies the author drew between the dogs and their people were hard to ignore. Citing Lockwood and Rindy's paper, Sager wrote,

"Though pit bulls constitute only two percent of the population, they have accounted in the last three years for fifteen of twenty-three reported dog bite fatalities." He then pointed out that Puerto Ricans like Tito and Chic "represent three percent of [Philadelphia's] population," yet "as a group, they have the lowest levels of education and income and the highest rates of teen pregnancy, infant mortality and criminal arrests in the city.

"Wherever there are boys who need something to be proud of and known for, there are boys who fight pit bulls," Sager explained.

> On the hard streets of the city, you are what you own: your moped, your blaster, your Adidas, your rap, your pit. Having a pit is not like having any other kind of dog. Pits do more than eat and shit and walk on a leash. They fight. And when they fight, they either win or die, and they do so, usually, with honor and style. They are perfect for ghettos like Little Puerto Rico—small enough to keep, tough enough to survive.

Juxtaposed with photographs of the boys walking their dogs on chains was a picture of a kindly white woman from the Women's Humane Society hugging a pit bull presumably rescued from the urban hell of street dogfighting. The street fighting of the 1980s was to professional dogfighting what a bar brawl is to professional boxing: relatively quick, very dirty, and completely disorganized. There was almost no training or know-how involved, with informal or "off the chain" fights held in abandoned buildings more for novelty or bragging rights than as part of larger operations.

And so it went, all summer long (the same summer, incidentally, that the Discovery Channel would air its first "Shark Week"). On July 6, *People* published "An Instinct for the Kill," which was followed on July 27 by a *Sports Illustrated* cover story and a feature in *Time* that called pit bulls "time bombs on legs." (All three publications are owned by the same company, Time Incorporated.) The *Sports Illustrated* layout featured photographs of menacing-looking dogs and dark-skinned men with grim, scowling faces set against a backdrop of graffiti and garbage, while *Time* observed, "Ferocious pit bulls can be seen any day with their drug-dealer owners on the corner of Ninth and Butler Streets in North Philadelphia. The dogs, with names like Murder, Hitler, and Scarface, wear metal-studded collars concealing crack and cocaine and the day's proceeds. They are equally visible on Chicago's West and South sides, where teenage boys have taken to brandishing their fierce pit bulls just as they would a switchblade or a gun. 'It's a macho thing, like carrying a weapon,' says Jane Alvaro of the Anti-Cruelty Society."

The cover of *Sports Illustrated,* July 27, 1987

In the 1980s, *People, Sports Illustrated,* and *Time* had a combined readership estimated at 120 million, approximately half of America's population, as well as large audiences overseas. The rash of coverage during the summer of 1987 left an indelible mark on public opinion. Now pit bulls that merely ran toward people or jumped on them were said to be "lunging" or "attacking." My fastest-growing file is the one filled with articles about pit bull "attacks" that resulted in no real injury; some did not even involve physical contact between the dog and its "target." Here is a small sampling of incidents that drew media attention:

- In 2006, a news station in Lubbock, Texas, reported that an elderly man was "attacked" by a pit bull that "lunged" at him but did not bite. "Fortunately, the victim was more shaken than injured," the reporter said.
- In 2009, *The Guardian* mourned the "thousands of urban trees mauled" by pit bulls in Great Britain.
- A pit bull "knocked down" a five-year-old child at a park in Tyler, Texas, and caused "minor scratches" in 2011.
- In Detroit in 2012, a woman reportedly "fended off" an "attacking" pit bull with nothing more than a bottle of hand sanitizer. She was neither bitten nor injured.
- This 2012 headline, from Nanaimo, British Columbia, speaks for itself: "Pit Bull in Life-or-Death Struggle with Giant Beaver."

In 1990, an Emmy-winning television reporter from Denver named Wendy Bergen was so eager to provide more sensational pit bull material for her news broadcast that she enlisted local pit bull breeders to stage a dogfight, then ominously told viewers that such fights were common in the Denver area. When it was discovered that her entire four-part series, "Blood Sport," was a massive fraud, Bergen was arrested and charged with multiple counts of dogfighting, conspiracy to commit dogfighting, and perjury. After a twelve-day trial, a judge sentenced her to pay $20,000 in fines.

Admirably, a few skeptical journalists noticed what was happening. Debra Saunders of the *San Francisco Chronicle* was one of the most prescient, and she was the first to notice that this conversation seemed to have little to do with dogs. On July 30, in a level-headed editorial, she wrote:

> This says some sad things about America. The pit bull scare illustrates how skittish we have become . . . skittish and ineffective, because this fad scare will do next to nothing to lessen dog attacks . . . Fad scares have been on the rise since we first learned about AIDS. Stations found that their ratings shot up whenever they ran an AIDS exposé or [a program on] the dangers of crack cocaine. They saw that scaring viewers sells . . . Pit bulls make for good local TV because they require no expertise. No need for facts, just get the best teeth shot.

There was something else at work here, according to Saunders, something that felt more than a little uncomfortable, and that was the easy allure of scapegoating. "There aren't a whole lot of pit bull owners to alienate," she wrote. "There are no pit bull advertisers. Fad scares scare and soothe at the same time. If we stop taking crack or get rid of a nearby pit bull, we're saved. Unlike the Middle East or acid rain, the pit bull problem is easy: Get rid of pit bulls. It won't upset an ecological food chain. No jobs will be lost. Most people won't be offended."

Saunders was wrong on that last point. There *were* many people who found this scapegoating both frightening and offensive, but they had neither the money nor the social capital to challenge it.

After the summer of 1987, the pit bull virtues most celebrated during the early twentieth century under the waving flag of American patriotism were now recast in a sinister light: "Never gives up" turned into "never lets go." "Tenacious" became "frenzied." "Athletic" and "spirited" became "monstrous" and "crazed." Instead of symbolizing the country's valor, pit bulls symbolized the brutality of its brokenness, and America very easily cast them aside.

In addition to being quoted in every major pit bull news story of the 1980s, Randy Lockwood once appeared on the Christian talk show *The 700 Club*; evidence in itself that no outlet could resist the story. Lockwood remembers that the segment included a rescued pit bull named Nemo ("Very nice dog," Lockwood recalls) and commenced with a prayer circle. "Dear Lord, thank you for Dr. Lockwood's expertise," the producers began. "And dear Lord, please help Nemo to relax."

For Lockwood and many of his colleagues, the "killer pit bull" narrative felt propelled forward by a giant engine of fear that no amount of reason could halt, and it plowed through the sciences almost as completely as it took over the press. Suddenly everyone just *knew* that pit bulls were dangerous, and researchers stopped being skeptical. Don Clifford, the Ohio veterinarian who toured the country lecturing about the dangers of pit bulls, once asked Lockwood for his file on pit-bull-related fatalities. "Of course I gave him what I had," Lockwood told me, "then I offered to give him the data on all the *other* breeds of dog involved in human deaths. He told me that he didn't want those. He wasn't interested." Clifford then went on to write a manual on "vicious" dogs for the American Veterinary Medical Association's Professional Liability Insurance Trust that was so misleading that the AVMA later pulled it from circulation.

But Lockwood's participation in the pit bull media bonanza would come back to haunt him when politicians and anti–pit bull activists used several of his public statements as ammunition for breed bans. "Fighting dogs lie all the time," Lockwood said during a 2004 law enforcement training video, the transcript of which was widely circulated by anti–pit bull groups. He went on to describe Hayward Turnipseed's dog offering a "play bow," then lunging at him, to illustrate that fighting dogs do not signal their intentions. (Some may not; there's anecdotal evidence of that being the case in specific subpopulations, but once again this claim has never been scientifically evaluated.) When I pressed Lockwood on this incident, however, he said that Turnipseed's dog was the only one—out of more than seven hundred fighting dogs—that he remembers as being a "liar" in the way he recounted. He never expected that this casual statement would be dissected under the harsh glare of the pit bull culture wars.

Lockwood explained that his intention was not to condemn the dogs, but to draw attention to the terrible things dogfighters were doing to animals for their own entertainment and financial gain. He was speaking from the perspective of a cruelty investigator, a man who had seen reflected in certain animals the darkest corners of the human psyche, not the easygoing companions curled up on couches across the country. The operatic sadism he encountered during dogfighting investigations—dogs beaten, hanged, electrocuted with car batteries—upset and angered him. Like police officers or emergency-room physicians or medical examiners, cruelty investigators operate inside a world full of things they never wanted to see and will never be able to forget. No one calls a cruelty expert to evaluate happy family pets. Unfortunately for many animal advocates of the 1980s, quotations don't change, even when opinions do.

In the mid-1990s, Lockwood partnered with a public health epidemiologist at the Centers for Disease Control and Prevention to examine the issue of fatal dog bite injuries more closely. Dr. Jeffrey Sacks at the CDC's National Center for Injury Prevention had already been looking into the issue of dog bites, which stood out to him as worthy of more scrutiny after he examined the results of the CDC's National Health Interview Survey years earlier. Of all the types of unintentional injuries, Sacks believed that dog bites were among the most preventable.

At the outset, Sacks did not focus on dog breeds. But as he looked over the data, several things stuck out. One, DBRF victims were not people who had been chased down and killed by strays; overwhelmingly, they were killed in the homes where the dog lived, sometimes by their own pets, and many of the victims were neonates or small children. He also noticed that certain breed descriptors kept popping up, especially in the news reports, and one of those was "pit bull." When Sacks and Lockwood began working together, both realized that the news reports were inconsistent in their breed descriptions and that the data were generally contradictory and incomplete. Moreover, the news accounts did not capture all dog bite deaths.

Sacks explained the process to me as a classic problem in research. "The data you have are not the data you want, the data you want are not the data you need, and the data you need are not the data you can get." But because Sacks felt that the data, as imperfect as they were, might contribute to a larger conversation about preventing dog-bite-related injuries, he published what he had, careful to add several pages of caveats about what could or could not be established via media reports.

Sacks joined up again with Lockwood and several other public health professionals for another paper, this one a retrospective analysis of all American DBRFs from 1979 to 1998, which was published in the *Journal of the American Veterinary Medical Association* in 2000. The numbers suggested that "pit bulls" (loosely defined) were responsible for more than half of human DBRFs over the twenty years;* however, they showed that from 1993 to 1998, Rottweilers had far and away replaced pit bulls as the most frequently reported breed. Despite the authors' repeated insistence that these tables *not* be used to infer breed-specific risk (and an accompanying letter from the American Veterinary Medical Association), the news that science had confirmed pit bulls were the "most dangerous" of all dogs spread far and wide.

It did not appear that many of the reporters had read the actual paper or considered the latest trends in breed popularity. The problem with tracking human fatalities by breed, the CDC researchers discovered, is that there will always be a "number one." It is unlikely that we will ever eliminate DBRFs altogether, as there will always be a small portion of any dog population that is unsound. If Americans owned only beagles, poodles, and Pekingese, such a breed-based ranking system would inevitably make one breed appear more "dangerous" than the others. For this reason, the CDC stopped tracking dog bite fatalities this way shortly after the twenty-year DBRF study was published, because epidemiologists did not believe it yielded any useful information.

"If we based our studies of homicide on serial killers," Lockwood explained, "we would basically assume that most murderers are white and middle-class, but serial killers are a very special case. Not only are fatal dog attacks a minuscule representation of the total dog population *and* of the total pit bull population, they're a minuscule representation of the *dog bite* population."

Sacks concurred. "Breed has been over focused on as the problem," he said. "It's a lot more complicated than that. Any bite involves a confluence of factors: a dog's genetics, the victim's behavior, the dog's socialization, the dog's medical history, the owner's training. You can't isolate one factor and say 'That's it!'"

---

* Other breed groups, such as terriers, hounds, and retrievers, were sorted into discrete breeds (Labrador versus golden, and so forth).

"People like to have or give the illusion of dealing with the situation," Lockwood told me. "They're looking for the quick fix. It's sort of like thinking we're going to stop terrorism by taking your shoes off at the airport, rather than investing money and upgrading ground control because on-ground near collisions are a much more serious threat to air travel than someone with a bomb in his underwear, or his shoes. For years politicians have placated fears with quick fixes [like breed bans] that are ultimately unsuccessful. Usually by the time they're proven to be unsuccessful, the politician is out of office and not held responsible for that decision."

Some months later, Lockwood elaborated in a follow-up e-mail: "My views (and the views of the humane community) have evolved as we gained more understanding of dogs, dog behavior, dog genetics, human impacts on dog welfare, etc.," he said.

> Prior to 1980 I had seen perhaps a dozen pits and half a dozen fighting dogs. In the intervening years I have worked with hundreds, perhaps thousands, and have interacted with every type of owner from the lowest scum to the most noble rescuers.
>
> If [the then presidential candidate Mitt] Romney can change his view every 25 minutes I think I am entitled to changing mine over 25 years.

Lockwood no longer believes in the concept of "dangerous breeds." He prefers to focus on how and why individual dogs become dangerous.

"When I started out," he told me, "I thought that if there *was* something going on with pit bulls, maybe I could do something that might save some kids. We didn't know all the things we know now. I had no idea what this would become."

CHAPTER 11

LOOKING WHERE THE LIGHT IS

The guiding motto in the life of every natural philosopher should be, seek simplicity and distrust it.

—ALFRED NORTH WHITEHEAD, *The Concept of Nature*

Karen Delise had worked in the records department of the Suffolk County Sheriff's Office on Long Island for fifteen years when, in 1993, a two-year-old boy in a neighboring county was killed by a 121-pound Rottweiler that lived next door to his family. By then, there wasn't much in the way of human depravity that Delise hadn't at some point filed away on the shelves. Assaults, rapes, stabbings, murders, horrific cases of child abuse—the reports came across her desk in a flutter of copy paper and manila folders every day. But the public seemed to respond much differently to the death of William "Billy" Sheppard than it did to victims of homicide. The story consumed the local news cycle for three straight days. *You can't trust those dogs,* people said. *They are vicious, and they will turn on you.*

Delise didn't have an opinion on Rottweilers one way or the other. She had lived with many types of dogs in her thirty-seven years (boxers were her passion), and it never occurred to her to think of any particular breed as dangerous. Now the people of Long Island seemed to be grabbing pitchforks and firing up their torches, high on anti-Rottweiler righteousness to an extent that made her uneasy. "Wasn't it a bit premature to generalize about an entire *breed* based on one incident?" she thought. If her neighbors wanted something to be truly outraged over, she could show them a whole roomful of files that painted a very grim picture of human nature. The "beloved family pet turned homicidal traitor" headlines also didn't make sense to her. Did these terrifying events really come "out of nowhere," as so many claimed, or could there be more going on than the media reported? Decades later, she would come to be regarded as something akin to the Erin Brockovich of dog bite deaths.

Delise faced a number of hurdles in the early stages of her research. Armed with only an associate's degree, she had no professional background in the disciplines that would have helped her. But she was smart, resourceful, intensely skeptical, and willing to track down the experts who could best explain what she needed to know. Years of working at a law enforcement agency had taught her that assumptions, even the most basic, can send investigations far into the weeds. She limited her opinions to what she saw in the data.

Unfortunately, she encountered the same gaping holes in the available science that vexed Randy Lockwood and Jeff Sacks years earlier. The United States had no central reporting system for dog bites or DBRFs; there was no comprehensive, accurate registry of American dogs by breed; and the widespread use of the Internet was still several years away. What few peer-reviewed papers there were on DBRFs still relied primarily on media accounts.

Using those news articles as a starting point, Delise first cross-referenced each case against the scientific literature. She then began to correspond with several epidemiologists, including Jeffrey Sacks at the CDC, who generously shared his data. As was customary at the CDC, she focused on human deaths in which the medical examiner had listed the mechanical trauma of a dog bite injury—lacerations (cuts), avulsions (removal of skin), crushed bones, or exsanguination—as the primary cause of death. A man who was hit by a car while running from a dog would not make the list, for example, nor would one who died of an infection after a bite.

Immediately, Delise was struck by how rare these events were when compared with other forms of unintentional-injury deaths. When she started, there had never been more than thirty-two DBRFs in the United States in any given year, despite a human population that was then approaching 260 million and a dog population that exceeded 55 million. The numbers appeared to zigzag up and down at random. In 1993, the year Billy Sheppard died, only twenty other Americans were killed by dogs, but the media coverage of these cases was so overblown that it led the public to believe that dog bite deaths happened all the time.

Without access to membership databases like LexisNexis, Delise began calling research librarians across the country, asking them to dig into old newspaper archives, microfilm, and public records to find accounts of dog-bite-related deaths for her. By doing this, Delise found many DBRFs that Pinckney and Kennedy, Lockwood and Rindy, HSUS, and the CDC had all missed. Nearly every one was a case that did *not* involve pit bulls. These were harder to locate because they did not receive the same level of media coverage as the pit bull incidents. This trend continues in the age of digital media. An Internet search for Mia Gibson, a little girl killed by two Shiba Inus in Ohio in 2013, yields 1,400 mentions on a little more than one hundred news outlets, while a search for Ryan Maxwell, a seven-year-old killed by a pit bull the same year, brings up nearly 5,000 on more than two hundred news sites. Liam Perk, a two-year-old killed by his family's Weimaraner in 2009, brings up only 571 mentions and five outlets, whereas Justin Clinton, killed by two pit bulls in 2009, brings up 2,880 on almost thirty news sites.

What appears in local newspapers and on local television is as much a reflection of the public's obsessions as it is of the media's agenda. "Often we get our news tips from viewers and from sources within local governments," one Colorado television news reporter replied when asked online about pit bull stories. "We don't hear . . . about the other severe bites involving other breeds because people don't tell us about them."

Delise then reexamined Lockwood and Rindy's data. From October 1983 to December 1984, the authors logged nine DBRFs, seven of which were reported to be pit bulls. Delise eventually located a total of *twenty-four* DBRFs from the same time period. Lockwood and Rindy appeared to have located all the "pit bull" fatalities but none of the additional fifteen, which involved other breeds and mixes. In 1985, the year Lockwood skipped, there were seventeen DBRFs, and only four were reportedly linked to pit bulls. All in all, Lockwood and Rindy listed twenty-one fatalities that occurred between October 1983 and December 1986 and concluded that fourteen of these (67 percent) were caused by pit bulls. Delise, on the other hand, confirmed fifty-six DBRFs for the years 1983 to 1986, and only one-third (eighteen, or roughly 32 percent) *might* have involved pit bulls. But even these percentages meant little, because the numbers were so low. "Why was everyone looking only at the pit bull incidents?" she wondered. "Why weren't they asking the same questions about the other two-thirds?"

Forty-four different breeds (including the dachshund, the West Highland terrier, and the Jack Russell terrier) have been linked to human deaths since 1979, the year the CDC began tracking fatalities by breed. If "fighting genes" had prompted pit bulls to kill humans, then what explained the deaths caused by German shepherds, Labrador retrievers, Chesapeake Bay retrievers, Siberian huskies, Brittany spaniels, Newfoundlands, coonhounds, collies, and sheepdogs, among others? Even worse than pit bulls being described as one breed was that research was generally divided between papers on "pit bulls" and papers on "normal dogs," which sent a strong message that the two terms were mutually exclusive.

"The media hyped [the panic] up," Delise said, "but the so-called experts are truly the guilty parties for providing 'evidence' that pit bulls were not 'normal' dogs."

When Delise compared the breeds involved in DBRFs with trends in the AKC's registration statistics, she saw that the dogs most represented on her growing list most often followed booms and busts in overall popularity, especially the popularity of the larger breeds traditionally used for guard and protection work. When registrations of German shepherds and Dobermans shot up during the 1970s, so did their rank on the DBRF list. When pit bulls assumed "fad bad" status in the mid-1980s, their involvement in DBRFs also increased. Later, when Rottweiler registrations surged during the late 1990s, it didn't surprise Delise that they took the top spot. As soon as Rottweiler registrations dropped off, so did the number of Rottweiler DBRFs. In Canada, the majority of DBRFs were caused by that country's most popular working dogs, medium-to-large husky types.

The annual total of DBRFs in the United States always hovered at roughly the same low number. To Delise, that complicated the issue of "killer breeds" quite a bit. As Lockwood had observed, if "killer breeds" or even "killer bloodlines" of breeds had truly emerged, they would have added to the number of total DBRFs from other breeds, not replaced them. These trends also suggested to Delise that what owners expected from a dog (guard work versus companion animal) and the manner in which the animal was maintained might hold more predictive power about its involvement in a DBRF than what specific breed (if any) it belonged to. Aggressive or erratic behavior from an animal acquired primarily for guarding is more likely to be tolerated than it is if the dog's "job" is to be a pet.

Because there never had been and never would be an accurate way of knowing how many members of each breed there were in the American dog population, there was no way to calculate "relative risk" by breed, and that is assuming that all the breed attributions reported in the media were correct, which Delise realized could not be true. Without a denominator, a numerator is just a number.

What's more, there were far too many potentially confounding variables in each DBRF case that could not be controlled for. In order to determine the relative "dangerousness" of a population of dogs in a scientifically credible way, you would have to know how the dogs in that group were bred, raised, socialized, trained, fed, handled, and housed. You would also have to study the dogs' neurological makeup, serotonin levels, hormonal balance, and general health, then take into consideration the behavior of the humans in the dogs' immediate environment. Until the role of those variables could be determined, it wouldn't matter if all dog bite deaths were caused by purebred APBTs or AmStaffs; the total number of incidents would simply be too small to be significant.

To her chagrin, Delise discovered that different media outlets often reported conflicting details about each fatality. Journalists on short deadlines were sometimes confused by what their sources told them or didn't independently verify the information, especially regarding the breed of the offending animal(s).* Press accounts also offered contradictory information about how the dogs were maintained by their owners. One outlet might report that a dog lived outdoors on a chain with no human contact, while another stated the same dog was a "beloved family pet." Dog owners who feared charges of animal cruelty, criminal negligence, child endangerment, or even manslaughter tended to downplay (or completely deny) that their animals had a history of aggressive behavior, swearing that the fatal injuries "came out of nowhere." In many cases, animal control records later showed that the dog had a documented history of aggression and the owner a documented history of mismanaging it.

In the infamous fatal attack on the lacrosse coach Diane Whipple in

---

* In a bizarre coincidence, I witnessed this phenomenon as I was sitting on Delise's living room floor in July 2012 taking interview notes on my laptop. My e-mail in-box chimed with a news alert that a forty-year-old man named Ronnel Brown had been killed by a dog in Avondale, Ohio. One news site described the dog as a "Bullmastiff." I clicked through related links and found that another outlet called it a "bulldog." A third called it a "pit bull–bull mastiff mix." As more information became available throughout the day, news stations relayed that the dog had in fact belonged to a rare and relatively new breed: the Alapaha blue-blood bulldog, which itself is a mishmash of multiple different breeds, including the Catahoula leopard dog, the Great Dane, and the American bulldog, as well as the AmStaff. To this day, multiple sites still report Brown's death as a "pit bull" fatality.

San Francisco in 2001, witness after witness testified in court that the two Presa Canarios being cared for by the lawyers Robert Noel and Marjorie Knoller had repeatedly bit or lunged at them in the hallways of their apartment building or out on the street, in addition to attacking other dogs in public, before they ultimately killed Whipple. Yet both Noel and Knoller swore at the time that the dogs had never caused any problems. Noel served two and a half years in prison for involuntary manslaughter, and Knoller was given a sentence of fifteen years to life for second-degree murder and manslaughter.

From 2000 to 2013, authorities located sufficient evidence to bring criminal charges against reckless/negligent owners or caretakers in more than a quarter of DBRFs. In almost 40 percent of DBRFs from 2000 to 2009, either the owner had been cited for mismanaging the dog or there was a record of complaints about the dog on file with animal control. "That tells me that these incidents did *not* come out of nowhere, and the dog did not simply snap all of a sudden," Delise said. "These were preventable tragedies."

To get more accurate data, Delise did what no other researcher before her had done: she personally interviewed the police officers, animal control officers, and medical examiners who had directly handled each case. They sent her copies of their official reports, photographs of the dogs, photographs of the victims, photographs of the scene, autopsy results, necropsy results, and any other relevant documents from investigations that lasted months, even years, after the camera crews scuttled away.

When Delise compared the information that officers provided her with what the press disseminated to the public, she told me she was "totally gobsmacked," especially by the photographic evidence. Years before science would confirm that visual breed identification of dogs is inaccurate, Delise noticed that the breed IDs in DBRFs often made no sense at all. Many of the "pit bulls" responsible for DBRFs appeared to be generic mutts. The same went for some of the "German shepherds" and "chows" and "Labrador retrievers." So, if the breed descriptions weren't reliable in many of these cases, then what else were the media getting wrong?

As it turned out, quite a bit.

On December 11, 2002, *The New York Times* reported that an eighty-year-old resident of North Bergen, New Jersey, named Julia Mazziotto had been killed at home by her daughter's two pit bulls in an attack that left bite and claw marks on 80 percent of Mazziotto's body. In characteristic tabloid fashion, the *New York Post* opted for a less

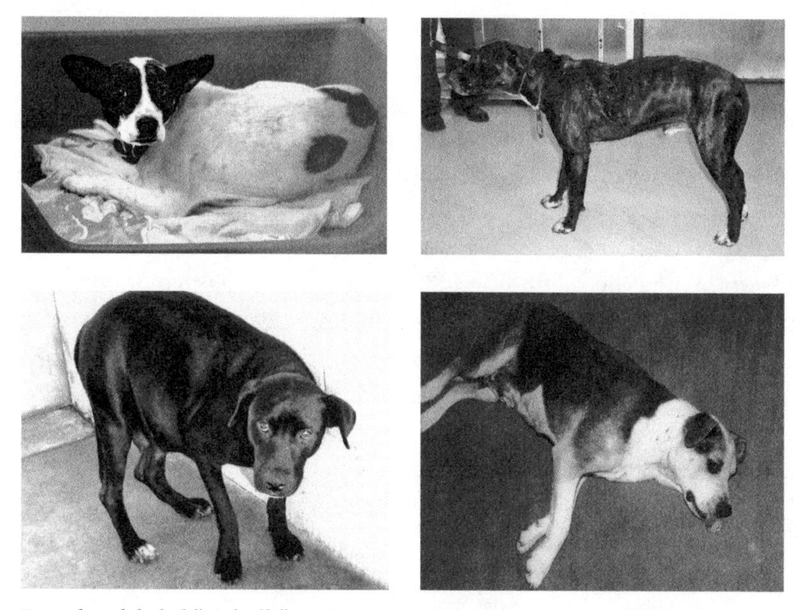

Four dogs labeled "pit bulls" in the media after involvement in DBRFs.

In news reports, the dog on the left was labeled an "Australian shepherd," and the dog on the right was labeled a "chow chow."

All six of the above dogs were reported to be purebreds, and all were linked to human deaths.

restrained approach, running the headline "Killer Pit Bulls Rip Granny to Shreds" and calling the dogs "bloodthirsty." A neighbor of Mazziotto's told *Newsday* that "the dogs were vicious," not because they had ever harmed anyone, but because "they barked a lot." Both pit bulls were impounded by the authorities, who planned to euthanize them, until Mazziotto's daughter, Terry Alaimo, stepped in. Alaimo couldn't believe that seven-year-old Shay and one-year-old Onyx would have harmed her mother at all, let alone killed her.

Alaimo contacted the Henry C. Lee Institute of Forensic Science in West Haven, Connecticut, and asked it to reevaluate the medical examiner's findings. The Lee Institute's forensic experts determined that Julia Mazziotto's wounds had been inflicted *postmortem* and that the elderly woman had actually died of a cardiac arrhythmia. The dogs had not been involved in her death, but after the fact they had pawed, scratched, and bitten her body.

Despite the experts' findings, Shay and Onyx remained impounded at the local animal control for eight months. A judge maintained that Onyx was "potentially dangerous" before releasing him back to Alaimo, making it nearly impossible for her to get homeowner's insurance afterward. Even in light of the new evidence, none of the papers that reported the incident as a pit bull fatality issued a correction.

In 2010, at least five news outlets reported that Ethel Horton, a sixty-five-year-old woman from South Carolina, had been killed by her nephew's pit bull. Dr. Janice Ross, the forensic pathologist who performed Horton's autopsy, found that Horton had indeed been bitten by the dog, but her bites were in no way fatal; Horton died from a heart attack after being bitten. Horton's most serious injuries, like Mazziotto's, were inflicted postmortem. James Chapple, whose 2007 death in Memphis was alleged by numerous Web sites to have been caused by pit bulls, was indeed severely *injured* by dogs during a gruesome incident in February, but he recovered and left the hospital. According to the Shelby County medical examiner and the Tennessee Department of Health, Mr. Chapple died months later of severe artherosclerosis exacerbated by hypertension, nephritis (kidney disease), alcoholism, hepatitis C, epilepsy, and "central cord syndrome" from a broken neck incurred when he fell down an embankment some years prior. In Katy, Texas, in 2012, multiple news agencies insisted that a ninety-six-year-old man named Juan Campos had been killed by his neighbor's pit bulls because there were bites to the man's lower legs, but the Harris County coroner could not determine if those wounds were pre- or postmortem. He ruled the cause of Campos's death "undetermined."

"If the coroner doesn't even know what killed the man," Delise said, "how would a local news reporter know?"

Even in cases where the breed ID was never questioned, and the dog responsible was a phenotypically distinct APBT, AmStaff, or American bully (no Staffordshire bull terrier has ever been linked to a human death in the United States), the circumstances surrounding these deaths, as in all DBRFs, were usually much more complicated than the press implied. In April 2005, a little girl named Cassidy Jeter was fatally mauled by two alleged pit bulls* in the backyard of her home in Hamtramck, Michigan, just outside Detroit. When the police arrived at the scene, they shot and killed both of the dogs. *The Detroit News* reported this as a straightforward "pit bull attack," saying that the child had "known the dogs since they were puppies" and that these well-cared-for "family pets" had suddenly turned on her as she walked through an alley to play on a set of swings. The community then clamored for pit bulls to be banned.

The police investigation, however, revealed that the two young dogs, an intact male and an intact female, were not "family pets" at all, at least not in the traditional sense of that term. They had belonged to Jeter's mother's recently deceased boyfriend, and Jeter's mother kept them isolated in the basement of a nearby abandoned house. One of the police officers told Delise that the male defecated a bright green ooze after it was shot, so its body was sent to the lab for a necropsy. The tests revealed that the dog had ingested a large amount of rat poison along with several carpentry nails and bits of cardboard. The stomach contents of the female were limited to cardboard and pieces of plastic. There was no food in the entire gastrointestinal tract of either dog, indicating that not only was one dying from poisoning but neither had eaten real food for days, if not weeks.

None of that changed the mind of Jeter's grandmother Brenda Ashford. "There's just something about that breed," Ashford told a reporter after the little girl's death. "There's a killer instinct that's in those dogs."

Delise showed me stacks of police photographs of emaciated dogs shivering in the snow or living in mud pits, in piles of trash, in crates or kennels too small for them to stand up in, dogs that had gnawed at their chains so long that they had filed their teeth all the way down, dogs with collars embedded in their necks or maggots crawling from their wounds, and females bred until their hips had given out.

In one DBRF that occurred on the South Side of Chicago in 2013, the owner had been cited by animal control so many times that the final report was ninety-six pages long. "No dog would have stood a chance in that house," Delise said. "How can we say this is about the dog's breed?"

Very few of the dogs that had killed people looked frightening. Many were smallish and terrified looking. "You know what really surprises me,

---

* Media reports offered conflicting breed descriptions.

though?" Delise went on. "How many dogs in America live like this and how few of them ever hurt anybody. It's just this tiny, *tiny* number. How come *that* isn't more indicative of their various 'breed traits'? Aren't the millions of dogs that *never* harm anyone more significant than the few that do?" She shook her head. "If any of these breeds were as vicious and unpredictable as some people make them out to be, then there should be blood running in the streets! Thousands and thousands of people would die from dog bites every year. And they don't."

Some animal lovers believe that any dog that kills a human *must* have been neglected or abused. That "there are no bad dogs, just bad owners." Delise did not agree with such a broad generalization. On that point, she defers to veterinarians and animal behaviorists, who for years have copiously documented chemical imbalances and neurological problems in dogs that do not result from abuse or training. As with humans, some dogs are simply not "all there," even if they have grown up in wonderful homes. The reverse is also true, as many animal rescuers can attest. Dogs raised in terribly deprived environments can be stable, loving family pets. It's not how the dog is "raised" that makes the largest difference in public safety but how the dog is maintained by its owner.

In the majority of cases, however, Delise did see a highly predictable—and therefore preventable—pattern. In 2011, after almost two decades of independent research, she partnered with the veterinary epidemiologist Dr. Gary Patronek at Tufts; Dr. Jeffrey Sacks, the public health epidemiologist at the CDC's Center for Injury Prevention; Dr. Amy Marder, an applied animal behaviorist at Boston's Center for Shelter Dogs; and Donald Cleary, one of her colleagues at the National Canine Research Council (the organization she founded to investigate DBRFs), to examine the coexisting factors in 256 dog bite deaths that occurred from 2000 to 2009. What they found was peer-reviewed and published in the *Journal of the American Veterinary Medical Association* in 2013.

Four or more significant variables related to the care, custody, or control of the attacking dog were present in the preponderance of DBRFs the authors studied. In 87 percent of the cases, the victim had been bitten when no able-bodied person was present to intervene. Three-quarters of these incidents occurred on or around the property where the dog lived (the dog's "turf"), and 85 percent of the time the victim was a visitor who did not know the dog or had no real relationship with it.

Because it is difficult for most dogs to kill a healthy human adult, people who are the most physically vulnerable, such as children under the age of nine and elderly adults, are the most likely to die from serious bite injuries. In the cases of child victims, lack of adult supervision was the most common recurring theme. Parents or babysitters had often left children alone with dogs inside the house or let them play alone with dogs out in the yard. In a number of cases, the child had wandered out of the house and approached a strange dog in the yard of a neighbor. To Delise and her co-authors, this underscored the need for *constant* supervision of children when animals are around, even if the dogs are small, just as injury experts recommend the constant supervision of children around swimming pools and other bodies of water, no matter how good a swimmer the child is.

Eighty-four percent of the dogs involved in DBRFs were not spayed or neutered, almost 80 percent weighed between fifty and one hundred pounds, and 60 percent were male. The dogs' level of socialization with humans played a critical role. Seventy-six percent of these cases involved dogs that were habitually isolated in some way, either chained in the backyard, kept in a small pen twenty-four hours a day, crated, or confined in a seldom-visited area of the house (such as a spare room or basement), and never allowed a normal social development. These Delise classified as "resident dogs," rather than family pets. Whether the dog lived inside or outside was less important than the amount of regular exercise, attention, and affection the animal received from its caretakers. If a resident dog remains outside, Delise stressed, it is imperative that it be properly contained so that it cannot escape and potentially harm a person or other pet.

In the cases of extreme animal neglect, certain scenarios—specifically, tethering—appeared to be more dangerous than others. Nearly one-third of DBRFs in Delise's files involved chained dogs. A 1994 case-control study of biting dogs co-authored by Jeffrey Sacks also identified chaining as a significant environmental factor. It is hard to know whether these dogs became dangerous because they were chained, or if they were chained because they were thought to be dangerous. One series of studies conducted in the Bahamas found that the negative media portrayals of pit bulls might contribute to owners assuming the dogs *need* to be chained. Whatever the reasoning, a neglected resident dog on a chain is more likely to become mentally frustrated from lack of exercise and fearful of strangers because it cannot flee to defend itself, while other people, dogs, and/or wildlife are free to approach, taunt, or harm it (in the Southwest, chained dogs are often called "coyote bait"). Unaltered males and unaltered females with puppies can become especially territorial in this scenario, the former even more so if they smell females in heat nearby. If the dog doesn't have proper shelter, then physical stress from lack of warmth in the winter or shade in the summer can also push the animal past its limits. A strange child wandering into its yard in these circumstances, especially if the dog is hungry or sick, is a recipe for disaster.

Finally, in more than 40 percent of DBRFs the owner had a record of prior mismanagement of the dog involved (such as citations for loose dogs or nuisance barking), and in a little more than 20 percent, he/she had a documented history of animal abuse or neglect.

If four or more of these important threads are woven into DBRFs so often, then the public's focus on breed as the only relevant factor in these rare but tragic incidents is in itself a dangerous miscalculation.

The majority of medical journal articles written about pit bulls were authored by physicians who were qualified to speak on subjects related to human health but not on matters of animal anatomy, health, or behavior. Over time, their research reflected the strong bias of the media, in some cases directly citing the popular press (rather than veterinary professionals) for "facts" on canine science. The claims made in these journals then fueled the same media myth machine they had drawn their information from in a sort of closed feedback loop, not unlike the dubious link between spitz dogs and rabies that prevailed among physicians in the 1870s.

After Pinckney and Kennedy's 1980 paper, the next two most cited medical journal articles about pit bulls are both case reports of dog bite injuries, or observations of individual patients, written up by surgeons. One was published in *Texas Medicine* in 1988, and the other was published in *The Journal of Trauma* in 1989. According to the five levels of evidence put forth by Oxford University's Centre for Evidence-Based Medicine, case reports and case series rank at number four (that is, next-to-least reliable in drawing scientific conclusions) because they are considered anecdotal. Anecdotes are used by scientists to form hypotheses, not to confirm them. Randomized controlled trials (RCTs) and systematic reviews of RCTs, on the other hand, are the most reliable. With injuries such as serious dog bites, there is a limit to what can be known from medical publications, because there will never be a randomized controlled trial in which dogs are allowed to bite and injure humans for research purposes.

In both of these case reports, the breeds of dog were identified based on whatever the patient or the patient's family member wrote on a medical intake form, and these were never verified for accuracy. One study that examined dog bites in Los Angeles found that from 1989 to 1996 only 138 of 1,109 dog bite victims, just 12 percent, were able to provide a breed description. "Most patients could only describe the size and colour of the dog that bit them," the authors wrote.

The authors of the 1988 case report opted to present serious dog bite injury as a pit-bull-specific problem based on severe injuries sustained by one man. "Most breeds do not repeatedly bite their victims," they wrote, "however, a pit bull attack has been compared to a shark attack and often results in multiple bites and extensive soft-tissue loss. Although the teeth of dogs are not very sharp, they can exert a force of 200 to 450 psi. Pit bulls inflict more serious bite wounds than do other dogs because they tend to attack the deep muscles, hold on, and shake." They continued with a brief history of the American pit bull terrier, noting, "The pit bull, it is said, is becoming the dog of choice for drug dealers."

It's generally frowned upon for scientists to speak outside their particular areas of expertise, but when they do, they are encouraged to base their claims on other peer-reviewed research. None of the above statements made by the authors, Steven Viegas, Jason Calhoun, and Jon Mader, are backed up elsewhere in the scientific literature. The authors cite one source for those statements, and that source is E. M. Swift's 1987 cover story for *Sports Illustrated.* (Benno Kroll's drug-addled dogfighting dispatch for *GEO* is also listed in the references.) For the specific information about the pit bull's bite, they cite an additional study, a 1980 paper on prophylactic antibiotics by Dr. Michael Callaham. Callaham's paper makes no reference to any breed of dog or any level of bite force. The sentence appears to have been lifted, almost verbatim, from another of Callaham's publications, one that does not mention dog breeds.

Because of the common claims about the unique nature of pit bull bite injuries, I asked to see the photographs Delise had received from medical examiners in DBRF cases. The images of crime scenes and autopsies were hard to stomach, but the injuries looked virtually identical, regardless of the type of dog involved. For a nonvenomous animal to bite and kill a human being, the injuries must be very serious, or the victim must be extraordinarily fragile. In 2012, a Siberian husky merely lifted a two-day-old infant out of a car seat by his head (the car seat had been placed on the living room floor, and the child's mother briefly left the infant unattended with the dog), and those injuries were serious enough to cause the baby's death.

A 2014 study of facial dog bites in the Denver area found "no significant association between the type of dog breed and the number of bite injuries." In addition, "there was no statistically significant association between wounds needing reconstruction versus direct repair according to dog breed." Forensic bite-mark specialists in Germany lament, "The differentiation of dog races is problematic . . . [In a previously cited case] it was impossible to differentiate between [the bites of] a Rottweiler and a German shepherd." The most recent edition of the American Board of Forensic Odontology's reference manual, the official textbook for experts in dental identification and bite-mark analysis, does not mention breed of dog at all.

When I consulted Dr. Kenneth Cohrn, a forensic odontologist in the Department of Pathology at the University of Florida who works with UF's veterinary forensics lab, he replied in an e-mail, "I would not look at a dog bite and say which breed is responsible. What you look for is individual variation, missing or malpostioned [*sic*] teeth, teeth spacing, chipped or broken teeth and the like."

In 1989, Dr. Bret Baack and his colleagues published another case report, this one documenting injuries sustained by one patient: a nine-year-old girl severely bitten by dogs in New Mexico. Like Viegas and his colleagues, the authors concluded that "a pit bull terrier injury tends to be more severe than that caused by other breeds because of the propensity of these dogs to bite deeply into the tissues and to hold and tear rather than merely snap and recoil. Pit bulls bite with greater force than most dogs (up to 1,800 lb/sq. in.)." Here we have a claim that is opposite the one made by Viegas, who wrote that the pit bull was uniquely dangerous because it "repeatedly bites" its victims. The source for Baack's bite force claim is another phantom reference, a 1983 paper on dog bites and penicillin that makes no mention of either pit bulls *or* bite force.

Baack and his co-authors continue, "Once [pit bulls] have their victim in a hold, they do not merely maintain the 'bite,' but continue to grind their premolars and molars into the tissue while the canine teeth stabilize the hold." It is physically impossible for a dog to hold on to something with its front teeth and simultaneously "grind" with its molars. This claim is sourced to a 1980 document on dogfighting published by the Humane Society of the United States. Baack's other sources are Swift's *Sports Illustrated* article and the 1987 feature from *Time*.

These misleading references pile up. In 2007, a doctor from the Wayne County, Michigan, Medical Examiner's Office, Cheryl Loewe, co-authored a similar paper, also a case report, on the injuries sustained by six children from the Detroit area who had been killed by dogs, specifically by pit bulls, over a period of almost twenty years. One of them was Cassidy Jeter, the six-year-old who died in Hamtramck in 2005, but the authors made no mention of the attacking dogs' starvation and poisoning. "Many of the canines involved in dog attacks," Loewe wrote, "can generate up to 1800 pounds of force per square inch with a bite, which is enough to penetrate sheet metal." The cited source is a 2001 paper that notes, "Because it is illegal to own a pit bull in the County of Denver, we rarely see injuries by this breed."

"When you actually go back and read through these articles, line by line," Delise told me, "you begin to ask yourself, why are medical professionals writing about animal behavior in the first place? If they want to write about antibiotic treatment for dog bites or stitching up bite wounds, that makes sense. But animal behavior? Would any of these journals allow veterinarians or behaviorists to write about human surgery?" She paused. "All my life, I trusted that what I read in places like this was accurate; that someone had checked it out. I assumed that doctors were careful people who know a lot more than I do. Then I find these glaring errors, and who am I? I'm nobody," she said. "How did an article that cites *Sports Illustrated* pass muster with a peer-review board of scientists?"

By far the most baffling example of flimsy citation in the medical literature occurred in a paper written by doctors at the University of Texas Health Science Center's Department of Surgery in San Antonio that was published in a 2011 issue of the *Annals of Surgery*. In an attempt to "prove" that pit bulls inflict more severe injuries on dog bite victims than other types of dogs, Drs. John Bini and Stephen Cohn and their co-authors examined 228 cases of dog bite injuries from their hospital's records. In only one-third of those incidents were breed descriptions of the biting dog available, and as always these were self-reported identifications that were not independently confirmed. Using that third as their data set (thus leaving out two-thirds of the injuries), the authors then relied on a grab bag of uncredentialed and unscientific sources for both data and context, including two dog rescue Web sites, a television show produced for entertainment purposes by the *National Geographic* channel, the Web site of a personal injury lawyer, and a YouTube video created by DogsBite.org, an activist group dedicated to banning pit bulls. John Bini, the paper's lead author, did not respond to my requests for comment about why he and his authors considered these sources reputable.

DogsBite.org was created by a Web designer named Colleen Lynn who was bitten on the arm twice "for approximately five seconds" by an unaltered male "pit bull mix" while jogging through a Seattle neighborhood in 2007. During the incident, Lynn's arm was fractured, either by the dog or from her subsequent fall to the ground. Lynn sustained seven puncture wounds (equivalent to two total bites), but her broken arm required a stabilizer bar and screws to repair, making her injury serious enough to warrant a dangerous dog investigation. The dog that bit Lynn was subsequently euthanized, and Lynn received a sizable payout from the owners' insurance company. She then dedicated herself to the promotion of breed-ban laws.

Lynn has no professional credentials in statistics, epidemiology, or animal behavior; neither do the sources she relies on most frequently. Before her bite injury, Lynn maintained the fortune-telling Web site Divine Lady.com, on which she referred to herself as "Divine Lady, Beholder of the Soul." In 2011, she self-published the third edition of *Divine Lady's Guide to the Runes*. (The original Divine Lady Web site now redirects to RuneCast.com, owned and operated by Lynn Media Group.) Lynn did not respond to my requests for an interview.

Like many other journalists, I initially believed that DogsBite.org was well sourced and evidence based because of its very polished and professional layout. To her credit, Lynn is an excellent Web designer. When I read the fine print, however, I noticed a number of warning signs. For one, DogsBite.org contradicts everything put forth by the groups most qualified to speak about animal science, animal behavior, and dog bite epidemiology: the American Veterinary Medical Association, the Centers for Disease Control and Prevention, the American College of Veterinary Behaviorists, the Animal Behavior Society, the National Animal Care and Control Association, the Association of Professional Dog Trainers, and almost every other animal welfare organization in the country other than PETA. According to DogsBite.org, these groups have been co-opted by the "pit bull lobby," a shady cabal that supporters of the site imply is financed by dogfighters. The site is also littered with childish ad hominems like "pit nutter" (an epithet for pit bull owners) and "science whore," a term used to describe the veterinarians and behaviorists who insist that there is no scientific basis for breed bans. Worse, commenters are allowed to express violent fantasies about killing dogs ("Pit Bulls should be used for target practice," one wrote in 2008). In its sidebar, DogsBite.org links to blogs filled with gratuitous photographs of pit bulls that have been shot or stabbed under headings such as "Fuck Pit Bulls and the Faggots Who Own Pit Bulls." While there is nothing at all wrong with dog bite victims providing support for one another, or with venting their outrage on the Internet in whatever manner they choose, what appears on this Web site should not be confused with credible scientific information.

Yet, in the quest for journalistic balance, many reporters have given DogsBite.org and the Centers for Disease Control and Prevention equal airtime, framing the issue as though two scientifically rigorous institutions just happen to disagree. In most cases, reporters don't even bother to consult the CDC (or the AVMA, or any other groups of professional scientists), preferring to pit DogsBite.org against animal rescuers or shelter workers, none of whom are qualified to speak on scientific matters of behavior, genetics, or public health.

Most of the information on DogsBite.org comes from one self-published paper on "dog attack deaths and maimings" by a man named Merritt Clifton, the former editor of the *Animal People* newsletter who runs an animal rights news blog called *Animals 24-7* and claims to have been bitten by a pit bull in 1982. Because Clifton's paper has been cited both in the mainstream press and in several court cases as evidence of the rational basis for breed-specific legislation, a thorough examination of it is necessary. Like Lynn, Clifton possesses no relevant credentials. His supporters call him an "award-winning journalist," but he readily admits that his research methods are limited to scanning media reports and classified ads rather than personally speaking with investigators or reviewing primary source documents. The award he received in 2010 was presented to him by the administrators of an infectious-disease-tracking Listserv, not a journalistic organization.

According to Dr. Rory Coker, a physics professor at the University of Texas at Austin who has taken great pains to outline the differences between empirical science and pseudoscience, the hallmark of pseudoscientists is their tendency to "clip newspaper reports, collect hearsay, cite other pseudoscience books," rather than making "an independent investigation to check their sources." Additionally, "pseudoscience always avoids putting its claims to a meaningful test," especially peer review, and "pseudoscientists are fond of imaginary conspiracies."

Clifton's dog attack "report" fulfills almost every criterion on Coker's list, particularly this one: "The emphasis [of pseudoscience] is not on meaningful, controlled, repeatable scientific experiments. Instead it is on unverifiable eyewitness testimony, stories and tall tales, hearsay, rumor, and dubious anecdotes. Genuine scientific literature is either ignored or misinterpreted."

Clifton's paper has never been peer-reviewed and it contains no citations, so the information therein cannot be verified. It does not draw upon government sources, public health records, or expert opinion. In defense of his reliance on news clips, Clifton writes, "Media coverage incorporates information from police reports, animal control reports, witness accounts, victim accounts in many instances, and hospital reports. Media coverage is, in short, multi-sourced, unlike reports from any single source." In other words, he is content to let others do his reporting for him.

Then there is the problem of his data. Clifton's paper purports to cover both the United States and Canada, but as of this writing it contains only 3,533 human victims over a thirty-year period. The Agency for Healthcare Research and Quality estimates that 9,500 Americans are hospitalized overnight for dog bite injuries *each year.* If his paper is to be believed, Clifton has captured only 2 percent of serious American dog bite injuries, an even smaller fraction of the total if you add in the Canadian data. ("The severity of the logged attacks appears to be at approximately the 1-bite-in-10,000 level," he writes. Unfortunately, neither he nor anyone else can determine the severity of the bite injuries the media does not cover.) Clifton is adamant that this tiny sample shows the sky-high "actuarial risk" of pit bull ownership in both countries. More worrisome is that he defines "dog attack deaths" as deaths in which dogs were only tangentially related, such as victims who ran from dogs and were hit by cars or, in one case, a child who was strangled by a scarf that a dog pulled on. Numerous deaths on Clifton's list are contradicted by official medical examiners' reports.

As far as disfiguring injuries go (another metric Clifton includes in his paper), the American Society of Plastic Surgeons reported that in just three years, 2011, 2012, and 2013, more than eighty-four thousand Americans received reconstructive surgery for dog bites. That number has *dropped* almost 40 percent since 2000, even as the dog population has increased. On top of those inconsistencies, Clifton includes breeds of dog in his data set that do not exist, such as the "East Highland Terrier."

"Only 15 fatal dog attacks are known to have occurred in the entire span from 1930 through 1960," Clifton asserted in 2014, "including nine by pit bulls, two by Dobermans, and four by unidentified mutts." Within a few hours at one university research library, however, I located seventy-two separate cases in the United States alone, the majority of which were attributed to German shepherds or "police dogs." Had I included Canada, as Clifton claims he does, or broadened my search to include deaths in which dogs were vaguely involved (cars, scarves), that number would have been much higher.

Despite telling the audience at a 2014 animal rights conference, "I have more than a hundred peer-reviewed publications," Clifton has authored only two papers that might generously be considered "peer-reviewed." Both were papers on rabies published in *Asian Biomedicine,* an open-access journal based in Bangkok. In one of them, he cites his own newsletter twelve times. The rest of the citations came mainly from Web sites, news reports, and press releases.

When I offered to travel to his home on Whidbey Island, Washington, to interview him, Clifton not only declined my request but replied, "You are not a journalist; you are an advocate," called me a "pit bull Holocaust denier" (!), and made a jab about my deceased father, whom I presume he learned about by looking me up on the Internet. I asked by e-mail if he would be willing to provide the sources for some of his more outrageous statements, but he did not respond.

The lack of a scientific foundation for the claims made about pit bulls in the press, in medical journals, and in courts of law is deeply disturbing. Jeffrey Sacks likened dog bite research to the well-worn joke about the drunk trying to find his keys in a darkened parking lot. Another man comes out, sees the first one crawling around on his knees, and asks what he is doing. The first man says, "I've lost my keys! I can't find them!"

The second says, "Well, where were you when you dropped them?"

The first points to the far side of the lot and says, "Over there."

"Over *there*?" the second man says. "Then what are you doing all the way over *here*?"

The first man points up to the streetlamp and replies, "I'm looking where the light is."

CHAPTER 12

"DON'T BELIEVE THE HYPE"

> The problem with stereotypes is not
> that they are untrue, but that they are incomplete.
> They make one story become the only story.
>
> —CHIMAMANDA NGOZI ADICHIE,
> "The Danger of a Single Story"

Dave Wilson remembers the exact moment he saw his first American pit bull terrier. It happened one morning in 1984 when he was in the ninth grade. He had missed the bus to high school and had to walk instead. A white dog tied to a tree in front of a neighborhood housing complex caught Wilson's attention immediately. "At first I was a little bit afraid of him, to be honest with you," Wilson told me. "He was just so . . . *confident.* But I was also completely infatuated with him, almost obsessed. He was different from anything I'd ever seen. Before long, I was missing the bus on purpose, just so that I could walk by and see him. His name was Cujo, but he wasn't a mean dog at all; he just *looked* tough."

Where Wilson came from, a "lower-middle class" area of Prince George's County, Maryland, just outside Washington, D.C., nothing was more important than looking tough. "When you're a young guy growing up in an environment where everyone is trying to be that way," he said, "you want to be cool and portray that image." The legend of the APBT as an "ultra dog" was just starting to reach the D.C. area, and its reputation as a reluctant but fierce defender greatly appealed to Wilson and the people he hung around with. Owning a muscular pit bull seemed to be a good way to complete the picture he wanted to project, but deep down, what he really wanted was a companion. The teenager saved up $50, which he promptly forked over to Cujo's owner for a female puppy he named Jinx.

At roughly the same time, the burgeoning hip-hop movement was crossing over into the mainstream and becoming a major cultural force, one that simultaneously reflected life on the streets and reinforced it. Within a few years of Wilson first seeing Cujo in Maryland, pit bull dogs would be taken up as that movement's unofficial mascot. The dogs made appearances in the lyrics of Big Daddy Kane and Salt-N-Pepa. Tone Lōc bragged to reporters that he owned a pit bull named Yesca (Chicano slang for marijuana), and the gangsta rapper Ice-T taunted on his *O.G. Original Gangster* album that he had a "dope" pit bull named Felony. Soon pit bulls were pictured alongside Snoop Dogg and Dr. Dre, as well as on album covers for DMX and Missy Elliott. Eve, the first female rap artist to join up with DMX's Ruff Ryders label, often called herself a "pit bull in a skirt." The video for one of Jay-Z's biggest radio hits, "99 Problems" (1999), featured grainy black-and-white footage of a staged dogfight right before an appearance by rock producer Rick Rubin. Were rap artists meditating on the reality they came from, or celebrating violence? It was often hard to tell.

The images of pit bulls in early hip-hop, like the images of expensive jewelry and fancy cars, were images of power, and they became most prevalent during the years when the majority of hip-hop consumers—the urban underclass—didn't have any. Thousands of unemployed young people roamed the streets of depopulated cities that had been left to rot. In the words of the music journalist and historian Jeff Chang, those young people "built codes, rules, and vocabularies for themselves to compensate for scarcity and lack. Their play was the organized chaos of the unseen and the unheard."

While designer labels connoted wealth, pit bulls connoted strength. But not only strength. They also symbolized rebellion, self-sufficiency, and a willingness to defend oneself at any cost—the qualities most necessary to survive life on the streets. That the mainstream media portrayed the dogs as terrifying and unpredictable only boosted their stock with rappers, which then made them even more desirable to disaffected youth, including gang members, but by no means limited to them.

Much like baggy jeans or oversized T-shirts, pit bulls eventually became another normal part of street style, and their significance was wide open to interpretation. In the pit bull, a young man could see whatever version of himself he wanted, from family defender to resourceful hustler to unbowed survivor. Never before had a generation of African-Americans, especially the urban poor, connected with a specific type of dog they felt represented their collective identity; in fact, throughout American history, their own dogs had systematically been taken away from them while whites used dogs to keep blacks in line. Among the most iconic images of the civil rights movement were those of Bull Connor's police officers setting their attack-trained German shepherds on peaceful demonstrators, including children, in the spring of 1963. The comedian Richard Pryor joked about being terrorized by police German shepherds in several of his routines. If the German police dog was the Apollonian enforcer of law and order, then the smaller, sleeker pit bull was its Dionysian counterpart.

"Black males, we are America's pit bull," the actor Michael B. Jordan said while promoting the film *Fruitvale Station* in 2013. "We're labeled vicious, inhumane and left to die on the street." In the film, Jordan plays the role of Oscar Grant, a young unarmed African-American man who was shot and killed by a white police officer in Oakland, California, in 2009. In one of the film's scenes, Grant attempts to rescue from the middle of the road a pit bull that has been hit by a car. "Oscar was [also] kind of left for dead," Jordan said. "So many of us young African-American males left for dead . . . We get branded a lot for being vicious, not human, so we wanted to show the humanity."

But the romance of the urban pit bull, like the romance of the "thug life" that early rap music glamorized, was punctured by the ugly and unglamorous reality of poverty, unemployment, and senseless violence that plagued American cities during that time. In one of his memoirs, *The Ice Opinion*, Ice-T described the New Jersey neighborhood where he grew up as "the compiled interest of lost hope" where "anybody weaker than the next man will be victimized by the stronger." Human life was cheap and getting cheaper when he wrote those words: almost twenty-five thousand Americans were murdered in 1993 (an all-time high), and black men were by far the most common victims. Once street dogfighting found its way into the most blighted corners of the country, where a kill-or-be-killed ethos permeated daily existence, a pit bull's life was the cheapest of all, as Mike Sager had described in his 1987 *Rolling Stone* feature.

Historically, professional dogfighters in America were white men from rural backgrounds. After the media coverage of fighting in the 1970s, that demographic shifted dramatically. "Many of the dogfighters at that time were from the South, and they were still racist," Joel Halverson, a former APBT conformation judge for the American Dog Breeders Association who has raised pit bulls since the 1970s, told me. "They wouldn't sell good dogs to [black people], and they wouldn't teach them how to go about things. So these guys got taken advantage of, and they just did whatever they wanted." The new brand of urban street fighter developed a reputation for being even more ruthless than his predecessors.*

Scenes of street dogfighting then began showing up in hip-hop iconography, most memorably in the work of the rapper Earl Simmons, known as DMX. In his 2002 memoir, Simmons described at length the loneliness and brutality he experienced growing up in the Yonkers housing projects and the constant rage he felt at the broken world around him. When not bouncing in and out of juvenile homes (he entered his first when he was ten years old), Simmons spent most of his time alone on the streets with neighborhood dogs. Because his mother would not allow him to keep a pet dog of his own, Simmons cared for strays on the roof of his apartment building, where he fed the animals table scraps and gave them old blankets to sleep on. He wrote that the dogs provided "the love and companionship I needed . . . I knew I would never be alone again."

But as it is with many young men who grow up in similar circumstances, love and violence were tightly interwoven for Simmons. After encountering multiple street dogfighters in his neighborhood, all of whom radiated a toughness that he did not, Simmons stole his own pit bull, Boomer, from a local junkyard. "Boomer was like a real person," Simmons recalled, "a nigga with his own unique personality and attitude. He was my companion, my ride. Anything I did, Boomer and me did together. Anything I got, Boomer and me shared. We were inseparable . . . [T]he dog was more like me than I ever imagined, and I was like my dog." One of Simmons's business partners later told a reporter for *Billboard,* "I think Boomer was literally [Simmons's] one and only best friend growing up."

Part of what Simmons enjoyed about Boomer was that other people in the neighborhood were afraid of him. "Nobody could fuck with me and my dog," he wrote. Left to his own devices, however, the disturbed young man's relationship with Boomer darkened. Before long, Simmons was setting the dog on alley cats and fighting him in basement projects.

* Whether or not that is actually true is hard to say, because both strains of cruelty were growing increasingly violent. By the mid-1990s, there was also a great deal of crossover between urban and rural dogfighting, especially in places like Texas, Ohio, and Missouri.

The victim had turned victimizer. Even after he became a major commercial success in the music industry, Simmons never left those early years in the projects far behind. He was arrested twice on animal cruelty charges when the bodies of three dead pit bulls were found buried in the yard of his Arizona home. (He would also be diagnosed with severe bipolar disorder.)

In his Maryland neighborhood, the young Dave Wilson felt the aftershocks of this growing phenomenon but wanted something different for himself and his dogs. Back in the 1990s, if you lived in a place like the Washington, D.C., area or the South Side of Chicago or South Central Los Angeles, there were very few places for dog lovers to gather unless they had a great deal of money. The vision of pet culture advertised in Purina commercials was thought to be the province of the rich, and pit bulls were summarily excluded from that picture.

"What we were told back then is that fighting was all these dogs could do and all they were good for," Wilson said. "But even though I knew people involved with that, I never could actually get myself to do it. Something conflicted internally. My dogs always wanted to please me, and I could never imagine putting them in harm's way." But he was still hanging with the wrong crowd, which led to a series of run-ins with the law, the specifics of which he didn't want to revisit. "I didn't finish school; I was sent to juvenile detention centers," he finally said. "My life was headed in a very bad direction." After being incarcerated for more than a year, Wilson vowed that he would turn his life around.

Knowing that Wilson was interested in pit bulls, a friend gifted him with a book about the American Staffordshire terrier, the AKC show breed derived from the APBT, which seemed to be free of the baggage Wilson wanted to avoid. "I didn't want to go back to the pit bull world I came from," he said. "It was just too negative. The AmStaffs had the same look of toughness that I liked, but they seemed more elegant and flashy and beautiful. They also seemed more laid-back." After Wilson resumed his normal life, he became a dog trainer and decided that devoting himself to the AmStaff provided a much better way forward. One AKC AmStaff in particular, Steeltown's Blue Monday, had a Weimaraner-ish silvery blue coat that Wilson found extraordinary. Within a few years, he saved up enough money to buy eleven acres in the Virginia countryside so that he could concentrate on raising AmStaffs and stay out of trouble. "There's no pizza delivery out there, that's for sure," he joked, "but the dogs helped me focus on doing something positive with my life."

The blue color that he had seen and loved in Blue Monday was a mutation of the allele for a black coat that is typically seen as a fault in American pit bull terriers but is celebrated in AmStaffs. Wilson found it in a few other lines and began breeding his own blue dogs. They were so "sharp" looking that he called his bloodline Razor's Edge, a name he and a friend came up with when they were still kids. Wilson then began entering his dogs in AKC conformation shows. "The [APBT] world had been mostly men, but the AmStaff world was me and a bunch of old ladies," Wilson said, laughing. "The good thing was, they were *single* old ladies, and I was a young guy, so they would sell me the dogs that I wanted."

Unlike the breeders of working APBTs, Wilson did not see any future in breeding for gameness. He consciously selected for mellower animals that didn't need an intense amount of exercise, because he felt that this attribute was more suitable for pets living in densely packed urban environments. He wanted his dogs to look imposing on the outside but have more layabout, marshmallow temperaments. This was a completely different goal from the high-octane athleticism that the old guard of APBT lovers was trying to attain.

As time passed, the Razor's Edge look grew stouter and more bulldog-like. The dogs had larger heads, broader chests, and shorter legs than other AKC AmStaffs, which made them stand out. Wilson nicknamed his dogs "American bullies." He then joined the UKC conformation circuit and registered his AmStaffs as APBTs (the UKC allows AKC dogs to be dual registered, because the two breeds started from the same stock), which gave him a toehold in the cultures of both clubs. Around 2002, Wilson partnered with a Los Angeles breeder named Richard Barajas, the founder of another "bully" line of AmStaffs called the Gottiline who had similar goals: refining the pit bulls that had become so popular in urban neighborhoods into a flashy show breed valued for its looks, not for fighting.

The stockier shape and easygoing reputation of the American bully caught on. Wilson soon gained a following of fans that came from backgrounds like his, people who wore ball caps and gold chains and would have otherwise felt uncomfortable in the tweedy world of dog shows. Though Wilson was white, most of his friends and followers were not, and he was overjoyed that his presence on the kennel-club circuit had inspired a younger, more diverse crowd to show an interest in canine conformation. Finally urban dog lovers would have a place to show off their pets in a healthy, fun way. He assumed the kennel clubs would appreciate this infusion of new blood, too. He was wrong.

According to Wilson, the UKC then changed its APBT breed standard to exclude American bullies from competition. (The UKC did not respond to my requests for comment about Wilson's statements.) After that, Wilson and his friends began to feel that they were not welcome at UKC shows, even if they stayed on the sidelines. In a 2013 Web post, he recounted a UKC event in California during which, he says, the hosts called the police and asked the American bully crowd to leave. "I went to speak to the host and the exact response they [*sic*] gave me was . . . they are gangsters and thugs and we don't want them here," Wilson wrote.

> I looked back and saw people with kids, people cooking food on grills, people smiling and just trying to get out and do something fun with their dogs . . . I understand if something is disruptive to the show, but this was not the case. This was discrimination, I realized first they discriminated against our style of dog, and now they are discriminating against people . . . People trying to do something positive with their dogs were treated as if they were criminals . . . When I came back I felt something needed to be done, these dogs and the people who supported them needed a home and a way to do something positive with the breed.

At the same time that this was happening in kennel clubs, vocal animal advocates were criticizing urban youth for breeding dogs, yet they also refused to adopt out pit bulls from their shelters. They then euthanized the dogs after telling the public that the animals were "unwanted." In 1999 alone, the Pennsylvania SPCA (PSPCA) put down more than *four thousand* pit bulls rather than put them up for adoption. When asked to explain this policy, Erik Hendricks, then the organization's executive director, pointed out the "very high percentage of inner-city kids with these pit bulls." He went so far as to support a breed ban in Philadelphia, saying, "There's a definite racial element here . . . The fact is that most of the people we're talking about are black and Hispanic kids in their teens and twenties. They see the dogs as macho and think of themselves as having the dog's characteristics: toughness, courage, power . . . because minority kids are involved, many politicians are reluctant to take a tough stand." The message this sent to the wider public was that pit bulls could not be pets and the people who wanted them didn't count as worthy pet owners.

In San Francisco in 1997, it was more difficult to adopt a pit bull from that city's SPCA than it was to purchase a firearm. The adoption process required not only that the potential owner be fingerprinted but also that he or she submit to a personal interview, a criminal background check, several home inspections, and interviews with both his veterinarian and his landlord. The adopter was then asked to submit *five* personal references, attend mandatory dog-training classes, and have the dog spayed/neutered, photographed, and microchipped. This gauntlet was so daunting that most potential owners didn't even consider running it. Why sign up to be treated with such suspicion when you could get a free puppy from a friend or neighbor?

At some shelters, the very fact of wanting a pit bull marked a potential adopter as an "unsuitable" candidate. The executive director of the Montgomery County SPCA in Norristown, Pennsylvania, said bluntly in 2000, "The people interested in [pit bulls] we will not adopt to." The president of the Central Westchester Humane Society in Elmsford, New York, supported this approach. "If someone comes here looking for a pit bull," she said, "I'm immediately suspicious." Stephanie Shain, the chief operating officer for the Washington Humane Society in Washington, D.C., recalled that a colleague from another organization once said to her, "You should see the people that come in wanting to adopt pit bulls. They are simply *disgusting.*"*

Like their owners, the dogs tended to be stereotyped as "criminals." Police officers in Harrisburg, Illinois, held stray pit bulls at gunpoint until animal control arrived to collect them. Once in shelters, these "perps" (as one *Newsday* reporter referred to them) were typically isolated from other animals out of fear that their "innate fighting instincts" might take over. This approach denied them any chance at developing healthy canine social skills, which, like a self-fulfilling prophecy, led them to be reactive with other dogs. Shelter staff at a number of facilities were prohibited from walking or handling pit bulls at all, not even for reasons of basic sanitation. Instead, the pit bulls' kennels were hosed down with the dogs inside. When these animals eventually went "kennel crazy," the dogs were labeled as "aggressive" and euthanized.

Then, of course, there was the baffling position of PETA, long considered the hardest of animal rights hard-liners; in fact, the organization believes that pet-keeping of any type is exploitative and should be

---

* The PSPCA and the Washington Humane Society, both of which once banned pit bull adoptions, have abandoned breed-specific policies and have seen dramatic increases in their adoption rates. Both shelters are now leaders in community outreach. The San Francisco SPCA has relaxed its adoption procedures as well.

abolished. Its anti–pit bull rhetoric was cleverly wrapped in the language of concern. But it was also highly inaccurate. "Pit bulls are perhaps the most abused dogs on the planet," Ingrid Newkirk, the group's founder and president, wrote in one widely circulated newspaper editorial. "These days, they are kept for protection by almost every drug dealer and pimp in every major city and beyond." Newkirk believed that this was reason enough not just to bar pit bull adoptions from shelters but to eradicate pit bulls from American communities. "People who genuinely care about dogs won't be affected by a ban on pits," she wrote. "They can go to the shelter and save one of the countless other breeds and lovable mutts sitting on death row through no fault of their own."

Given this climate, Wilson began his own registry, the American Bully Kennel Club, in 2004. "The UKC told us that our dogs didn't meet their standard," Hector Lopez, an American bully breeder from Long Beach, California, told me at an ABKC show in Atlanta. "Really, though, I think it was because they didn't want people like us around." Shawn Mullins, a breeder from Oklahoma, echoed Lopez's experience: "Since the UKC didn't seem to want us, we began putting together our own events to celebrate our dogs."

Lopez, Mullins, and others like them were drawn to the American bully for the same reason that originally motivated Wilson. They loved the flashy appearance of large pit bulls, and they loved seeing their dogs in thick studded collars and other "tough" accessories, but this, they told me, was pure pageantry. Underneath it, they wanted what Lopez called "maximum couch-potato" companions.

American bully breeders took it upon themselves to create an entirely different kind of dog show, a family-oriented event with food and music (much like a neighborhood block party) rather than a competitive, buttoned-up affair. They called these gatherings "bully barbecues" and invited dog rescue groups to attend as well. They also added a Save-a-Bully category to their formal competitions so that dogs adopted from rescues and shelters could compete, despite not having pedigrees. "After we held the first few bully barbecues in Los Angeles," Shawn Mullins said, "guys from all different neighborhoods—guys who thought they hated each other—would come together and have a good time, just because they loved their dogs. They put their issues aside and focused on what they had in common." When I met Mullins at a 2013 ABKC event, one of his dogs was decked out in a spiffy dress collar and sat on a literal pedestal in front of him, unfazed by the swirling crowd of animal lovers. Several children and their parents stopped to take photos with the slobbery giant. Mullins smiled and scratched his dog's ears. "*That's* what these dogs are about. *That's* what these dogs can do. They bring people *together.*"

The culture of the ABKC presented the pit bull as both a highly prized companion and a luxury investment, something to be cherished and protected, not abused. The value of the dog was no longer attached to a devastating culture of violence and death. Now a shiny coat and bright eyes were more esteemed than visible scars or a scrappy temperament.

Over time, the American bully replaced the APBT as a favorite with the next generation of hip-hoppers and professional athletes. Few of today's American bully breeders sell pups for less than $800, and many charge closer to $2,500, sometimes more than twice that much. Antwan Patton, better known as the hip-hop megastar Big Boi of Outkast, runs an American bully kennel in Atlanta with his brother, James, called Pitfall. Puppies from Pitfall Kennels (whose bloodlines are mostly Razor's Edge) have sold for more than $6,000 apiece to a roster of celebrities that includes the tennis champion Serena Williams, the rapper 50 Cent, and the music producer turned mogul Jermaine Dupri. Thanks to such high-profile proponents of the breed, the ABKC now registers more than 100,000 dogs a year from thirty different breeds and has a presence on five of seven continents. In 2013, the UKC recognized that the American bully was here to stay and accepted the bully into its registry as a separate breed.

For any dog breed, popularity comes with problems, especially when money is involved, and overbreeding is one of the largest. Since the early 1990s, a number of American bully breeders have played fast and loose with their dogs' pedigrees, crossing in other types of bulldogs and various mastiffs to make their dogs heavier. These dogs can weigh well over one hundred pounds—three times the size of the average American pit bull terrier. They are getting bigger every year, which once again raises the age-old question of just how far the term "pit bull" can be stretched. It is not uncommon in the American bully world for breeders to advertise their stud dogs based on the measurements of their head size, with the largest heads being considered the most impressive.

Wilson is particularly pained by the breeders who disregard their animals' health in the quest for a more extreme, oversized look; as a result, "exotic" American bullies have begun to suffer from the same hereditary health defects as English bulldogs (breathing problems, hip and joint problems, and difficulty whelping naturally, among them), which is why Wilson has banned exotics from ABKC competition. Likewise, some of the young upstarts who want a piece of the "bully game" have no knowledge or experience in animal breeding, and they don't care to learn from their elders.

When he first started the ABKC, unscrupulous breeders were not Wilson's only problem. Many critics looked at the club's predominantly urban demographic and incorrectly assumed that it endorsed dogfighting. A 2006 editorial in a Fredericksburg, Virginia, newspaper politely referred to an upcoming ABKC show as "a questionable source of entertainment." A columnist for the same paper wrote, "Not everything primitive walks on all fours. Pit bulls have become a macho status symbol among ghetto toughs and rural rednecks, who meet at an improbable crossroads of intimidation and swagger."

In 2009, a lobbyist for the Oklahoma Animal Protection Association mounted an unsuccessful campaign to have ABKC events banned across the state, saying that the group existed only to promote dogfighting. "It's outrageous," the lobbyist told a reporter. "It's the equivalent of allowing a Klan rally and saying it's a social dialog summit." But he seemed unaware that ABKC dogs are much too fat and slow to have ever been used for that purpose, as anyone who has seen them up close would know. What's more, the ABKC prohibits any discussion or promotion of illegal activities at its events. A member who is suspected of harming his dogs in any way is promptly banned for life.

But the stereotypes are difficult to shatter, and as in all groups there are some who live up to the worst of them. American bullies remain outcasts in the already cast-out world of pit bull culture. "Anything that portrays pit bulls in a tough or macho way is not helpful," one humane activist said when asked about the ABKC in 2008. "These dogs in too many cases are being kept as status symbols for their owners. At the end of the day, dogs are better off with people who care about them."

Inside the Meadowlands Exposition Center in Secaucus, New Jersey, three main things separated the American Bully Kennel Club's national dog show from its AKC, UKC, and Westminster brethren: the amount of syncopated bass vibrating up the walls from the deejay's central command post (the hip-hop superstars Lil Jon and Usher were shaking the speakers when I arrived), the presence of a full cash bar, and the number of parents pushing strollers through the expo hall. Having anticipated this, the show's organizers had set up a large children's play area near the back of the room and filled it with coloring books and boxes of crayons. Every time the deejay threw a Taylor Swift song into the rap rotation, the toddlers camped out in the kiddie corner bounced and clapped, shaking their diapered rumps to the beat.

It is estimated that preparing a show dog for Westminster costs between $50,000 and $100,000. Approximately a quarter of that goes toward hiring a professional handler to make the dog look good in the ring. That was not the case here. ABKC members bred, trained, groomed, and handled their own dogs. The WASPy sense of kennel-club entitlement that Christopher Guest parodied in his 2000 dog-show send-up, *Best in Show* ("We were so lucky to have been raised on catalogs!") was replaced by full-throated street swagger. The kennels' names—Thug Blood, Outlaw, Cash Money—invoked brash hustler machismo, while *Atomic Dogg*, the lifestyle magazine dedicated to the "bully game," featured photographs of greased-up bikini models and shiny cars with twenty-inch rims ("dubs") next to dogs with giant heads.

Nothing made a more powerful visual statement than the merchandise tables, which displayed hundreds of spiked collars that glinted in the fluorescent lights. There were spikes for every taste and occasion: spikes with Louis Vuitton logos on them, spikes encrusted with rhinestones, even spikes made from bullet casings stood on end. None of this actually came across as intimidating—far from it—but there were plenty of stereotypes to go around, if that's what you were looking for.

That was not, however, what Kim Wolf was looking for. With business cards in hand and a camera slung around her neck, the thirty-two-year-old Wolf looked a bit like a sorority girl who had walked through the wrong door. (When she was a student at Colgate University, she told me, she wore pearls with her Lilly Pulitzer dresses.) Yet, as she made her way through the aisles of tables, Wolf blended in easily. Her training as a geriatric social worker was evident in her warmth and genuine curiosity about the people she met. She snapped photographs of dog owners and asked them to tell her an interesting fact or story about their pets, which she wrote down in a notebook that she carried with her. One man introduced his tricolor bully, which was draped in gold hardware, as "Mr. Guccianni." "I just love animals," he told her. "All animals. I've had animals all my life." The thought of fighting a dog repulsed him. "I like dogs that look good," he said, laughing. "Why would I want to get them all banged up? My dogs are gorgeous!"

. . .

Inspired by Brandon Stanton's Humans of New York street photography project, which has been celebrated (and sometimes criticized) for telling the stories of everyday New Yorkers, Wolf decided in July 2013 to attempt something similar, but with New Yorkers and their dogs. By highlighting "the universality of the human-canine bond," Wolf hoped that Dogs of New York would challenge the stereotypes around pet ownership that are still evident in the broader culture while reminding viewers that every human-dog relationship is unique. It began as nothing more than a Facebook page and some low-resolution iPhone photographs, but soon it grew to have more than twenty thousand fans. Buoyed by the public support, Wolf began a community outreach program called Ruff Riders to provide food and pet care supplies for dog and cat owners who are struggling to make ends meet. Both projects are housed under Wolf's nonprofit, Beyond Breed.

Wolf's experiences on the streets of New York City, like mine back in North Carolina, did not support the idea that urban dog owners acquire pit bulls only for "status" or intimidation. One of her first subjects, in fact, was a man named Marion whom Wolf met near her apartment in Bushwick, Brooklyn. Wolf noticed that Marion's pit bull, Lady, had a piece of carpet draped over her back that was tied in place with shoestrings. Marion told Wolf that he had recently taken Lady in after finding her on the street, and he was afraid she might get cold on their daily walks. He could not afford to buy Lady a sweater, so he made her a "jacket" out of the materials he had on hand. "Obviously, Marion loved Lady very much and cared for her because he was worried that she had been mistreated," Wolf said. "He didn't get a pit bull to be 'macho' or 'tough.' "

Another man, Nick, had tried to adopt a dog through his local animal shelter but feared he would be ineligible because the shelter performed criminal background checks and Nick had a drug possession charge on his record (he mimed smoking a joint). Certain he would be turned away at an adoption facility, Nick bought his pit bull, Snow, from a breeder in his neighborhood. "Snow was my first dog that was all mine, you know?" he told Wolf. "I've had him since he was a puppy. He follows me everywhere. I'm on the couch, he's on the couch. I'm in the kitchen, he's in the kitchen. I'm sleeping on the bed, he's in the bed. I spoil him. I've thought about having another dog, but I don't want Snow to be jealous. Like I said, he's my first dog that was all mine. I want to give him everything."

So what did unite the people Wolf met in their love of pit bulls? The research on pet selection is instructive. In 2012, Dr. Emily Weiss, the vice president of shelter research and development at the ASPCA, and her colleagues surveyed approximately fifteen hundred pet owners about why they chose the animal they now have. They found that pet selection is much like dating: An animal's physical appearance makes a big difference in the early stages of a relationship but is less important over time. This is especially true of dogs. More than a quarter of dog owners listed appearance as the number one factor that influenced their choice of pet, whereas behavior/personality was primary for only 15.8 percent. When the owners were asked in slightly different terms if the dog's appearance was important to them, more than 75 percent said yes. What each person finds physically appealing in a dog, as in another human, is different, and it is influenced by many factors, including personal experiences and cultural trends. As Hal Herzog discovered in his study of AKC trends, social contagion—wanting something because people close to you want it—is a big one.

The same goes for what we see in the dogs of other people. Our assumptions about the person standing *next* to the dog play a significant role in how we perceive that dog's temperament, even if we are looking only at still images. In 2013, researchers in California asked 228 test subjects to view photographs of a Labrador retriever, a border collie, and an American bully (labeled a "pit bull"), then complete a questionnaire about the dog's probable levels of friendliness, approachability, intelligence, and aggressiveness, among other traits. Most of the subjects ranked the pit bull higher than the Labrador and the border collie in the aggression category. Then the researchers asked their subjects to view photographs of the same dogs posed with a series of five human handlers including a male child, an elderly woman, and what was described as a "rough male." The photographs of the pit bull next to the elderly woman and the male child greatly *decreased* the subjects' perception of the pit bull's aggressiveness, while the photograph of the dog with the "rough male" *increased* it. In other words, context is key.

Two studies conducted in 2006 and 2009 attempted to establish whether or not there is any significant relationship between owning a certain type of dog and being a certain type of person, specifically, if the desire for a so-called aggressive breed marks the owner as a social deviant. Both papers are riddled with errors. The first examined animal control records from 355 dog owners in Hamilton County, Ohio, between 2000 and 2002. It divided them into two groups: owners of "low-risk" breeds and owners of "high-risk" breeds. Five of the "breeds" listed ("hounds," "terriers," "spaniels," "pit bulls," and "mixed breeds") are not breeds at all, but breed *groups,* one is a hybrid ("Terripoo"), and one appears to be made up ("Ahra"). The "high-risk" group contained only four "breeds": pit bulls (lumped together), Akitas, Rottweilers, and chows. (For unexplained reasons, the authors did not include German shepherds or Dobermans in the data set.) Only eight of the dogs in the study were labeled "mixed breeds," even though more than half of American dogs are not purebred. "Pit bulls" were also by far the dogs most represented in the sample (153), almost seven times more common than the next most popular category, "terriers" (24).

From there, each owner group was further split into those who had been cited for a failure to comply with animal control laws and those who had licensed their dogs, but no explanation was given as to how many licenses might have arisen from citations. The authors then compared the numbers of the dog owners' criminal convictions or "citations" (which included minor traffic tickets) in each group, concluding that "ownership of a 'high-risk' (vicious) dog can be a significant marker of general deviance."

Given the glaring problems above, this conclusion makes absolutely no sense. Until 2012, Ohio was a state in which all dogs that resembled "pit bulls" were automatically considered "vicious" and subject to heavy regulation, so the residents of Hamilton County who chose to keep pit bulls in 2000 did so in a much different context from other pet owners. The authors themselves point out that the legislation "was difficult to understand and lacked the educational resources to make registration readily available to [pit bull] owners." That doesn't tell us anything about the distribution of dogs and owners in a place where such ordinances do not exist. As we know from almost a century's worth of AKC data and other cultural trends, a significant portion of the people who seek out "imposing" breeds are more frightened than frightening.

The second study, published in 2009 by psychologists at West Virginia University, was an anonymous online survey of 758 college students. These authors concluded that "vicious dog ownership may be a simple marker of broader social deviance" after they sorted the survey responses into categories based on self-reported acts of "deviance." Once again, the authors labeled breeds in question as "vicious" based on reputation, not behavior, and the dogs included in this category were "pit bulls," chows,

Rottweilers, and Akitas, along with wolf hybrids and Dobermans. German shepherds were once again omitted.

Almost 70 percent of the "vicious" dog owners in this sample turned out to be women, and the most common type of "crime" they committed was "status offense," such as running away from home. The results were well within the margin of error (approximately twice the standard deviation). But the most outrageous part of the paper was the battery of "psychopathology subscales" that the authors included at the end. They assert that owners of dogs from breeds stereotyped by insurance agents as being "vicious" bear scrutiny as possible psychopaths when many of their anonymous teenage subjects did nothing more than run away from home. There are so many problems with this conclusion that it's hard to know where to begin.

While it *is* likely that highly antisocial people seek out dogs they believe project a scary image, the logic doesn't work in reverse; many people in prison have tattoos, but tattoos have become so mainstream that having one hardly means you have been to prison. Even among avowed social "deviants," the human-canine bond is much more complicated than media stereotypes suggest. In 2011, researchers in the U.K. interviewed more than two dozen self-identified members of gangs and other "street-based youth groups" about their attitudes toward animals. The top reason that young people on the street acquired so-called status dogs was the same reason that the well-to-do and elderly acquired their pets: they were looking for companionship and social connection. Many of the kids came from backgrounds where both were difficult to come by. (In America, a child is more likely to grow up with a pet than with a father who lives in the home.) "She's my best friend," a teenager named Jack said of his American bulldog. "I've had a hard time, and she's the one I want to spend time with. She's always with me." For most of the participants, personal protection and ego enhancement were secondary concerns. The researchers concluded that the British media's insistence on a narrative of violence ("feral children and their feral pets") was inaccurate. Such a fear-based narrative, they said, drove a disproportionate government response (breed-specific legislation) that not only misrepresented youth street culture but also "prevents [urban youth] from reaping the many rewards of the human-animal bond."

"People [in the U.K.] talk about 'weapon dogs' without needing anything as coarse as evidence that the dog might be used as a weapon," wrote the columnist Zoe Williams in *The Guardian.* "All that really means is a burly Staffordshire bull terrier with brass chest furniture in the company of young, ideally black, men . . . It would never be OK to say: 'I'm afraid of young men, especially large groups of them, especially the ones without much money'—so in order to articulate that, these people are broken down into their constituent parts."

The sociologist Arnold Arluke has noted that dogs, specifically pit bulls, can provide a convenient scrim behind which people can voice negative comments about other humans. "When vented indirectly toward animals," he wrote of anti-pit-bull sentiment, "hostility toward human groups may not raise public ridicule because these inappropriate remarks are one step away from us." Arluke also documented the collaboration between police and humane groups in an unnamed "major metropolitan area" that used pit bull ownership as a means of criminally profiling black men. "Members of this task force carried out joint 'sweeps' in suspected inner-city neighborhoods to spot 'suspicious' dog owners and 'disarm' them by taking their animals," he wrote. "Driving through certain high-risk neighborhoods allowed for opportunistic spotting of African Americans walking with pit bulls on sidewalks or sitting on stoops with their animals, the assumption being that these dogs were not pets but illegal and dangerous weapons."

If the owners of these dogs were unable to provide proof of licensing when asked, the animal control officers seized the dogs and then required the owners to provide copious amounts of personal information in order to get them back, which gifted the police with names, addresses, and phone numbers the department would not otherwise have had. Likewise, in early 2015, a Republican representative from Mississippi proposed a house bill that would declare all pit-bull-type dogs across the state "dangerous" and allow law enforcement officers to search a residence *without a warrant,* provided they "suspected" a pit bull was present. In both cases, the dogs were used as pretexts for end runs around the Constitution.

Many people who believe that pit bulls are uniquely dangerous to humans maintain that their feelings are based solely on the actions of the dogs and not racial or cultural animus toward their owners, and for a number of them that is undoubtedly true. But the loud reverberation of racialized language, especially the word "thug," in their criticisms is deafening. In much the same way that pit bulls have been systematically "de-caninized," their owners have been dehumanized. One writer from Louisiana slammed "the animals who own the animals . . . [who are]

generally young thugs who see their pit bull as an expression of their own toughness." Another, from Schenectady, New York, complained about "the other kind of pit bull owner, maybe a drug dealer or some other kind of street thug, who keeps an animal that is designed to intimidate . . . and represents a hazard to innocent strangers minding their own business, trying to walk down the street." A third wrote an "open letter to pit bull defenders" in Shawnee, Oklahoma: "Your pit bull—complete with studded collar—does not impress. It does not scream macho or gangsta. All it says is . . . I can't afford homeowners [*sic*] insurance because of this risky dog. No neighborhood will allow me in. And I owe thousands in animal control fines. That is how pit bull owners should be perceived."

The most telling statement of all, however, came from a resident of Sterling Heights, Michigan, who, when asked by a reporter why he supported a city-wide ban on pit bulls in 2010, had this to say: "We have inner-city people who bought homes here . . . They don't need to bring their pit bulls here."

CHAPTER 13

TRAINING THE DOG

Whoever fights monsters should see to it
that he does not become a monster.

—FRIEDRICH NIETZSCHE, *Beyond Good and Evil*

No one is more dangerous than he who imagines himself
pure in heart: for his purity, by definition, is unassailable.

—JAMES BALDWIN, *Nobody Knows My Name*

Tara FitzGerald had lived in New York City with her five-year-old dog, Biko, for four months when she crouched down to scratch his belly in the hallway of her apartment. As usual, the dog loved the attention, but then FitzGerald noticed a subtle, almost imperceptible shift in his mood. "Did you just growl at me, Beeks?" she joked. Later, she wrote,

> The words were barely out of my mouth when his stocky frame came flying toward my face and knocked me backwards to the floor. His teeth were bared in a tight grimace, and he was snarling in a voice of wild fury I had never heard before. He seemed to be powered by a savage, primal rage. I still can't quite reconstruct how it happened so quickly, how he went from prone to airborne in a split second, how [he] could flip and fly in one smooth movement, gaining such speed and trajectory that I never stood a chance of raising my arm in time to shield my face.

FitzGerald felt a "soft tearing sensation, like slow ripping through silk. Then hot blood gushed from my mouth, pumping out in heavy spurts."

Biko had torn through her face, removing "a large amount of soft tissue from lip and cheek," according to her medical report from Bellevue Hospital. She would eventually require the expertise of several plastic surgeons, as well as that of a therapist who helped her work through her post-traumatic stress disorder. FitzGerald would never look—or be—the same. "What does it mean when your face is no longer the face you recognize?" she wrote.

When her veterinarian recommended that she euthanize Biko, FitzGerald felt conflicted. "I was horrified by the idea, and also a little frightened of the possibility that my decision might be coming from a vengeful place. I didn't hate [my dog], but I did feel betrayed by him." In the end, she decided that euthanasia was the best option because of something she did not originally disclose to medical personnel: Biko had bitten her before.

FitzGerald's dog was a basset hound.

The pain and grief that FitzGerald felt over her disfigurement is evident in her writing, but she never blamed Biko's actions on his breed, nor did she indict all other basset hounds. Rather, she accepted that after five years of love and companionship something incomprehensible had gone very wrong with her dog. Her story did not make the news.

On a sweltering afternoon in the summer of 2012, I met with Tony Solesky, a painting contractor who had recently become a strident, highly visible anti-pit-bull activist and a fixture in the local press in the four years since his son, Dominic, was almost killed by a pit bull. Ten-year-old Dominic had been playing Nerf tag with friends in the alley behind his family's row home in Towson, Maryland, when an intact male dog belonging to a man named Thomas O'Halloran escaped from the makeshift five-by-five chain-link pen it shared with an intact female (who had just given birth to a litter of pups) and first bit one of Dominic's friends, then savagely punctured Dominic's femoral artery. The boy's injuries were so severe that they necessitated multiple blood transfusions and surgeries, a seventeen-day stay at Johns Hopkins Hospital, and months of physical therapy before he was again able to play the sports he loved. The consequences of O'Halloran's profound negligence had been so grave that most people following the case assumed he would serve jail time for reckless endangerment. He didn't. By pleading guilty to the charges, O'Halloran got off with two years' probation and a $500 fine.

The Soleskys went after O'Halloran in civil court, but he declared bankruptcy. Undeterred, the family sued O'Halloran's landlord, Dorothy Tracey, for damages. (Her insurance company, State Farm, settled with the Soleskys out of court.) Tracey then countersued, and the case fell into the meat grinder of the legal system for the next several years. Until 2012, Maryland was one of several states in which dog bite liability was governed by the state's common law, which holds that an owner is liable for injuries caused by his dog only if it can be proven that the owner knew (or should have known) that his dog had aggressive tendencies. If the owner insisted that the dog never behaved in that manner before, the burden fell on the plaintiff to present contrary evidence. This legal loophole is often referred to as a "one free bite" rule, and it provides precious little relief for bite victims, even when their injuries are life-threatening.

That changed in April 2012 when the state's court of appeals not only ruled in favor of the Soleskys but proceeded one step further. It said that the *type* of dogs O'Halloran owned—pit bulls—established a prima facie case that the animals were vicious. "Because of it's [*sic*] aggressive and vicious nature and its capability to inflict serious and sometimes fatal injuries," the ruling stated, "pit bulls and cross-bred pit bulls are inherently dangerous."

The opinion relied on previous court rulings that upheld breed bans in other jurisdictions, which in turn relied on the various academic studies based on media reports. The ruling also did not articulate how "pit bull" or "cross-bred pit bull" should be defined or who was qualified to make such a determination. This greatly troubled the dissenting judge, who wrote, "Succumbing to the allure of bad facts leads inevitably to the development of bad law."

Bypassing the legislature, the court then changed the common law from "one free bite" to a standard called "strict liability" for owners of pit bulls, meaning that they would be automatically liable for any damages caused by their pets, no matter the circumstances. Owners of all other dogs (even large guardian breeds such as German shepherds, Rottweilers, Dobermans, and various mastiffs) would still be protected under the original "one free bite" law.

This decision had the potential to affect the lives of many Maryland families, because the verdict extended to landlords who rented to pit bull owners, and thus influenced the insurance companies who covered those rental properties. If the insurance companies decided to stiffen their policies or landlords raised rents in response to the ruling—which they did—tenants across Maryland (whose dogs had not posed any problems) could now be forced to either give up their pets or give up their homes. Legally, these renters were not entitled to a grace period in which to make other arrangements.

Those at the bottom of the economic ladder were by far the most vulnerable. On August 10, the president of Armistead Homes Corporation sent a letter to the fifteen hundred residents of Armistead Gardens, a low-income housing cooperative in northeast Baltimore, informing them that if owners of "pit bulls or cross-bred pit bull mixes" did not get rid of the dogs "immediately," the board would pursue legal action ("including termination") against the leaseholders. According to Charles Edwards, the attorney for one resident who subsequently sued the housing corporation, as many as five hundred Armistead Gardens residents—one in three—owned a dog that someone might consider a "pit bull."

"It's been heartbreaking at times, parents with children, everybody crying, including our staff members and other customers," said Jennifer Brause, the executive director of the Baltimore Animal Rescue and Care Shelter, when describing pet surrenders after the court decision. "This has been hard for everyone. We even had a purebred boxer come in—a dog that obviously isn't a pit bull—because the landlord thought it was a pit bull and suddenly got nervous after the ruling. The tenant didn't have the resources to fight and, in the end, was afraid of retaliation by the landlord."

Some Marylanders felt that this law disproportionately punished the urban poor. "Over time," a man named Lawrence Grandpre wrote in *The Baltimore Sun,*

> it seems that "pit bull" has become a synonym for "black," and thus a similar bias seems to be at play here. As a black person raised in Baltimore, pit bulls were a central part of the social fabric of my life. The best dog I ever had was a pit bull, and he was the sweetest thing I have ever met. I am confident that if you were to ask the vast majority of pit bull owners in this city, they will tell you the same thing. For black folks like me who grew up with them, we love them because when we were born into a violent world not of our choosing, they protected us.

That families who were barely hanging on could be yanked around in this way because of one person's criminal recklessness seemed excessive. Prince George's County, Maryland, had enforced a ban on pit bulls for years (at a cost of $250,000 a year), yet the chief of Prince George's County Animal Management Division, Rodney Taylor, did not believe it improved the safety of the community. In fact, he repeatedly said that it was a waste of money and manpower. Taylor spoke out against the ruling in *Tracey v. Solesky*, telling members of the legislative task force, "One of the most difficult challenges we have as an organization is going to someone's house, knocking on their door, and seeing their American pit bull terrier sitting in their living room watching television with the family, and [we] have to take it out. Where the dog has done nothing wrong, no problems, but just because [of] its breed, he has to be removed."

While I could only imagine the Soleskys' immense grief over what happened to Dominic, I knew all too well what it felt like to want justice and not get it. In the spring of 2006, my mother, then fifty-five, survived a violent home invasion in Savannah, Georgia, during which she was beaten and left unconscious while the intruder(s) slipped back into the night. Fingerprint analysis and DNA tests found no matches in the state's criminal database, and no one was ever charged with the crime.

Critics of the Maryland ruling dismissed Solesky as someone who used his son's trauma to raise his own profile, but as a person who had almost lost a loved one because of a stranger's horrendous choices, I felt they were being unfair. How could they not sympathize with a parent who had nearly lost a child? What happened to Dominic Solesky should not have happened, and our existing laws are pitifully weak when it comes to holding owners responsible. The case was an inescapable reminder that if not properly maintained, dogs are physically capable of inflicting serious harm, and the concerns of those harmed by them deserve respect and consideration. I hoped that the elder Solesky could help me understand the toll that this had taken on him and why he felt that the eradication of pit bulls was the best way to ease that burden.

At Solesky's request, I joined him at his "yacht club" in east Baltimore, which turned out to be little more than a ranch-type house adjacent to a few boat slips and outbuildings. Ushering me onto the deck of his thirty-one-foot-long cabin cruiser, Solesky offered me a cold drink from the fridge, for which I was exceptionally grateful in the heat. He struck me as a gracious, affable person.

Before we got into the specifics of that ordeal, however, Solesky wanted to explain something. In fact, he wanted to explain many things. Reclining in bare feet, Solesky expounded on the finer points of golf, hunting field trials, boater safety, and his tenure as block captain of his neighborhood association, but mainly he wanted me to understand how much he knew about dogs. This, he said, was based on his own experiences as a hunter (Brittany spaniels are his breed of choice) and "deductive reasoning."

Dog breeds exist for a reason, Solesky told me, and those breeds have preinstalled instincts. No amount of training or environmental influence can make a dog go against what nature has programmed on its genetic software, he said. If pointers point and retrievers retrieve (in what was becoming a familiar refrain), then dogs from "fighting breeds" are destined to maim and kill. All that training can do is force the dog to submit to your dominance. This was important. Solesky believed dominance was the key to understanding most things in life. Mental toughness. "Intestinal fortitude." He prized those virtues above all others.

Solesky then shared a personal anecdote that he hoped would erase any doubts I might have had about certain breeds of dog being instinctively violent.

When he was growing up in the densely packed row houses of Baltimore, Solesky and his family lived next to a man who raised beagles and Akitas, the latter of which were, in his words, "bad-ass fighting dogs." One day a loose Siberian husky attacked the Soleskys' dog, a mutt that had also gotten loose, and "really mauled the hell out of it." According to him, this husky had been a problem in the neighborhood for a while, even nipping Solesky's brother, so the neighbor warned the husky's owner that she had better keep her dog restrained. Not satisfied with the response he got, the neighbor took thirteen-year-old Solesky aside, grabbed one of his Akitas, and said, "Let me show you something." When the husky again escaped from its yard, the neighbor turned the Akita loose.

"The Akita brutalized [the husky]," Solesky recalled, the corners of his mouth turning upward slightly. "The dog was squealing and screaming. It was really tough to break up both of them. It was unbelievable. I had never seen anything like it." By the end of the fight, he said, the husky was just "laying there," and the Akita's white fur was soaked with blood.

It must have been horrible—even traumatic—to see something so cruel, I thought, especially as a thirteen-year-old child. I asked Solesky how he reacted. With satisfaction, he crossed his arms and smiled. "I was glad that husky was getting its due, as brutal as that sounds. I guess [its owner] took him to the vet or whatever, but that dog never came out and did that again. That dog got trained."

In the run-up to a new political landscape of "broken windows" policing, being tough on pit bulls during the 1980s was an essential part of being tough on crime. New York City's Democratic mayor, Ed Koch, was one of the first leaders of a large city to decide that pit bulls had to go. Koch had nothing against potentially dangerous dogs, per se. In 1981, he called for unsupervised German shepherds to patrol the perimeter fences of subway storage yards to harass would-be graffiti artists, saying, "If I had my way, I wouldn't put in dogs, but wolves." Koch did this with the enthusiastic support of the Metropolitan Transportation Authority, whose president beamed over a litter of German shepherd pups. "I hope they will turn into vicious attack puppies," he said. "The more the merrier."

But on the other side of the tracks, as it were, Koch decided that pit bulls were a problem because of who was thought to own them. Noting the link between pit bulls and drug dealers during a press conference at city hall, the mayor said, "I perceive the pit bull in the same league as the great white shark, on occasion." He added, "There are dogs, dangerous dogs, and then there are pit bulls." Pit bulls were banned as part of the city's health code in 1989; the ban was lifted two years later after a series of due process and equal protection challenges, but the New York City Housing Authority passed a number of regulations that made it impossible to keep pit bulls in public housing.

After the "killer dog" media circus of 1987, one incident was all it took to unleash anti-pit-bull fury in any city. Miami-Dade County banned pit bulls in the spring of 1989, after the press seized on news of a dog attack that badly scarred the face of an eight-year-old girl named Melissa Moreira. In Denver, a similarly high-profile attack on an evangelical pastor named Wilbur Billingsley in May prompted the city to enact restrictions on pit bull ownership so prohibitive (like a mandatory $300,000 insurance policy) that they amounted to an all-out ban, because few insurance companies would write the policies and the ones that did charged high prices. Editorials in both the *Miami Herald* and the *Rocky Mountain News* applauded their local governments' attempts to eradicate this "urban menace" from communities.

On paper, the breed-specific laws in Miami and Denver were sold as a gentle way of letting the dogs gradually disappear from those cities without having to confiscate them aggressively. Existing owners of American pit bull terriers, American Staffordshire terriers, or Staffordshire bull terriers, as well as owners of mutts that resembled those breeds, would be "grandfathered" in, so long as they retained the requisite insurance and had their dogs sterilized, tattooed, and registered with the city by a given deadline. If an owner failed to comply with these measures, however, his or her dog would be seized by animal control and destroyed. After the deadline, anyone bringing a pit bull into the city or "harboring" a pit bull would be subject to legal action.

But in the cases of families who could not secure the necessary insurance, who passed the deadline unaware of the laws, or who moved to these areas after the passing of the ordinances, breed-specific laws turned into an assault on pet owners whose dogs hadn't harmed anyone. The lengths to which Denver law enforcement in particular went to enforce its ordinance spawned several major lawsuits, as well as a great deal of national criticism.

One of the more disturbing episodes took place in 2002, when an army staff sergeant named Heidi Tufto, who had just moved to Denver and was unaware of the city's pit bull ban, took her dog for a walk in a local park. According to her account, police officers patrolling the area ordered her to surrender her pet on the spot, then held her at gunpoint while animal control snared the dog with a catch-pole. Traumatized, Tufto arranged to have her dog flown out of the city on a military transport plane and then moved away. Others were not so lucky. Karen Breslin, who represented several plaintiffs in lawsuits against the City of Denver, told me that several of her clients were forced to pick up their dogs from the municipal shelter in body bags.

"There appears to be a racial end of this," a Denver-area veterinarian told the *Rocky Mountain News* in 2005. "Look at the dogs that have been impounded, and the surnames of their owners . . . They aren't killing dogs from Cherry Creek [a wealthy Denver suburb]. They pick on the easiest people to pick on, the ones who give up easiest."

Breslin saw a similar trend. "Bans like this stay in place because the people who are on the receiving end of it are people who don't have the resources or the kind of life circumstances that position you to really fight when bad things happen," she said. "The people who have contacted me are by and large people without money for legal representa-

Pit bulls euthanized in Denver, 2006

tion. When something like this happens, if you're in that position, you think there's not much you can do except take it."

When I asked Bob Rohde, the president and CEO of Denver's Dumb Friends League, an animal shelter with an annual operating budget of $14 million, for his thoughts on breed-specific legislation, he said that he did not support it but then added, "Do people want pit bulls? How many families with children want pit bulls?"

Between 1992 and 2009, Denver impounded almost 5,300 "pit bulls" and killed an estimated 3,497 of them. Photographs from the Denver Animal Shelter (which operates the county's animal control services) obtained by *Westword* magazine in 2009 show a mountain of the dogs' corpses, all arranged in rows, like sacks of flour. When I visited the shelter in 2012, no one there was willing to speak with me about the law. Kory Nelson, the assistant city attorney who represented Denver in its most recent breed-ban lawsuit, initially responded to my request for an interview, then ceased communication. All this made the sign on the wall of the Denver Animal Shelter, which quoted Mahatma Gandhi, much more sobering: "The greatness of a nation and its moral progress can be judged by the way its animals are treated."

Those who feared that breed-specific policies were tinged with human prejudice had reason for concern, not because the architects of these policies (or their supporters) consciously set out to discriminate against human groups—in all likelihood, they did not—but because the use of racial stereotypes and coded language ("thug," "dealer") to justify the necessity of such laws perpetuated a long, troubling tradition.

Over the course of history, the dogs most often portrayed as "dangerous" and subjected to the highest penalties have belonged to people with the least political power. In England during the twelfth-century reign of Henry II, the king demanded that all mastiffs kept by commoners be seized and either killed or maimed by a game warden, who was ordered to cut out the center of the dog's foot with a hammer and chisel. This process was called "expeditation," and it rendered the dog permanently lame and effectively useless as a working animal. Henry hoped this would deter poaching in the royal forests (as mastiffs were used for catching large game), but he also saw it as a way of weakening his subjects psychologically. Once the mastiffs were taken away, commoners had one less method of self-defense. Some scholars believe that once the king went so far as to order a "mastiff massacre" to quickly levy taxes from villagers who scrambled for exemptions. In some areas of England, the law of expeditation stayed on the books well into the eighteenth century.

In antebellum America, the least powerful were African slaves. Among the planter class, it was accepted as a foregone conclusion that dogs owned by slaves were more likely to attack sheep and cause general mayhem around the estate. "It is not for any good purpose Negros raise, or keep dogs, but to aid them in their night robberies," George Washington wrote to a friend in 1790. Any of Washington's slaves who was found with a dog was severely whipped, the dog hanged. Thomas Jefferson also fretted over the potential for slave-dog attacks on sheep flocks. He informed his Monticello overseer in 1801 that "to secure wool enough, the negroes' dogs must all be killed. Do not spare a single one."

Over time, the justifications for banning slaves' dogs shifted into the moral realm. In an 1833 treatise called *Detail of a Plan for the Moral Improvement of Negroes on Plantations*, the Georgia rice planter Thomas Savage Clay contended that dogs must be outlawed in slave quarters because blacks were not morally sophisticated enough to care for animals properly. "One of the greatest pests on a plantation, is a pack of mangy, starving curs," Clay wrote. "They steal to escape starvation, and are then most unmercifully beaten: the children, seeing the cruelty of their parents, soon learn to imitate them, and both children and parents vent upon the poor animals, that passion which has been excited by some object beyond their power." Even free blacks in parts of Virginia, South Carolina, Maryland, and Arkansas were legally prohibited from owning dogs because of the fear that the dogs would somehow become "weaponized."

For Charles Ball, a slave who escaped from a South Carolina plantation around 1812, the love of a dog provided the only sense of comfort he knew. The rest of his daily existence had been engineered to strip him of his humanity. The devotion that Ball felt for his "constant companion," whom he poignantly named Trueman, grew so strong that it made the man's final escape almost impossible. Knowing that he could not afford to have the dog's bark give away his location once he had broken free, Ball recalled in his memoirs,

> I then turned to take a last farewell of my poor dog, that stood by the tree to which he was bound, looking wistfully at me. When I approached him, he licked my hands, and then rising on his hind feet and placing his fore paws on my breast, he uttered a long howl, which thrilled through my heart, as if he had said, "My master, do not leave me behind you." All the affection that the poor animal had testified for me in the course of his life, now rose fresh in my memory. I recollected that he had always been ready to lay down his life for me; that when I was tied and bound to the tree to be whipped, they were forced to compel me to order my dog to be quiet, to prevent him from attacking my executioner in my defence.

Any emotional tie that strengthened a slave's sense of self threatened his master's hold on him. For this reason, one cotton magnate stated in 1855 that dogs owned by slaves had caused "greater injury to the people of South Carolina than all the abolitionists in the world." Four years later, the South Carolina legislature officially wrote a slave-dog ban into state law. All this while police forces and lynch mobs used bloodhounds to track (and in some cases, kill) African-Americans until at least World War I. Carl Sandburg, who worked as a reporter before becoming a poet, described an event in Vicksburg, Mississippi, in which "a colored man accused of an assault on a white woman was placed in a hole that came to his shoulders. Earth was tamped around his neck, only his head being left above ground. A steel cage five feet square then was put over the head of the victim and a bulldog was put inside the cage. Around the dog's head was tied a paper bag filled with red pepper to inflame his nostrils and eyes. The dog immediately lunged at the victim's head. Further details are too gruesome to print."

James K. Vardaman, Mississippi's governor from 1904 to 1908, believed that the Constitution "did not apply to wild animals and niggers" and enjoyed setting black convicts loose in the woods so that he could "hunt" them with bloodhounds.

Once Darwin's theories of evolution and heredity made their way into the class-conscious Victorian dog fancy, white Americans increasingly fixated on matters of canine innateness, which they related directly back to the "inborn" tendencies of the dog's owner. Race inevitably seeped into the conversation. "You will find just as much difference between the southern negro's dog and the white man's dog, as there is between a canine owned by a negro and one which comes when a Senator snaps his fingers," wrote one commenter in an 1884 edition of South Carolina's *Anderson Intelligencer.* "The negro's dog is a slouch. He is built on the wrong principles to begin with . . . The master may be ever so independent, but the dog has no business on earth. He is an interloper." In 1891, an unsigned editorial in a Louisville newspaper declared, "The nigger dog is, by instinct, a thief, and he is the only inheritance for future generations of niggers. He is at heart as black as the ace of spades."

In townships around the South, whites demanded new laws to contend with the "dog evil." "Before the war no legislation was needed; for the whites kept the number of dogs within safe bounds," wrote one angry farmer in 1901. But afterward, he said, "the half-starved negro dog" destroyed sheep flocks and robbed hens' nests at will. No one was safe, he said. Something needed to be done.

It is hardly surprising that when the German Nazis passed a battery of animal rights laws in 1938, the codes were intended, at least in part, to elevate the legal and moral status of animals while decreasing that of ethnic minorities. Kosher butchery was, of course, the first thing to be outlawed, but in 1942 Hitler also banned Jews from keeping pets under the pretense of "saving" the animals from Jewish "cruelty." Counter to the professed aims of the law, pets belonging to Jewish families were seized and euthanized. The Nazis feared the "impure" blood of "Jewish dogs" would taint the gene pool of their own animals.

After the bans in Denver and Miami, the rush to ban "dangerous" breeds of dog quickly spread to other countries. In 1989, panic swept across the U.K. after two Rottweilers killed an eleven-year-old girl in Scotland. For the next two years, the English tabloids fixated on the Rottweiler as a savage predator. According to Dr. Anthony Podberscek of the Animal Welfare and Human-Animal Interactions Group at the University of Cambridge, 162 dog attacks were reported in the U.K. press from 1988 to 1992, and "although the German shepherd was the attacker in many cases, it did not receive as much publicity as the Rottweiler," with the latter commonly described as a "'devil,' 'killer,' or 'war' dog." "Rottweiler" appeared in more than 70 percent of the headlines that mentioned the breed of dog involved; "German shepherd" in only about 6 percent. Animal wardens in the U.K. grew so afraid of potentially encountering "killer dogs" that they began wearing "riot shields, chain mail, and ultrasonic stun guns."

Man and dog in Philadelphia, ca. 1890

In response to this crush of newspaper articles, British politicians looked to a legislative solution, but the one they settled on didn't involve German shepherds, the U.K.'s most frequent biters, or Rottweilers, the subjects of so much tabloid terror. Incredibly, the home secretary, Kenneth Baker, who had been chairman of the Conservative Party under Margaret Thatcher, decided that the best course of action would be to ban pit bulls, of which there were an estimated ten thousand in the U.K. at the time.

Baker later wrote that he selected pit bulls in part because they were "being bought by criminals such as drug dealers," though he admitted that "the issue was made more complicated by the fact that the largest number of reported bitings was caused by Alsatians [the British term for German shepherds] and other domestic breeds whose owners would never have regarded their pets as dangerous." He noted that social class and "optics" played into his decision a great deal. Banning pit bulls was a safe way to look as though he were being tough on crime without angering his political base, the wealthier men and women who patronized kennel clubs and enjoyed outdoor activities like foxhunting, skeet shooting, and other sports that required "green welly" boots. "There was a danger of over-reaction," Baker wrote, "with demands to have all dogs muzzled and to put Rottweilers, Dobermans, and Alsatians in the same category as pit bulls. This would have infuriated the 'green welly' brigade. However, the 'pit bull lobby' came to my aid by appearing in front of TV cameras with owners usually sporting tattoos and earrings while extolling the allegedly gentle nature of their dogs."

One dog "expert" supposedly told the home secretary that "unlike other breeds, [pit bulls] were unpredictable and could not be reliably trained," after which Baker felt that he had made the correct decision. Rather than "promising the slaughter of the ten thousand," which is what he wanted to do, Baker grudgingly created a long list of requirements, including public muzzling, mandatory sterilization, and tattoos. "This led to humorous exchanges about . . . whether the dog's tattoo should match that of the owner," Baker wrote, alluding to the character Radio Raheem from Spike Lee's 1989 film, *Do the Right Thing,* who is choked to death by a police officer. "Would pit bulls have 'love' and 'hate' inscribed on each knuckle?"

In 1991, Parliament signed into law the Dangerous Dogs Act (DDA), which banned "fighting breeds" like the American pit bull terrier, the American Staffordshire terrier, the Tosa Inu, the Dogo Argentino, and the Fila Brasileiro, though not Britain's own Staffordshire bull terrier, the original prototype for the APBT, or other breeds with "fighting heritage" like the boxer, the Akita, the shar-pei, and the Boston terrier. German shepherds, Rottweilers, and Dobermans were also left unregulated. The DDA was passed in spite of massive public opposition that has only grown with the passage of time. Not only did the DDA ban the import or possession of dogs that shared "characteristics" with the above breeds; it included in its list of punishable offenses the ownership of any dog that "behaves in a way that causes fear that an injury might occur." This means that if you live in the U.K. and your neighbor is frightened by your dog for whatever reason, the local government has the right to seize your dog and investigate you.

The DDA then led to a wave of pit bull bans in parts of Germany, Spain, Italy, Belgium, and the Netherlands, among other countries. All were geared toward keeping out the specter of American inner-city culture rather than addressing animal-related problems. (The Netherlands repealed its ban in 2008, and Italy followed in 2009.) To date, there is no central database of how many dogs have been seized and destroyed under the U.K.'s ban, but in almost every county polled by the U.K.'s Press Association (including in Scotland and Wales), the number of dogs killed increases each year. Dog bite injuries and hospitalizations are higher than ever, especially in the most economically depressed areas.

. . .

As far as Tony Solesky was concerned, pit bulls, like lead-based paint and faulty children's car seats, were defective merchandise. "Would you breed this dog if it didn't exist?" he asked me. "If you'd say yeah, then something's wrong with you. And I want to be quoted on that." Solesky said that other dog breeds, like golden retrievers, don't have the same problems as pit bulls because they have "built-in factors that made them innocuous. They had air bags and safety belts in them."

But try telling that to the parents of Aiden McGrew, a two-month-old baby killed by his family's "golden retriever/Labrador mix" in South Carolina in 2012. Or Kimberly Demaree, a woman from Chesterfield, Virginia, who was hospitalized after being severely injured by a golden retriever that jumped its five-foot-high fence in 2013. Or Kelly Lawrence, who sustained more than seventy puncture wounds and a cracked wrist when her golden retriever latched onto her arm and shook it "like a rag doll" in 2014. Or Isabelle Dinoire, a Frenchwoman who received the world's first transplanted face after her Labrador retriever chewed hers off following a suicide attempt in 2005. Most news outlets portrayed Dinoire's dog as a hero companion that tried desperately to "rouse" her owner, and they could very well be right. But in what light might that course of events have been cast if the dog had been a pit bull?

There is no breed of dog that is incapable of causing injury, and dogs are not toys or consumer goods. The human-canine bond is based entirely on the fact that a dog is a sentient creature with an emotional life. The relationship between a dog and its person grows and changes over time. It is reductive (and potentially dangerous) to assume that dogs of any type are either "safe" or "unsafe."

To use a more accurate car metaphor, the Honda Civic, like the pit bull, is small in size, fairly generic in appearance, inexpensive, and easy to acquire. These four characteristics make it one of the best-selling cars of all time. For those exact same reasons, the Civic is also the leading car bought, sold, and modified for purposes of street drag racing, a highly dangerous and illegal practice that kills approximately one hundred Americans every year (three times as many as are killed by all types of dogs combined). Yet no legislator has ever proposed a ban on the Honda Civic in order to correct errant human behavior by a small number of people. If the Civic were completely removed from the market, drag racers would move on to low-end Mitsubishis or Toyotas; in some places, they already have. In this equation, the car isn't the problem. The person racing it illegally is.

I asked Solesky why he didn't trust the large cache of expertise that researchers in public health and the animal sciences brought to bear on this subject, not to mention the position statements made by the American Bar Association and the White House, both of which maintain that focusing on breed is ineffective and that breed bans are bad policy. Nineteen states have adopted preemptive laws against the passing of breed-specific legislation.

He replied that the opinions of these groups could not be trusted because, to him, either their "vocational prejudice" biased them too heavily in favor of animals at the expense of humans or they had been overwhelmed by the "pit bull lobby."

For Solesky, as for many people who feel strongly about pit bulls, it was a zero-sum game: Either you were with him or you were with the pit-bull-owning terrorists. Either you believed in protecting humans or you believed in protecting animals, one or the other. "[Animal welfare] is primarily a Caucasian female thing," he told me. He found that problematic, not because it revealed a lack of diversity, but because women's "empathy and nurturing instincts debilitate their ability to be reasonable."

What about the owners who hadn't sought out pit bulls but had simply taken in strays? Or the owners of boxers and mastiffs and mutts that might be mistaken for pit bulls? Knowing what we know about the subjective nature of visual breed identification, who was equipped to make a decision that might leave someone else homeless?

Solesky shrugged. "How do you determine what a pit bull is? Here's how you do that: If somebody thinks it is, it is. If you can't prove it isn't, then it is."

Dr. Alan Beck is the director of the Center for the Human-Animal Bond at Purdue University, and his roots in the field of animal welfare run deep. From 1974 to 1979, he directed the New York City Department of Public Health's Bureau of Animal Affairs, he has served on the peer-review boards of several highly respected journals, and the research he conducted on the ecology of stray dogs in St. Louis during the 1970s (as well as his work on the therapeutic value of pets for patients in hospitals and nursing homes) is widely cited. His current area of study focuses on human interactions with AIBO, the robot dog manufactured by Sony. Beck is considered a pioneer in the field of human-animal relationships. He is also the only credentialed animal professional in America not employed by PETA who believes that breed bans work.

In Beck's view, pit bulls present two problems, one biological and one social. On the biological front, he believes that by virtue of the APBT's historical function, dogs that resemble it have what amounts to a congenital mental defect—explosive aggression—that is now spreading throughout the dog population in a disease-like manner. Beck framed it this way:

> Imagine a dog breed that is relatively rare, less than 10 percent in most places and never more than 20 percent, has an inherited disease such that it inflicts pain, lameness and even death to other dogs fare [*sic*] out of proportion to their place in the community. In addition human beings, especially children, are also susceptible. The disease is such that the owners tend to relinquish their dogs to animal shelters long before their dog's natural lifespan, which is inhumane and a burden to animal societies trying to service many animals. Should we either aggressively breed away from the disease or at least discourage ownership?

The problem with this formulation, as I mentioned to Beck when I spoke with him, is that many breeders *have* been breeding away from fighting traits since at least the 1930s. What's more, the current epidemiological research into pet relinquishment indicates that dog owners most often surrender their pets because of personal difficulties, not because of their animals' behavior. Housing issues and lack of income are at the top of the list. Divorces, domestic violence, unemployment, and terminal illnesses can also be significant factors. Taking these obstacles into account, as well as the difficulty in arriving at accurate breed numbers, the huge problem with breed identification, the unknowns surrounding population genetics, our incomplete understanding of aggression, and the large percentage of mixed-breed dogs in the United States, it is hard to answer yes to Beck's rhetorical question. As many public health professionals have discovered, the wild-card variables and "dark figures" of dog bites are too numerous.

Keeping in mind those inherent limitations, what do the existing numbers tell us about the effectiveness of breed bans? According to a 2009 report from Colorado's Coalition for Living Safely with Dogs, the rate of dog bite injury hospitalizations (DBIH) in Denver County is much higher than that of similarly sized counties with no breed-specific laws. The Colorado Department of Public Health and the Environment recorded 367 DBIH in Denver County (2010 population: 600,158) between 1995 and 2011, while breed-neutral El Paso County (2010 population: 622,263) had only 189 during the same time period. The promised result of the breed law—a decrease in serious dog bite injuries—did not occur after its implementation, nor did the law prevent dog bite fatalities. In 1998, a little boy named Austin Cussins was killed by a "Rottweiler mix" while visiting his grandmother in Denver.

In 2010, Drs. Gary Patronek and Amy Marder (both of whom would later serve as Karen Delise's co-authors) used a theoretical model from public health called the "number needed to treat" to explore how many dogs would need to be eliminated in order to prevent a single dog-bite-related emergency room visit or hospitalization, which they called the "number needed to ban," or NNB. The figures they came up with were startling. In Colorado, 8,333 dogs would have to be banned (most of which would be destroyed) in order to prevent a single injurious bite. The number increased dramatically with the severity of injuries. Using national data from the American Society of Plastic Surgeons, Patronek and Marder calculated the NNB at 30,663 dogs—that is, dogs removed from their homes and either relocated or euthanized—"to prevent a single reconstructive procedure each year." That would be the equivalent of wiping out the entire human population of Long Branch, New Jersey, to prevent one nasty but nonfatal stabbing.

Most American municipal governments are now repealing pit bull bans rather than proposing them, as more evidence mounts that breed-specific laws are expensive and difficult to enforce, whereas well-resourced animal control and bite-prevention education is easier, cheaper, and universally supported by scientists. Between January 2012 and May 2014, 163 cities and towns either repealed their breed ordinances or rejected them, while only twenty-one passed bans. With the exceptions of Miami and Denver, most of the towns where breed-specific laws are still embraced tend to be small and relatively hermetic; as of this writing, less than 5 percent of Americans live in areas with breed-specific ordinances.

When Miami considered repealing its ban in 2012, an advance poll conducted by ClearView Research revealed that voters were unlikely to support a repeal *even if confronted with evidence that the ban did not work.* The idea that pit bulls were "different" had been hammered home for so many years that official reports and statistics to the contrary held no power.

Studies conducted in the U.K., Canada, Belgium, Spain, Italy, and the Netherlands have concluded that breed bans do not protect people from serious dog bite injuries. The main problem, researchers say, is that they are too shortsighted. Even if a city, county, state, or country were successfully able to rid itself of all "pit bulls" (however defined), what happens when the next "scary" dog comes along? The human population is still uneducated about proper animal husbandry and dog safety. It becomes a repeating game of whack-a-mole while public funds are drained. We are seeing this now, to a certain extent, as the much larger mastiff breeds, specifically the Presa Canario and the Cane Corso, grow in popularity.

In order to "prove" that these bans work, advocates of these laws often move the goalpost. Colleen Lynn, the founder of DogsBite.org, told *USA Today* in 2014, "[BSL]'s not designed to reduce all dog bites. It's breed-specific and meant to reduce pit bull maulings and fatalities." While it's true that you cannot be hurt by something that does not exist, that's the same as saying a ban on Honda Civics is designed to reduce Honda Civic injuries and deaths but refusing to acknowledge that reckless-driving laws would change the human behaviors *causing* those deaths, thus saving many more lives.

When I asked Beck about the data on which he was basing his assumptions, he pivoted to the social problem. "I know I would not want to live next to a pit bull family," he said.

In his research on stray dogs in St. Louis during the 1970s, Beck noted that strays were more numerous in under-resourced neighborhoods where there were few animal control officers, and in those neighborhoods bites were more likely to occur because the population density was high and children often ran around unsupervised. Did he now think that these issues might be worthy of more examination?

He laughed and said, "It would be nice to ban poor people, but we can't do that, so you have to use these surrogate measures." But what if other factors are contributing to these incidents? What if owner behavior plays a larger role than previously thought?

"I'm not so sure that we should keep working so hard to try to keep rationalizing these things," Beck said. "Maybe it was the owner, maybe the dog wants attention, maybe the owner's a minority, maybe the owner's stupid, maybe the owner is poor . . . We could keep torturing ourselves and trying to find reasons why it happened rather than say that maybe this particular kind of animal is not appropriate in urban America."

I asked Beck how many pit bull dogs he had ever handled or evaluated.

"And tell me why that's important?" he asked.

Surely it's necessary to have some direct experience with the animals he had talked, written, and testified about in court? He sighed. "[That's] like saying, 'You studied cancer, how much cancer did you have?' It's nonsense. I'm sorry, it just is."

Not everyone who has lost someone in a dog bite incident champions breed bans. In 2009, a two-year-old boy named Liam Perk was killed in Cape Coral, Florida, by an eight-year-old Weimaraner named Lloyd that his parents had owned without incident since he was a puppy. After consulting with experts about what happened, Joseph and Carrie Perk realized that they had failed to notice how much stress Lloyd displayed when he was around small children. Had they known more about dog behavior and body language, they could have kept Lloyd away from Liam while still attending to his needs for affection and exercise. To honor Liam's memory, the Perks partnered with other humane safety groups and established the Liam J. Perk Foundation, the goal of which is to educate families about dog bite prevention and safe child-dog interactions.

About a month after I met Tony Solesky, a man named Chris Mitchell walked into his job as the front desk manager for the LifeLine Animal Project, a nonprofit that provides low- or no-cost veterinary services in Atlanta, to the most devastating news he had ever received: one of his closest friends, a twenty-three-year-old dog rescuer and fellow LifeLine employee named Rebecca Carey, had died from a serious dog bite wound to her neck while caring for five dogs in her home. The dogs were reported as being two Presa Canarios (one she owned, and one she was looking after for a friend), two pit bulls, and a boxer mix.

When Jackie Cira, the owner of the other Presa Canario, found Carey's body in her home, she thought at first that Carey had possibly fallen and hit her head. The dogs did not seem "crazed" or dangerous. There was very little blood. It was not clear which had been responsible, but all five dogs were immediately euthanized by animal control before a determination could be made. "I don't know who did what, but I can say with certainty who did not," Cira told reporters. One of the pit bulls was so small, for example, that Carey had named him Napoleon, while each of the Presa Canarios topped one hundred pounds. One of the Presas was being nursed back to health after having been found dramatically malnourished. It's possible, Mitchell said, that two of the dogs got into a fight over food and Rebecca was inadvertently caught in the middle.

"Rebecca was a beautiful, creative, kind, thoughtful person," Chris Mitchell told me later. "And she was also one of the most skilled dog handlers I have ever known. It's not as though she was some inexperienced little girl who didn't know what she was doing. Every time we needed to get an aggressive dog out of its kennel for medical reasons, Rebecca was always the one who did it. She was very realistic and practical about the risks of working with animals, and if she believed a dog was behaving aggressively, she would *never* have taken that dog into her own home or placed it in anyone else's. She just wasn't like that. But she also realized that sometimes, unforeseen things like bad bites just . . . happen. That's the risk you take."

The most difficult part of losing such a close friend, Mitchell said, was reading the comments posted on the Internet from people in anti-pit-bull groups, who seemed to gloat over the tragedy. They did not seem to him to be "victims' advocates" at all. Some speculated that Carey must have suffered from a "lion-tamer complex" and used her death as further evidence that pit bulls should be banned. None of this was accurate, Mitchell told me. He believes that everyone is entitled to his or her own opinion, but online he characterized their comments as cruel and said that it was inappropriate to bash her, her dogs, and her way of life.

The commenters then turned on Mitchell, saying that he and his fellow "rescue angels" were to blame for Carey's death. "What I want to say to these people is," Mitchell said, "what happened to Rebecca was an accident. A terrible, terrible *accident.* Why don't you spend some time with these dogs you want to kill first? Why don't you learn a little about them before you make up your mind?"

When talking with Tony Solesky on the boat, I had asked him how he felt about the possibility that people might lose their homes over the court of appeals ruling. He shrugged. "I don't have much respect for somebody who would get that kind of dog and find out that they don't have the other things that are important in life in order. So I empathize very deeply with their pain, but I don't sympathize whatsoever with them. I think there's a very particular reason why I'm not in their situation and they're in their situation."

Solesky and I looked at this issue through two very different lenses that were ground into focus from our own personal experiences, and neither of us was, or ever could be, totally free from bias. When he thought about pit bull owners, he saw not only Thomas O'Halloran, whose blithe carelessness and refusal to take responsibility for an animal in his care had almost killed a young boy, but also a faceless legion of violent criminals who seem to care nothing about human suffering. When I thought about pit bull owners, I saw Doris in her wheelchair, sitting on the porch with Pretty Girl, and the hundreds of others like her I met while working with the Coalition to Unchain Dogs. I saw Lori Hensley toting bales of hay to her clients every winter, Diane Jessup waltzing her dogs through an obedience routine, the owners of American bullies and their blinged-out canine beefcakes. There was a vast sea of dog owners who were neither villainous nor virtuous, who were simply trying to get on with their lives.

The sociologist Bruce A. Jacobs once made a highly astute observation about these types of personal biases using the example of a white woman from suburban Cleveland and a black man from Baltimore having a discussion about the presence of police officers in their respective communities. To the black man, the police are to be feared and run away from, but law enforcement officers make the white woman feel safer. Each has experienced opposite realities, and the existence of one does not negate the other. "The problem isn't that the white woman from Cleveland or the black guy from L.A. or Baltimore is lying," Jacobs said. "The problem is that they are *both telling the truth* . . . That's got to be the point of departure for how we learn to face one another."

A little over a year after we talked, Tony Solesky ran as a write-in candidate for Baltimore county executive, using his experience in anti-pit-bull activism as the foundation of his political platform. He was not elected.

CHAPTER 14

DIFFERENT IS DEAD

You great star, what would your happiness be
had you not those for whom you shine?
—FRIEDRICH NIETZSCHE, *Thus Spake Zarathustra*

When the first meteorological maps of a large hurricane gathering strength in the Gulf of Mexico indicated that the storm was headed directly for New Orleans in August 2005, there was much discussion at the local, state, and national levels about what might befall the Crescent City's human residents but virtually none about what a major natural disaster might do to the animals that belonged to them. On August 28, the mayor of New Orleans, Ray Nagin, ordered a mandatory evacuation for all 500,000 New Orleanians, but many pet owners were horrified to discover that most of the hotels outside the city and none of the emergency shelters—save one, in Baton Rouge—would allow pets.

Among the tens of thousands of people who risked their lives to stay home, fear of abandoning their animals was a primary reason for that decision. But even after the levees broke and thousands of men, women, and children were stranded on their rooftops, sometimes for days, rescuers did not allow the displaced to bring their pets into emergency facilities. Some residents had to be forced at gunpoint to leave their companions behind. Just as evacuees were beginning to return to the city several weeks later, Hurricane Rita slammed into the Gulf, causing even more damage. An estimated 200,000 pets were displaced by both storms, and despite the truly heroic efforts of hundreds of volunteers, only about 15,000 of those animals were eventually rescued.

Original caption, 2005: "Lonzo Cutler, 34, who doesn't want to leave his pit bull behind, cradles the dog in front of his flooded home in the Ninth Ward neighborhood of New Orleans, La., as the rest of his family (in background) waits for rescuers to help them escape the barely-habitable area. As the Big Easy evacuates, already traumatized victims of Hurricane Katrina are making a choice: Head for safety or stay behind with a beloved pet."

"There is a class issue involved here," Karen Dawn, an animal advocate, wrote in *The Washington Post.* "While Marriott hotels welcomed the pets of Katrina evacuees as 'part of the family,' people who had to rely on the Red Cross for shelter were forced to abandon that part of the family or attempt to ride out the storm. It cannot be denied that many poor people are dead as a result of 'no pets' policies."

The rescuers who worked long hours in the swampy, post-Katrina wasteland would later recall that the majority of the dogs left behind were either pit bulls or mixed breeds that resembled pit bulls. This created problems for a number of the animal welfare agencies that flocked to the city to assist with re-homing lost pets. Some volunteers had never handled or worked with pit bulls and were frightened of them. For their part, the dogs were also terrified, some acted aggressively, and because they had been living for so long in contaminated water, many were close to death. Workers made their way through the aisles of kennels assembled at the Lamar Dixon Expo Center each morning, carrying out the bodies of dogs that had died overnight.

Not all of the approximately five hundred animal shelters that opened their doors to Katrina pets offered to take pit bulls out of the city. Some filled their transport vehicles with more "adoptable" dogs, while others couldn't take in pit bulls because of long-standing shelter policies. The idea that pit bulls were, in the words of Philadelphia's Fourth District councilman (and eventually the city's mayor), Michael Nutter, "nasty, despicable animals used primarily to scare and intimidate people, or to protect those involved in illegal activity," had spread throughout animal welfare.

A pit bull swims to a rescue boat during the aftermath of Hurricane Katrina

"There's no question that the pit bulls were separated out and treated differently [during the rescue effort]," a woman named Molly Gibb, who traveled to the Gulf to help lost pets after the hurricane, told me. "The media has done a very good job of 'de-dogging' the pit bull." Gibb would later foster a shy red-and-white pit bull that had been found tied to her dead owner, who had drowned. When the dog was picked up, she was wearing a fancy leather dress collar with metal studs on it, something Gibb recognized must have cost a good amount of money. "That man gave his *life* to save his dog," Gibb said, her voice trembling, "and other rescuers were complaining about what kind of collar she had on." (Later, the dog would be adopted by a pastor who took her to church with him on Sundays. He named her Faith.)

In her powerful National Book Award–winning novel, *Salvage the Bones* (2011), Jesmyn Ward presents an intricate and emotionally devastating portrait of the relationship between two Mississippi children, Esch and Skeetah, and their white pit bull, China, in the days before the storm. Because their own mother died in childbirth and their father numbs his grief with alcohol, Esch and her brother come to view China as the family matriarch. The dog provides both physical protection for them in a landscape full of wildness and desperation and, as a fighting dog, some semblance of an income when there is very little money to be had. When China is swept away by Katrina's floodwaters, it is clear that Esch and Skeetah have lost the force that held their family together. It is with China in mind that Esch fights to survive.

But even Gulf Coast residents who miraculously managed to stay with their pets throughout the entire ordeal were met with an ugly surprise at the end of it: pit bulls (as well as dogs from several other large breeds) would not be allowed in Federal Emergency Management Agency trailers. People who had already lost everything then lost their companions as well.

In the wake of the hurricane's destruction, pet custody battles raged between the animals' original owners (who were predominantly

African-American residents of the poor neighborhoods most devastated by the storm) and the rescuers that took them home (who were mainly white), which was extensively documented in a heart-wrenching 2009 film called *Mine.* In the cases of dogs who were found sick or chained, some rescuers admitted that they deliberately refrained from documenting information that might have helped their original owners find their animals, as they believed that it was in the dogs' best interest to live elsewhere. "The message is, 'You're poor, and we can take care of these dogs a lot better than you can,'" an attorney named Steven Wise observed.

Other rescuers, including a group called Stealth Volunteers, took the opposite approach, tirelessly dedicating themselves to reuniting lost pets with their owners, during those first few weeks and for months—even years—afterward, making hundreds of phone calls, arranging transports, and even putting up their own money for legal fees.

One custody case involved a woman named Kara Keyes, a resident of New Orleans's Seventh Ward. Keyes and her husband, Ronald, a forklift operator, became separated from their black-and-white pit bull, Crown, when they were forced to evacuate just before Katrina. Months later, a group that had formed to reunite Katrina's lost pets with their owners informed Keyes that Crown was now living in Marin County, California, with a clinical psychologist named Dr. Pia Salk, a pet columnist and frequent guest on *The Martha Stewart Show.* When Keyes contacted her, Salk refused to give the dog back, saying that she did not believe that Keyes would be able to treat the dog for heartworms. Instead, Salk offered to pay Keyes $1,000 for the dog. "No," Keyes told the *Los Angeles Times.* "The love that I have for Crown could never amount to any money."

"People who let their dogs suffer with hearworm [*sic*] and other diseases don't deserve to have pets," wrote one angry Internet commenter about the Katrina custody issue. "They had plenty of warning about the approaching storm and the responsibility to make a plan, but instead they discarded their pets like garbage and now expect to get them back after they've received free healthcare. If you can't afford to take care of them properly, you shouldn't have them. They are much better off wherever they are now."

"These people were victims of Katrina's wrath," argued another. "There was no plan to help the poor evacuate . . . [One evacuee] had a wheelchair-bound mother and young children to evacuate, without any help and no shelters allowed dogs, esp [*sic*] large dogs. Was she supposed to stay behind and strap her mom to anything that floats? We weren't there. I was there after and it's not the same thing. These people lost everything they owned, the right thing to do here is help them just that little bit [to] get back one part of normal to them."

In the end, less than 5 percent of the pets rescued during Katrina were ever united with their original owners.

The plight of animals during Hurricane Katrina was overwhelmingly tragic, but it did cast pit bulls in an extremely sympathetic light. Many of the volunteers who had never handled pit bulls came away feeling very warmly toward them, and the old shelter policies barring pit bull adoptions now looked antiquated, if not irrational. For the first time, the animal protection movement shifted from fearing or merely tolerating the dogs to actively advocating for them. So did the media. Best Friends Animal Society, a no-kill shelter, animal sanctuary, and advocacy organization based in Kanab, Utah, began a focused pit bull initiative, the goal of which was to promote the dogs as family pets. HSUS also started Pets for Life (PFL), a community-based outreach program devoted to providing pet care resources for the residents of underserved areas, which benefited hundreds of pit bull owners. Public donations poured in to help the pit bulls find homes, and new rescue groups devoted specifically to pit bulls multiplied.

One of the rescuers who traveled to New Orleans specifically to help pit bulls was Jane Berkey, a New York literary agent and the founder of Animal Farm Foundation (AFF), the most well resourced of the pit bull advocacy groups that emerged in the late 1980s and the 1990s to counter the widespread fear of pit bulls that was costing many thousands of dogs their lives. Spearheaded by AFF, Bay Area Dog Owners Responsible About Pit Bulls (BAD RAP), and Pit Bull Rescue Central (PBRC), the pit bull advocacy movement was unique among animal welfare campaigns in that it was more of a social justice crusade for equal treatment than a simple consciousness-raising mission. Early pit bull advocates spent much of their time fighting breed-discriminatory laws, as well as anti-pit-bull policies in animal shelters.

When Berkey heard that pit bulls were being left behind in New Orleans because they were thought to be "unadoptable," she decided that enough was enough. From her farm in New York's Hudson Valley, she had one of her horse trailers retrofitted to accommodate dog crates, then she and one of her newly hired dog trainers, Bernice Clifford, drove it to New Orleans. For several days, they pitched in at Lamar Dixon, where most of the animals that had been plucked out of the fetid floodwaters were being housed. "It was like downtown Mogadishu," Berkey recalled, "complete and utter chaos." Fortunately, she saw several pets happily reunited with their owners. When she and Clifford left the city, they took as many pit bulls (whose owners hadn't returned) back to New York as they could fit in the trailer. "They were getting the short end of the stick again," she said, "and I couldn't stand to see that any longer."

The e-mail I received from Stacey Coleman, the executive director of Animal Farm Foundation, recommended that I pack a sweatshirt and shoes that I wouldn't mind getting muddy. "Please remember that we are a working farm," Coleman wrote, adding, "You should probably leave your fancy clothes at home." AFF was located in a remote part of the Hudson Valley far away from any motels or rooms for rent—there was barely even cell phone reception—so in order to learn how the organization carried out its mission, I would need to stay on the farm itself.

The property's official name was Little Farm. It sounded so quaint, so rustic. I had visions of mucking out stables, or toting bags of grain to the chicken coop, or sidling up to a sweet old dairy cow, but upon my arrival I saw how wrong I had been. For one thing, Little Farm was not little. It was also not a "farm" in the sense that people I grew up with used that term, with tractors and hay bales and barefoot children running hoops along the road with sticks. Instead, it was four hundred acres of lush, rolling, impeccably landscaped estate, dotted with enough outbuildings to form a small village. A white colonial manor house stood atop the largest hill, gazing down like an all-seeing eye upon multiple barns, paddocks full of horses made of shiny muscle, and a lake framed by softly swaying willow trees. Maintenance workers zipped past in golf carts. The grass looked like raked velvet.

This was a farm in the way horse people used the word. I would soon discover that Jane Berkey was just such a horse person—dressage, to be exact, a highly formal type of "horse ballet." Because of her wealth and influence, Berkey was the woman Internet conspiracy theorists believed funded the all-powerful "pit bull lobby."

The small clapboard house that functioned as AFF's primary office building hummed like a sunlit beehive, with ringing telephones and rhythmically droning photocopiers and the soft clatter of keyboards in the background. The walls of the main conference room were covered in antique photographs of pit bulls from the nineteenth and early twentieth centuries, including the cover of a 1918 Red Cross magazine featuring a heroic soldier and his uniformed pit bull braving enemy fire on a World War I battlefield. A set of floor-to-ceiling bookshelves proudly displayed hundreds of sepia-toned portraits of people and their pit bulls through the ages: tuxedo-clad men in spectacles, flappers in finger waves, athletes in knee-length basketball jerseys, a girl in an elaborately embroidered sari, Canadian Mounties, several men in Russian Cossack hats.

Over the past ten years, a great deal has changed in animal welfare, thanks in part to the work of pit bull advocates. There has also been a significant amount of turnover at the country's largest humane organizations, with new leaders bringing new perspectives to bear. Most shelters have abandoned their strict pit bull euthanasia policies, but an estimated three-quarters of a million dogs labeled as pit bulls are still put down in American shelters every year. "When I show [clients] all our adoptable dogs," Kelly D'Agostino, the animal care supervisor for the Massachusetts SPCA in Boston, told me, "they say, 'These are just pit bulls. Where are your real dogs?' "

AFF, along with many other rescue and advocacy groups, has worked to change this type of thinking by funding public education campaigns and seminars, internships, and outreach programs. "If we don't have faith that people are going to do the right things with their dogs," Stacey Coleman said, "then why are we even in this business? Why is there so much gloom and doom? If you really believe that pit bulls are so different that they need special handling and special policies, then why *wouldn't* you support breed bans? Because you are sending the public the same message. Policies that make it impossible for pit bulls to find loving homes are made based on fear, not fact."

AFF's small shelter and dog-training center housed roughly twenty dogs taken in from local animal controls, cruelty cases, or fight busts. A few of the tenants stayed in traditional large wire kennels, but most had their own "rooms," complete with furniture, in order to acclimate them to living in a home environment. A staff of six trainers and a veterinary technician rotated through every day, exercising the dogs in outdoor play groups during the morning, then working on basic obedience training in the afternoon. The property contained several spacious play yards, a converted paddock filled with professional agility equipment for the dogs to climb around on, and a large pond with a dog-friendly diving dock. Shelter staff regularly took their charges on car trips into the nearest town and tried to expose them to as many new people and experiences as possible.

"People believe that we have great dogs because we handpick them in some way," Coleman said, "but that's not true. We pull dogs from all backgrounds, and we usually don't know a lot about them before they arrive, only that someone, somewhere labeled them as 'pit bulls.' We try to give them as much socialization and training and enrichment as we can while they are here, and that makes a big difference. A lot of them have never had the opportunity to learn how to be dogs, especially if they've been isolated or cooped up in a facility for a long time. What we've found is most of them don't really need to be 'rehabilitated'; they need to be *hab*ilitated, if that makes sense."

Coleman and I drove up a hill lined with apple trees until we arrived at the big white house where Jane Berkey lived with her two dogs. Beside it was a small guesthouse where I'd be sleeping. "Dinner is at Jane's house at six," Coleman said. "After dinner, someone will bring your dog to you."

"My dog?"

"Yes," she said, smiling. "Your roommate. When we have visitors, we try to pair them with dogs from the shelter so that the dogs enjoy the company and you can tell us more about how they do on their overnight stays. A member of the shelter staff will drop a dog off each night and pick him or her up each morning. They'll bring you instructions and a questionnaire."

Jane Berkey, who was in her early sixties when I met her, exuded forthrightness. Her voice had the slightest rasp of the ex-smoker in it, and her tone wasn't formal, necessarily, but it wasn't informal, either. When she fetched me from the guesthouse later that evening, she chummily hooked her arm into my elbow. Walking with swift, purposeful strides, she guided me past the pool, under an arbor of grapevines, and through the hydrangea-framed patio into her house, whose cavernous rooms were filled with antique pit bull memorabilia, including original prints of Wallace Robinson's World War I posters.

Peeking into one doorway, I noticed a photograph of a young woman with soft, cascading curls and a cigarette dangling from her varnished fingertips. She bore an uncanny resemblance to the actress Bridget Fonda. "That was me," Jane said. "Wedding number two."

We sat down to talk in a formal living room lit by a giant crystal chandelier, underneath which sat two ornately carved carousel pit bulls. The statues were modeled on Jane's two most beloved dogs, Petal and Joshua, both recently deceased. They had been commissioned from the California artist Tim Racer, who, along with his partner, Donna Reynolds, founded BAD RAP, the rescue that had successfully re-homed many of the fighting dogs seized from Michael Vick's property in 2007. I remembered something that Tim Racer had told me when I visited BAD RAP in Oakland, California, some months prior. "It took me fourteen years of working with these dogs before I truly stopped believing all the things people said about them," Racer said. "That is how powerful those messages can be."

"What you have to understand," Jane said, "is that we are now at the point at which the very discussion of the problem advances the problem. The term 'pit bull' now has so much coin that *everyone* gets something out of it. Musicians, rappers, politicians, reporters, whoever. Everybody gains something from that term. Except the dogs. They just end up dead."

Although she grew up in an affluent Long Island suburb, there was always something about Berkey that cut against the grain, and it showed itself fairly early. As an avowed child of the 1960s, she married her first husband, John Rotrosen, at eighteen, then drove out to San Francisco on a BMW R50 motorcycle to soak up the flower power still trailing behind the Summer of Love. When she came back east to attend New York University (where her husband attended medical school), the newlyweds lived for a time in a quasi-commune on Varick Street in New York City's West Village.

Jane joined the fledgling animal rights movement in the early 1970s, after her husband laid out for her in graphic detail the plight of animals kept in research laboratories. Over the next few years, she attended animal research protests and became more involved with specific campaigns, heavily influenced, as many were, by the ideas of the Australian moral philosopher Peter Singer. "It wasn't so much about the individual animals back then as it was about the larger idea," Berkey said. "It was much more of an ethical issue about justice, fairness, and decency, versus arrogance and stupidity. Our species has a bit of an arrogance problem."

Fast-forward twenty-five years, past the establishment of her highly successful Jane Rotrosen Agency, with its stable of best-selling commercial fiction authors, past two more husbands and two more divorces. Needing a break from the hectic pace of Manhattan, Jane found herself retreating to her horse farm in rural Dutchess County more frequently, which she had always planned to turn into a horse rescue. While there, she volunteered four days a week to walk dogs at the local public shelter. Some of the other shelter workers cringed around dogs labeled as pit bulls, so the task of walking them usually fell to Berkey. "They never seemed any different to me," she recalled, "only that they carried a heavier burden."

Then she saw a photograph that would change the course of her life, a quick snap of a tricolored female pit bull with an inquisitive, tender face. During the winter months, which Jane spent at her house in Palm Beach, Florida, she customarily asked her house manager, Beth, to take pictures of the Dutchess County shelter dogs and post them on local bulletin boards to increase the dogs' chances of adoption in Jane's absence. It was her way of feeling a bit less guilty for not being there. When Jane saw one of the photographs Beth had taken, the one of the tricolored dog, Jane told me, "Right then, I just knew."

The dog was covered in dirt and feces, and "her soul was just gone," she remembered. By the time Berkey finally returned to the Dutchess County shelter, the dog "had completely given up." Jane snapped a leash on her, walked her over to a grooming tub, and painstakingly cleaned her off. "She looked like a monkey in a laboratory to me," Jane said. "She looked up at me, and there was just no question. She and I were going to travel together." Berkey's voice caught in her throat. "I can experience it right now. It's like she's here with me."

Jane took the dog home and named her Petal. At first, Berkey lavished resources on Petal and on her second pit bull, Joshua, going so far as to fly them around the country on private planes whenever she traveled. "I wanted to show that yes, pit bulls can be owned by someone who will do this for them," she said. "That these dogs don't always have to be given less." (Now, she says, she would much rather spend that money on homeless dogs, who need it more.) Her own animals brought her so much joy that it seemed unconscionable that other shelters refused to put pit bulls up for adoption and, even worse, that in some parts of the country families could have their companions taken away because of what they looked like.

The profound connection that Jane felt with Petal turned into a personal mission. She would devote all her time, energy, and considerable resources to "getting pit bulls back on the bus." Rather than rescuing horses and farm animals, as Berkey originally planned, Animal Farm Foundation would "restore the image of the American Pit Bull Terrier and protect him from discrimination and cruelty," according to its earliest mission statement. That changed when Jane discovered that the amoebic term "pit bull" encompassed a much larger group of animals than the APBT.

"First we damned the dogs called pit bulls with all this scary, incorrect information," she said. "Then we attached that label to mixed-breed dogs that aren't even pit bulls. At this point, a pit bull is *not even a dog;* it's a social construct! The term has a life of its own. But once you put that label on a dog, then the context of everything it *does* changes. And if something goes wrong, God forbid, it's the shot heard round the world."

The scapegoating of pit bulls that grew out of the inflammatory rhetoric of the 1980s shocked Berkey, as did the vindictiveness that came with it. In early 2006, Kathy McIntyre, the editor of the *Gateway News* in Commerce City, Colorado, published the names and addresses of every registered pit bull in the city—all twenty-four of them—as a sort of canine "blacklist." "My philosophy and my reason for doing this is if we know where the pit bulls in Commerce City live," McIntyre said, "we can take measures to protect ourselves."

Later that year, a mixed-breed puppy was tortured and baked to death in an oven by two Atlanta teenagers. Dr. Melinda Merck, the forensic veterinarian who consulted on the case, recalls that there was a discussion about which kind of puppy to use in court as part of a live demonstration. When the possibility of using a pit bull puppy came up, the prosecutor decided against it. He feared that the jury might not sympathize with the imagined suffering of a pit bull, so a German shepherd puppy was used instead.

In May 2009, a two-year-old female pit bull was doused with gasoline and set on fire in Baltimore. She survived the initial attack, but her burn injuries, which covered 95 percent of her body, were so severe that she had to be euthanized four days later. Her case became a rallying cry against breed prejudice. Despite these crimes, however, some Americans still could not bring themselves to think of pit bulls with the same consideration they gave to other breeds. "Pit bulls should be boiled alive like lobsters and fed to their idiot owners," the author and gay-rights activist Dan Savage wrote in a 2007 column for *The Stranger.*

Unlike most dog rescuers, who must scrimp and save, even go into debt, to help the dogs in their care, Jane's financial privilege put her in a unique position (AFF's 2013 assets were listed at more than $1.5 million). She began attending veterinary conferences, consulting with behavior experts, and arming herself with mountains of scientific literature. In an effort to correct for the second-class treatment pit bulls typically received, AFF offered grants to rescues and shelters that provided pit-bull-specific services, such as training programs or spay/neuter discounts. Jane then learned about Karen Delise's research group, the National Canine Research Council, and offered to fund it so that the public would benefit from comprehensive information about dog-bite-fatality cases.

I asked Delise whether Berkey or AFF, an avowed advocacy organization, had ever pressured her to alter her data or make her reports more favorable to pit bulls, as many anti-pit-bull activists had alleged. "Absolutely not," Delise said. "Never. Jane may not want to know the details of the pit-bull-related incidents, but I do. I want every scrap of information I can find about these cases. There is nothing I hate more than being wrong." This is one of the reasons, she said, that she had consulted with epidemiologists and put her data through the peer-review process.

But new data and changed minds could not fully repair the damage that the old scientific blunders had caused, nor could they soothe the fears of many people in animal sheltering. For a time, the ASPCA's official presentation, "The Care of Pit Bulls in a Shelter Environment," included such talking points as "install a panic button in rooms housing pit bulls along with other restraint equipment." It also described "cases of experienced handlers . . . being attacked without warning or provocation" and included several slides on "bait dogs." Many of these points sounded as though they had been taken directly from the media scare stories of the 1980s, and it was clear that the authors of the presentation did not consider pit bulls "normal" dogs. Shelters across the country, which looked to the ASPCA as an example, took these messages to heart.*

For a pit bull to be adopted into a private home during those first years of the "turnaround," it had to be *perfect.* Every behavior easily forgotten in dogs from other breeds, something as slight as a growl or a bared tooth, was a potential headline, which pit bull advocates simply could not afford. As Diane Jessup put it, "We were fighting for our lives." If a pit bull

---

* The ASPCA no longer uses this presentation as part of its shelter education program.

tolerated some situations but not others—if it could not properly be considered an "ambassador" for its breed, in all places, with all people and pets, at all times—then it had to be euthanized, even at AFF. Jane remembers with haunting clarity all the dogs she put down out of fear of what the animal "might" do. "In trying to make them perfect," she said, "we deprived them of being dogs. In the end, the things I loved, I made less."

It was right around 2000 that Jane called up Diane Jessup for the first time. She wanted to learn more about the American pit bull terrier, and there were few people more qualified to teach her than Jessup was. Despite the obvious differences in their backgrounds and deportment, the two women are incredibly—almost eerily—similar: tough, smart, opinionated, and stubborn, but also tenderhearted and generous with those they know well. Their shared passion for animals kindled a fast friendship. Jane sold Diane's novel, *The Dog Who Spoke with Gods,* to St. Martin's Press, and it was Jane who financed Diane's now-defunct Law Dogs USA program, which trained shelter pit bulls for jobs in law enforcement. Jane named her favorite dressage horse after Dread, Diane's first APBT. Berkey had immense respect for Jessup as a trainer, and she admired the deep connection she shared with her animals.

But on the issue of how best to advocate for pit bulls, the two eventually grew apart. Diane wanted to preserve the one breed that she loved so much, which she believed was inherently different from all others, and she felt that the existence of so many pit bulls in shelters (not unlike a market flooded with knockoffs) damaged the reputation and the future of that breed. To her, the "generic" pit bulls—the knockoffs—had to go. "Jane thinks that there should be a pit bull in every home," Jessup told me. "And I don't."

Jane, on the other hand, was much more concerned about the dogs in shelters—the dogs of today. They came from scattershot backgrounds that did not adhere to any one breed profile. These dogs were dying by the thousands. "Ride the horse you're on" was one of her favorite sayings, and she believed in the "rising tide raises all boats" theory of animal advocacy. To Jane, a dog was a dog, and, more important, a life was a life. "I do *not* believe there should be a pit bull in every home," she insisted. "That's just not true. I don't even believe there should be a dog in every home. I believe people should have the dogs they want, if they want them, and that we shouldn't put up so many barriers to adoption."

Diane invoked her years of practical training experience with purebred APBTs and her career as an animal control officer. The dogs she raised displayed traits she worked hard to cultivate. Jane marshaled her years of experience with shelter dogs and the opinions of behavior experts. The dogs she worked with were incredibly diverse. On this issue, Berkey and Jessup simply could not agree.

"What I tried to convey to Diane," Jane told me, "is that *different is dead.* What she says scares people. And it's not scientifically accurate. What she is pursuing with all the tough talk, given the stigma that is attached to these dogs, will actually be the end of the dog. What I am pursuing, which I believe is more accurate to the dogs we have in most of America, is what will save the dogs that she loves." Each woman had devoted her life to pit bulls, but their proposed solutions were almost photographic negatives of each other. There was no single policy that fixed the problem. There probably never would be. Each dog was different.

Others involved in re-homing shelter dogs also retreated to these same poles. Marcy Setter of Pit Bull Rescue Central, a large adoption network started by a woman named Veronique Chesser in the 1990s, agreed with Jessup: pit bulls require a special kind of owner. "With other breeds," Setter told me, "you don't walk down the street and have to put up with people yelling 'baby killer' at you. You don't have to worry about the government passing a law and taking your family member away. And you don't have to worry about going to the dog park, knowing that if there is *any* kind of incident with another dog, regardless of *your* dog's intentions—whether it was frightened or defending itself—it won't matter because you've just created a media event. So if you're telling me that a golden retriever owner has to have the same level of responsibility, you're crazy." PBRC even recommended that pit bull owners keep a "break stick" (a tool that resembles a large wooden or plastic letter opener) handy to pry their pet's jaws apart if it got into a fight.

Aimee Sadler, a longtime animal trainer who has worked with horses and marine mammals as well as dogs, disagreed. "I've worked with hundreds, if not thousands, of pit bulls in shelters," she said, "and you can tell me until you are blue in the face that they are more aggressive toward other dogs, but that is simply not what I have seen. Many times what people assume is a, quote, inborn aggression toward other dogs is simply a lack of socialization and experience with other dogs." To remedy this, Sadler pioneered the Dogs Playing for Life program, which facilitates play groups for shelter dogs in order to give them exercise, help them learn canine social skills, and increase their chances for adoption.

Berkey continued, "I always believed that language reflects thought. Then a rhetorician friend of mine corrected me: no, language reflects *habit.* And we have so many bad habits with respect to dogs, things we say we love so much." This is why she preferred to place "pit bull" in quotation marks. Berkey felt that what we believe about the meaning of those two words shapes our perceptions and expectations of the dogs we encounter in everyday life.

But Diane Jessup's dogs were not theoretical, least of all to her. They were not constructs; they had pedigrees. As convinced as Diane was that pit bulls were "different," Berkey was equally convinced that they were not. This is why she leaned so heavily on science, which presents a more complex, nuanced picture of genes and behavior. "Without that," she said, "it's all dueling anecdotes." Berkey's goal was not to preserve a historic breed; her goal was to help pit bulls in shelters not die. "I just want to get them back on the bus," Jane said. "I want them to have the same opportunities that any other dog would have. That's all I am trying to do."

Back in the guesthouse, I opened the door to hear the tapping of claws on kennel wire, followed by a low, snorting whine. Inside a crate near the bed was a tan-colored puppy with a Roman nose and giant gremlin ears that sprouted from his head. I had almost forgotten about my "roommate," who had arrived with a leash, a few toys, and some paperwork that was fanned out on the desk. "Hello!" it read. "My name is Pinto Bean!" According to the papers, Pinto was not really a puppy; he was nearly full grown at almost a year of age and a Lilliputian twenty-five pounds.

"Well, hello, Pinto," I said, unlatching the crate. He sprang into my arms and we played for an hour or so. First we went for a short walk, during which he barked at a few shadows. Then we worked on the commands he had learned during training ("sit," "stay," "wait"), all of which were nicely executed. After that we played some fetch and tug, and he entertained me with some unbearably earnest, yet still puppy-sized, growls and play barks. Finally, I put him back in the kennel with his blanket and toys. "Night, buddy," I called over to him as I crawled into bed.

Pinto sighed and lay down with his paws over his snout. His ears drooped. I turned away and tried not to look at him. When I finally rolled over, I could feel him blinking at me. He looked like the Count of

Monte Cristo in there. I did not know then that Pinto, who seemed very gentlemanly and well adjusted, had spent the first six months of his life isolated and confined at an animal control facility.

"Aw, hell," I said, tossing back the covers. "Fine."

He slept with his head on my shoulder the whole night, lightly snoring into my ear.

By far, the most significant turning point in the lives of American pit bulls was the April 2007 arrest of Michael Vick, star quarterback for the Atlanta Falcons football team, on federal felony charges related to the operation of a professional interstate dogfighting ring that was run out of his property in rural Surry County, Virginia. Based on forensic evidence collected on Vick's property and the testimony of several witnesses, dogs from Vick's Bad Newz Kennels were discovered to have been drowned, hanged, electrocuted, and beaten to death in addition to the daily pain and suffering they experienced as victims of dogfighting. Several, including one that had been fatally beaten, were killed at Vick's own hands.

"I may have become more dedicated to the deep study of dogs than I was to my Falcons playbook," Vick wrote in his 2012 memoir, *Finally Free*. "I became better at reading dogs than reading defenses." He recalled thinking, "It ain't that bad. It's wrong for the dogs, but this is what these dogs like to do. This is why they're bred." Having grown up in a rough section of Newport News, Virginia, he said that he saw his first dogfight at the age of eight.

In a deeply ironic turn of events, Vick found out about the charges against him while he was playing golf. He pleaded guilty to conspiracy to commit dogfighting and was sentenced to three years in federal prison. He served twenty-three months.

"You might say, 'come on Mike, how could you do those things to those dogs?'" he wrote after being released from prison in 2009. "And you're right . . . I ask myself those questions every day. What kind of person does this? How does a human-being [*sic*] treat dogs or any animal with pain and cruelty? The hard part for me is the answer to these questions. Because the answer is ME. I am trying so hard to be a better person, because who I was, I am ashamed of. 'Cause see, my whole life has been numb." Vick made several apologetic public statements, but for animal lovers, his contrition fell on deaf ears. He always seemed to be apologizing to the people in his life, not to the dogs he had tortured and killed. Why should he be allowed to return to a career in professional sports, one in which he would make millions of dollars and be held up as a role model for thousands of young people? His critics were adamant that he shouldn't.

Forty-nine dogs were eventually seized from Vick's property through a massive effort coordinated by the ASPCA. Because of the athlete's financial assets, the courts required him to pay almost $1 million for the care of the dogs long-term, which allowed humane workers a unique opportunity to show the world it was possible for ex–fighting dogs to rebound from a lifetime of abuse. After thorough behavioral evaluations, two of the dogs were euthanized (one for health reasons), twenty-two were given sanctuary by Best Friends Animal Society,* ten went to BAD RAP, and the remaining fifteen went to other shelters around the country. Several ended up at Animal Farm Foundation.

The rescue and rehabilitation of the surviving "Vicktory Dogs," as beautifully narrated in *The Lost Dogs* by Jim Gorant, was a powerful moment in animal welfare. For the first time, fighting pit bulls were seen as victims of a terrible crime, not co-conspirators in it. The groundswell of support for their rehabilitation was enormous, with the media actively cheering them on. *Sports Illustrated,* which had played a central role in the panic of 1987, ran a moving feature by Gorant that included one of Vick's decidedly unfrightening pit bulls on the magazine's cover. "Finally," Jane recalled, "the public hated something worse than it hated pit bulls, and that was Michael Vick."

Others in animal rights were skeptical of the rehabilitation efforts. A spokesperson for PETA said, "These dogs are a ticking time bomb. Rehabilitating fighting dogs is not in the cards. It's widely accepted that euthanasia is the most humane thing for them." Wayne Pacelle, the president and chief executive officer of the Humane Society of the United States, told *The New York Times,* "Officials from our organization have examined some of these dogs and, generally speaking, they are some of the most aggressively trained pit bulls in the country. Hundreds of thousands of less-violent pit bulls, who are better candidates to be rehabilitated, are being put down. The fate of these dogs will be up to the government, but we have recommended to them, and believe they will eventually be

* Of these twenty-two, twelve have been fully rehabilitated and placed in homes, some acting as service or therapy dogs. Two had to be euthanized after arriving at Best Friends because of injuries sustained in dogfighting at Bad Newz Kennels, two were ordered by the court to remain at Best Friends sanctuary for life, and six—it is hoped—will soon be ready for adoption.

put down." In 2009, however, HSUS changed its official policy about the euthanasia of dogs from fighting busts, and has since taken an active role in working to end breed-based legislation and breed-based housing policies. "Pit bulls are special not because of their breed, but because they are dogs," Pacelle wrote in 2014. "And like any other dog, they deserve to be kept safe with loving families." (For a time, HSUS also partnered with Vick in a controversial anti-dogfighting campaign.)

The fact that thirty-five fighting dogs would go on to be successfully placed in pet homes should have been an unalloyed triumph for animal-loving Americans, and for most it was. But still, the ghosts of race and class clouded the media coverage, as they always did in high-profile stories about pit bulls. Several African-American celebrities, including the actor Jamie Foxx and the actress and television personality Whoopi Goldberg, spoke out in Vick's defense before learning the full extent of his crimes. Pointing out what he believed was media hypocrisy, the NBA star Stephon Marbury said, "We don't say anything about people who shoot deer or shoot other animals." The comedian Chris Rock was equally puzzled. "What the hell did Michael Vick do, man?" he said. "A dog? A pit bull ain't even a real dog."

Statements like these then prompted their own backlash, and the rhetoric escalated. Critics called for Vick to be "neutered," electrocuted, or torn apart by dogs. Cartoonists portrayed him as an animal. PETA demanded that he receive a brain scan to test for possible psychopathy before being allowed to return to football. Threats were made against Vick's family members, specifically his children. In 2010, the conservative television commentator Tucker Carlson said, "I'm a Christian, I've made mistakes myself, I believe fervently in second chances, but Michael Vick killed dogs, and he did [it] in a heartless and cruel way. And I think personally he should have been executed for that. He wasn't."

Responding to Carlson and others who heaped opprobrium on Vick and called for his "execution" in a manner she felt was disproportionate to his crimes (especially given that other prominent figures in sports had been convicted of murder, rape, domestic violence, and child abuse), the Princeton professor and television host Melissa Harris-Perry wrote in *The Nation,* "Not only have animals been used as weapons against black people, but many African Americans feel that the suffering of animals evokes more empathy and concern among whites than does the suffering of black people . . . the decision to demonize Vick seems motivated by something more pernicious than concern for animal welfare. It seems to be about race."

The author and ethicist Dr. Kathy Rudy, who teaches several classes on animal rights and animal welfare at Duke University, also found the threats of violence against Vick and the wholesale condemnation of hip-hop troubling. "I am not saying dogfighting is acceptable," Rudy stated, "but rather that Vick should be publicly criticized for that activity, not for his participation in hip-hop subculture . . . We need to find ways to condemn dogfighting without denigrating black culture with it."

Once again, the dogs themselves were used as a pretext for a much more contentious war over human values and culture. No pundit mentioned that in 1969, when dogfighting was illegal but not a felony, *Sports Illustrated* ran a feature on the New Orleans Saints that focused on Doug Atkins, the Saints' defensive end, and his dog, a pit bull named Rebel that Atkins openly admitted to fighting. "Matched him with a Doberman last night and the Doberman gave him fits for four or five minutes, but ol' Rebel never quit," Atkins told the reporter. "Why, he can fight at full speed for 35, 40 minutes and he finally wore that Doberman right down. Got him down and probably would have killed him, but ol' Rebel ain't got any teeth. Had to gum the Doberman until he quit." Atkins, who is white, was inducted into the Pro Football Hall of Fame in 1982.

Almost all these commentators agreed that intentional animal cruelty is morally reprehensible and that it must always be prosecuted, no matter who the perpetrator is or what background he comes from. No one deserves a "free pass" on cruelty. What seemed to raise some eyebrows was not that Vick was charged, convicted, and sentenced for sickening offenses, but that, because of his prominence and legal loopholes in certain jurisdictions, his sentence was much stiffer than those of many professional fighters who had perpetrated the same acts against more animals over a much longer period of time, most of whom were white. Ed Faron, the so-called godfather of dogfighting, had been at it for forty years by the time he was arrested in Wilkes County, North Carolina, in 2008. He received only a ten-month sentence.

Were critics like Melissa Harris-Perry correct? Did Americans seem to care more about the suffering of animals than about the suffering of black people, or even about the suffering of human beings in general? It often appears that way, but the question, like so many others, is more complex than a simple either/or. When presented with fictional "news" items about the alleged beatings of a puppy, an adult dog, a human infant, or a human adult, college students in one sociological study conducted at Northeastern University registered their levels of emotional distress as being highest after reading about the abuse of the human child (which was followed in descending order by the puppy, the adult dog, and the human adult). In this scenario, the perceived helplessness and vulnerability of the victim was paramount; the adult human was assumed to be more capable of "fending for himself" than the victims in the remaining three categories. Other research indicates that a person's emotional proximity to the victim (whether or not an imperiled person or dog is a stranger to you) plays more of a role than species: subjects were theoretically more likely to save the life of their own dog than that of a foreign tourist, but they were more likely to save the life of a foreign tourist than that of a dog they did not know.

In each of these scenarios, however, what seems to be the most powerful factor in generating empathy is a focus on the individual human or animal (one with a name and a story) rather than on a faceless, remote group or statistic. Because of Michael Vick's celebrity status, his former dogs' names, personalities, and long paths to recovery were exhaustively chronicled by the press. Too often in crime reporting, human victims don't receive that level of attention.

Fortunately for them, the dogs themselves were oblivious to such matters. The Vicktory Dogs proved that all former fighting dogs are not lost causes. When the public embraced them, it advanced the notion that victims of cruelty deserve a fair shot. "Vick showed the worst of us, our bloodlust," wrote Donna Reynolds, one of the cofounders of BAD RAP, "but this [rescue effort] showed the best. I don't think any of us thought it was possible—the government, the rescuers, the people involved. We like to think we have life figured out, and it's nice that it can still surprise us, that sometimes we can accomplish things we had only dreamed of. We've moved our evolution forward."

The protocols for handling dogs from fight busts also changed. When more than five hundred pit bulls were seized from multiple fighting operations during a multistate sting operation in 2009, each was given a full behavior evaluation by the ASPCA, as well as time to decompress and recover. Within a few weeks, more than half were deemed suitable for placement in pet homes. "You can't save them all," Randy Lockwood said, "but we do have an obligation to the victims of cruelty to undo some of the harm that humans have done. The spectrum of behaviors, even among fighting dogs, is enormous. These animals have been exposed to the worst combination of genetics and environment, and many can still be good, loving pets."

In the wake of the Vick scandal, dozens more rescue organizations and advocacy groups were started around the country, and the tide of breed prejudice finally began to ebb. National Geographic Channel aired the first pit-bull-centric reality show, *DogTown,* in 2008, which was then followed by two Animal Planet shows, *Pit Bulls and Parolees* (2009) and *Pit Boss* (2010). Almost overnight there were pit bull parades and "pit crews." Nonprofits like The Unexpected Pit Bull (founded in 2004) and Pinups for Pitbulls (founded in 2005) quickly drew in crowds of fans with positive messaging and fun themed calendars. The lives of countless thousands of dogs that would otherwise have been condemned to euthanasia were saved as the combined result of these efforts.

And yet, in attempting to solve one set of problems, pit bull advocates inadvertently stumbled on a different set. "If pit bulls are so difficult to identify," Dr. James Serpell, the director of the Center for the Interaction of Animals and Society at the University of Pennsylvania's School of Veterinary Medicine, asked me, "then how do these advocacy groups know what they are rescuing? You can't have it both ways." Others wondered why one type of dog needed an entire social movement in the first place. Weren't breed-specific incentives, parades, and moral crusades the flip side of breed-specific laws? Didn't it feed into the same sort of tribalism? "I once knew someone who said that pit bull pride is worth the prejudice," Berkey told me. "But trust me, it isn't."

Some of the old negative inaccuracies about locking jaws and supernatural strength were replaced with positive untruths, such as the trope that pit bulls were referred to as "nanny dogs" in the early twentieth century. Just as the twenty-four-hour news cycle amplified 1980s fearmongering, so, too, did the rise of social media make it possible for reams of feel-good myths to circulate. Now anyone who had ever seen a pit bull could declare himself an authority and insist that "it's all how you raise them." Additionally, some animal advocates displayed a cringe-worthy lack of cultural sensitivity by first equating breedism with human racism, then using coded racial language to condemn certain pit bull owners. A popular T-shirt read "Pit Bulls Are for Hugs, Not Thugs." This type of marketing implied that pit bulls were only acceptable pets when they belonged to upper- and middle-class white people. It allowed the specter of the "sinister other" to lurk in the background.

As Berkey had noted, there was profit in pit bulls for virtually everyone, and after 2007 they were also profitable for advocates. Because of its profound emotional gravity, the cruelty narrative—specifically the dogfighting narrative—was now an effective way of generating donations. In an almost carbon-copy replication of 1974, the news was again flooded with stories about dogfighting, exposés about dogfighting, "expert" commentary about dogfighting, and opinion polls about dogfighting. The inaccurate message to the public—that "pit bull" equals "dogfighting"—had not changed, nor had the inflated estimates of its scope. One widely circulated figure, that there are currently forty thousand professional dogfighters in the United States, was based on a crude guess about magazine subscriptions, not actual data. Investigation and arrest records, as well as the testimony of multiple confidential informants in dogfighting cases, indicate that the real numbers are probably not even a third of that estimate. Instances of the crime have dropped lower than they have been in three decades. Yet in 2014 the ASPCA declared April 8 National Dog Fighting Awareness Day, which was accompanied by a major publicity campaign. After Michael Vick, the last thing that America's pit bulls needed was more "awareness" about dogfighting. What they needed was much less of it.

"Overstating the problem of dog fighting, directing it at young black youths, and flashing America with images of dogs engaged in battle might work from a marketing perspective," Donna Reynolds of BAD RAP wrote in a 2011 blog post, "but it can actually work against the cause by deepening fear and stereotypes. This is especially true in communities that have [breed-specific legislation]. From a big picture perspective, the larger threat to the dogs is much more insidious and mainstream than the guy in the ghetto: It's the unforgivable prejudice that forces families to surrender their dogs to a shelter because no property owner will rent to them."

This was a keen observation. As of 2014, 35 percent of Americans lived in rental housing, and nearly 40 percent of renters own dogs, but only 24 percent of American apartments allow dogs on the premises. Of those 24 percent, the number that will rent to owners of dogs perceived as being "pit bulls" is infinitesimal. Forty-nine of the fifty apartment complexes closest to my North Carolina home will not rent to owners of pit bulls. The fiftieth rental company, which is twenty-five miles away, does not rent units for less than $850 a month—significantly higher than many working Americans can afford. My only other option, should I ever need to find rental housing with my dogs, would be to put myself at the mercy of a local slumlord. It is likely that a large number of renters are either hiding their dogs, moving to less safe properties in order to keep them, or facing eviction.

Landlords most often implement dog breed or weight restrictions because their insurance carriers demand it, and insurance companies are all too happy to highlight the millions they pay out each year in dog bite cases as justification for these policies. It's not personal, they say; it's just a matter of "actuarial risk."

In 2013, the Insurance Information Institute (III) reported that costs related to dog bite injuries made up one-third of all homeowner's insurance liability claims. Members of the media then misrepresented this (at least in headlines) as being one-third of *all* homeowner's insurance claims. The insurance industry did pay out $483 million for 17,359 bite injury cases that year, which sounds like a staggering figure until you put it next to the other data. According to the III, only 4.8 percent of insured homeowners filed *any* kind of claim in 2013, and most of those claims (97 percent) sought compensation for property damage incurred in fires, floods, hurricanes, and burglaries. Of all the claims filed that year, only *3 percent* were liability claims. Only 2.7 percent related to bodily injury of any type. Divided into thirds, that means dog bites accounted for less than 1 percent of homeowner's insurance payouts—0.9 percent, to be exact. Those injuries were inflicted by dogs from many different breeds. The costs of these settlements have increased, not because the injuries are growing more severe, but because the cost of health care is climbing dramatically (health expenditures in the United States have increased more than 800 percent since 1960). What's more, the estimated insurance payouts for dog bite claims have *dropped* more than 50 percent since 1994, when they topped $1 billion nationwide. In terms of sheer numbers, deer again pose a larger threat; American insurance companies paid out $4 billion for damage caused by deer-vehicle collisions in 2012.

Most of us accept that if an insurance underwriter tells us that a condition, behavior, or personal choice represents an "actuarial risk," then he or she must have solid empirical evidence to back that up. But the scholar Brian J. Glenn (who trained as an insurance underwriter) points out that the entire discipline of actuarial research originally grew out of the insurers' desire to make decisions based on the perceived morality of a customer without the customer (or the public) knowing that's what was being done. In a remarkable feat of reverse engineering, the insurance industry devised algorithms to fit the story it wanted to tell about who was a legitimate member of "mainstream" American society and who was not. This process was called character underwriting, and it was still a key part of insurance training when Glenn went through the process in the 1990s.

To that end, insurers have historically gathered a tremendous amount of mathematical data to justify denying coverage to African-Americans, Jews, gays, unmarried women, single mothers, and blue-collar workers because they represented, in the parlance of insurance underwriting, "moral hazards," not physical ones. Until the Fair Housing Act of 1968 made it illegal for landlords to discriminate based on race, insurance companies often invoked actuarial risk when denying coverage to homeowners in urban neighborhoods, a process we now call redlining because of the actual red line insurers drew on maps that separated the "good" and insurable from the "bad" and "risky" areas of American cities. Residents of the "risky" communities typically had lower incomes, their homes were older and in need of repair, and their neighborhoods experienced more burglaries, so the insurance companies felt justified denying them coverage. It was simply about the numbers, they said. As the journalist Ta-Nehisi Coates elucidated in a thorough investigation of redlining for *The Atlantic,* these practices erected the economic equivalent of the Berlin Wall between American haves and have-nots. Despite these policies being illegal, they continued well past 1968.

As overt race and class discrimination became less socially acceptable, Brian Glenn explains, "The rhetoric of insurance exclusion changed to appear more scientifically based, keeping the underlying narratives about groups buried and undisclosed." Even today, he says, "the numbers, data, and forms merely hide the fact that applicants are still judged according to their standing in society." A family who keeps a car on blocks in the front yard may very well be denied coverage, while a family who keeps a boat in the front yard may not, despite the fact that both represent the same physical hazards to children playing in or around them. Glenn includes pit bulls in the category of moral status markers. "There are two stories being told," he writes. "The one about hazards is told to the outside world. The one about class and lifestyles is not."

"Our job should be to put ourselves out of business," Jane Berkey once told me. Yet the paradox of "different is dead" makes this task infinitely more difficult. AFF continues to fund breed-specific initiatives in shelters around the country, she said, because "the damage done to these dogs was so great that some kind of extra help was necessary." But doing so, she acknowledged, necessitates treating the dogs "differently." That is the central conundrum of the pit bull movement, for Berkey and others: How do you know when your mission has been accomplished? And what will become of pit bull advocates when that happens? Is there a future for National Love Your Average Pet Day?

"All I want is for this to be over," Berkey said. "I am ready to move on to other things." As for the charges of "pit bull lobbying" leveled against her by proponents of breed bans, she laughed. "If it were possible to simply buy a solution to this problem, I would have done it," she said. "But trust me, if it were that easy, this would have been over years ago." Her greatest hope is that the day will come when pit bull advocacy is no longer necessary, when America relinquishes its tight grip on the image and identity of the pit bull and simply lets them be dogs. "It will happen one dog at a time, one mind at a time," Jane said. "The *dogs* will open people's hearts. All we have to do is get out of their way."

CHAPTER 15

FOR LIFE

> The whole point of learning about the human race presumably is to give it mercy.
>
> —REYNOLDS PRICE, interview with *The Georgia Review*

Bam-bam-bam-bam-bam!

"Stop doing that," Stephen Parker said to his co-worker Ashley Mutch as the two stood on the front porch of a row home in North Philadelphia. Both carried clipboards and stacks of flyers, and both were dressed in black hooded sweatshirts bearing the Humane Society of the United States logo. A hard April wind whipped down Ontario Street, and Parker blew into his hands.

Mutch, a former humane law enforcement officer who had once starred in the Animal Planet reality show *Animal Cops: Philadelphia,* looked over her aviator sunglasses at him and raised her eyebrows. "Stop doing what?"

"Knocking like a cop, with the back of your fist. No one is gonna come to the door if you do that."

Mutch laughed. "I can't help it!" She tightened her black ponytail and bounced on the balls of her feet to stay warm. A lifelong athlete, she worked as a CrossFit coach in her off-hours. "Old habits and whatnot." After waiting a few seconds, she tried a light, one-knuckled "Shave and a Haircut" instead. The door opened and an owl-eyed elderly woman peeked out.

"Hi, there," Mutch said, extending her hand. "I'm Ashley, and this is Stephen, and we work for the Humane Society of the United States. Do you know anyone who has a pet?" The woman took her hand and shook it slowly, glancing over the HSUS logo on Mutch's sweatshirt with obvious skepticism. From a window a few doors down, Latin rap music poured into the street.

"I have a cat," the woman finally said, opening the door an inch wider. Mutch then explained why she and Parker were there: They were promoting HSUS's nationwide community outreach initiative, Pets for Life, and informing neighborhood residents about the program's upcoming Community Pet Day, an all-day outdoor event at which pet owners could get their animals vaccinated, sign up for spay and neuter appointments, enroll in dog-training classes, and pick up pet supplies including leashes, collars, and bags of food. Her pitch was cheerful and seamless, with a big *ta-da* ending: Every one of the services and supplies offered, including the spay and neuter surgeries, was free, no strings attached.

"Are you *sure* it's free?" the woman said, scanning the flyer as if looking for the fine print. "All of it?"

"Yes ma'am. Everything we offer is one hundred percent free."

"Is it only for dogs?"

"Nope. Cats, too."

"Will I need to show ID? Paycheck stubs?"

"Nope. Just show up with your pet."

"How many animals can I bring?"

"As many as you want."

"Is this for real?"

"Completely."

"God bless you! Do you have any more of those flyers?"

This short interaction is what the PFL outreach team referred to as a "touch," or a face-to-face conversation with a member of the community. Each staff member's goal was to rack up as many touches per week as possible, which were then tracked as official data. No other form of communication (phone, e-mail) counted as a touch, though calls and e-mails were tracked as well. I had been walking the streets of North Philly with the outreach team for three full days, and in that time I had observed dozens of touches. The original stack of two thousand bilingual flyers had thinned to a modest few hundred. The chat usually went the same way, shifting through the low gear of suspicion, past cautious approval, and through incredulity before cruising into relaxed gratitude, which is where it always seemed to land. In the days I had been out with the team, none of the strangers we met had said, "No, thank you," "Not interested," or "Go away."

Ashley Mutch speaking to a Pets for Life client in Philadelphia, 2013

"If you ever want to see how badly a service is needed somewhere," Amanda Arrington, the national director of Pets for Life (and founder of the Coalition to Unchain Dogs), told me, "watch what happens when you give it away for free. If you're giving away something great for free and nobody wants it, then you are doing something wrong." Because Philadelphia had been portrayed in the media as such a hotbed of animal cruelty and urban blight, I was especially curious about how PFL's Philly program might affect pit bulls and the people who loved them.

If successful, Pets for Life could achieve a watershed moment in the field of animal welfare. Like the Coalition, PFL's mission was to provide pet care supplies and veterinary services to residents of zip codes where neither were available. Instead of having to react to problems caused by cruelty, neglect, dog bites, and pet overpopulation, PFL staff felt that they might be able to prevent those problems by making the right information and resources available to all members of the community. This not only would foster goodwill and help close a very large gap in animal welfare but also could potentially save humane organizations a great deal of time and money down the line. By and large, PFL staff members came from the neighborhoods the program aimed to serve, making connections much easier to facilitate.

Before settling on the PFL approach, HSUS had first tried an awareness-raising and educational program called End Dogfighting in Atlanta, Chicago, Philadelphia, and Detroit. It didn't work. The program's title focused on a crime that most pet owners in the community did not want to be associated with, just as most pregnant mothers would not want to be seen taking parenting classes from a program called End Child Abuse. By zeroing in specifically on pit bulls, End Dogfighting made potential clients suspicious. "No one likes to be singled out or feel targeted," Arrington said. "Many people with pit-bull-type dogs are fearful of getting into trouble, facing prejudice, dealing with housing issues and insurance, or even losing their dogs. They were afraid the program was some type of bait and switch. Most people with pit bulls don't want to be segmented from the larger community of pet ownership. They want their dogs to be seen as just that, dogs."

A Pets for Life client and her dogs, Philadelphia

Eighty-eight percent of PFL's clients had never contacted a shelter or animal control. Seventy-seven percent had never taken their pet to a veterinarian. Doing so was very difficult for them, especially if they were elderly or did not own cars. "I grew up right here," Stephen Parker told me, "and when I was younger, we wouldn't take our animals to the PSPCA, because we assumed that they would try and take our pets away. When we saw the red animal control trucks, we'd hide our dogs." When PFL cast a wider net, included all companion animals, and put a positive spin on the program, however, the number of pit bulls serviced almost doubled, including the number of dogs being spayed and neutered. The staff also learned that the issues faced by pet owners were changing. Only a quarter of its clients had pit bulls, while almost half had dogs from toy breeds. Only 9 percent had gotten dogs from breeders, while 55 percent got them from friends or family members.

I asked Parker if he had ever encountered dogfighting in his neighborhood as a kid. "I could count on one hand the times I ever even *heard* about dogfighting," he said, "let alone met anyone who had ever done it. What you see on the news is just not reality."

"Dog fighting . . . we're not saying it doesn't exist," Kenny Lamberti, HSUS's Philadelphia program manager before Ashley Mutch, explained to a local reporter, "but through our work and what we see in these communities across the country, it's really a minimal problem. It exists, but it's really under the radar and in the shadows. Focusing on that, we really weren't touching enough people or enough pets."

Lamberti, who is now PFL's national outreach director, originally began his work with animals as a dog trainer in Dorchester, Massachusetts, one who was specifically interested in working with pit bulls. His calves are scrawled with pit bull tattoos, and for a time he even worked in pit bull advocacy as a trainer for Animal Farm Foundation. But then, he told me, he saw that he could be a more effective advocate if he broadened his efforts to all people and all animals rather than devoting his energy to a select few. "The best thing that could happen for pit bulls," he said, "would be for everyone to stop talking about them."

In Arrington's experience, first with the Coalition and now with HSUS, most of the thousands of people she has come across are trying to do the right thing for their animals; they just might not have access to the right tools. "Our clients *don't need to be told* to love their pets," she said. "They don't need to be told *not* to fight their dogs. What they need is very simple. They need nonjudgmental information about pet care and pet health. They need access to resources. They need a collar and a leash."

Mutch felt strongly that community-focused outreach was infinitely more effective than punitive laws and citations, and she often used her former job as a humane law enforcement officer as an example. Law enforcement was an important component of building safe and humane communities, she said, but it didn't need to be the *only* component. "Back then, everything was confrontational," she told me. "It was all about taking people's dogs away if they looked sick, rather than helping them get treatment for the dog. I wore my bulletproof vest and my gun like it was the *holy frickin' grail.* Now we're just trying to help people. When I knock on someone's door now, they sometimes still think I'm there to bust them, but I'm not. I'm there to offer them all this free stuff! It's amazing how much ground you can cover just by being nice. Not being preachy or judging, just being nice."

PFL Philadelphia's unofficial headquarters was located inside the PSPCA, an organization that, like so many others over the past decade, has overhauled its policies and now embraces a breed-neutral approach to pet adoption. There, in addition to Ashley Mutch and Stephen Parker, I met Janice Poleon, PFL Philly's outreach coordinator, and Devell Brookins, part of the outreach team who also taught the program's dog-training classes.

Years ago, Poleon planned to go into the hospitality industry, but a chance meeting with a friend's pit bull steered her more toward animal welfare. Soon Poleon's interest in animals broadened into community-based social work, of which pets were only one piece. After working with Pets for Life for more than a year, she, too, had experienced not-so-subtle human prejudice where animals were concerned, and that only inspired her to advocate more for the people she served.

"The things people say to me on the phone, not realizing that I am black, are shocking," she told me as she thumbed through her spay/neuter paperwork for the week. "Truly shocking. They make comments about 'thugs' and 'those people' in the 'ghetto.' They think everyone's a dogfighter. But they don't know anything about our clients. Sometimes our clients can't even feed themselves, but they find a way to feed their pets."

Stephen Parker, who was sitting in the chair next to her, nodded knowingly. Parker played drums in a gospel band and had once served in the navy. "Devell and I got stopped by the cops once when we were training our dogs for weight pull," he said. "We get disapproving looks from law enforcement all the time. It's even worse in the winter, when we're wearing our hoodies. People look at me, and they make assumptions. They don't know that I've traveled around the world, or that I make music, or that I work for the Humane Society. They just see a black guy in a hoodie with a pit bull."

Devell Brookins then joined us. In addition to training dogs as part of the outreach effort, Brookins enjoyed parsing the finer points of crop rotation (he had attended Philadelphia's W. B. Saul agricultural high school), vegetable gardening, and cake decorating. For a time, he admitted, he bred pit bulls for extra cash. "I didn't have a job," he explained. "I needed the money." Both he and Parker, who grew up in the same neighborhood, joined PFL after attending the free training classes with their pit bulls, Ace and Red. During the ill-fated End Dogfighting program, Brookins was one of the only people who showed up.

The members of that night's training class, which Brookins and Parker were leading, then began filtering in. One of them was Jonathan Bricker, a twenty-four-year-old PFL client turned volunteer, who walked in with his eight-year-old son, Nyrese, and his two absentminded pit bulls, Danger and Envy. Mutch sat Nyrese down at the table with a pencil and a stack of paper for drawing. "Dude," she said to him, winking, "you have to help me figure out the schedule for Saturday. Can you do that?"

Brookins passed out bags of liver treats to everyone and then took the class through the "touch" and "leave it" commands. At first, Danger and Envy bounced off each other like gobs of Jell-O, but eventually they settled down and got to work. Envy turned out to be a smart, conscientious student, but Danger had a shorter attention span. Whenever I looked

Devell Brookins and Ace

over, Envy was sitting politely on the sidelines, while Danger faced the wrong direction or solicited head scratches. Bricker laughed, just shook his head, and looked at the floor. "You're killin' me, Danger," he said. "You're just killin' me, man."

After a few minutes, everyone took a time-out while Brookins and Parker set up some agility equipment in the middle of the room. Bricker told me that he had recently served two years in prison on drug-related charges. Because of his record, he said, finding a job had been virtually impossible, which worried him a great deal because, in addition to Nyrese, he and his wife had an eight-month-old son. Not content to sit around, Bricker spent most of his time volunteering with PFL. "I hope they'll hire me one day," he told me. "I would love to work with animals."

Bricker learned about PFL a year earlier through its street outreach team. His first dog, a mixed-breed named Sunny, had developed a skin allergy and lost a large amount of fur, and Bricker didn't know how to help her. "Then I met Miss Janice"—he tilted his head in Poleon's direction—"and she helped me get Sunny treated." Soon after that, a friend gave his wife two pit bull puppies—Danger and Envy—for her birthday. He brought them to one of PFL's community events to get them their shots, which is where he learned about the training classes, something he had always been interested in.

"I started coming to the classes and I just kept coming," Bricker said. He stood up and smiled, revealing a small gap between his teeth, and went over to give Poleon a hug. "Someone helped my dogs, so I want to help other people's dogs," he said. "I want to pay it back."

I did not sleep at all the night before the big community event. This new way of thinking about human-animal relationships seemed wonderful, and it felt good to believe that what I had seen Lori Hensley accomplish with the Coalition would work on a national scale. It sounded like the beginning of a peaceful revolution. But what if, like End Dogfighting, this program sounded good on paper but fizzled out in the real world? What if the residents of North Philly didn't want what PFL was offering?

Janice Poleon and Stephen Parker meet with Pets for Life clients in Philadelphia

The problems of pet overpopulation, animal cruelty, and neglect are incredibly complicated, but I agreed with Mutch that dividing the community into "good" and "bad" pet owners and "safe" or "dangerous" dog breeds was not the best way to tackle those issues. No one responds well to having their emotional relationships dictated by someone else telling them what shapes of dogs they can and cannot have. The best chance that those who care about animals have to reduce these problems seemed to come from keeping animals in the homes they already have and supporting positive, healthy relationships between people and their canine companions. As Lori so often said, that involves caring as much about the person as you do about the pet.

"As an animal welfare worker, I was the one who for so long was guilty of neglect," Mutch told me. "I neglected the underserved community because I judged them—judged them for being poor and trying really hard just to survive. I judged them and took away family members, and pushed them even farther away from the animal welfare organization and the people who should actually be the ones to help them. I judged the house they lived in, the block they lived on, and the car they drove—which for some reason I believed told me clues about how good or bad of a pet owner they were."

I couldn't help but wonder what might have changed for Michael Vick and others like him if, instead of the culture of violence and despair that took hold in Newport News when he was young, members of some Pets for Life–type program had been positive, visible members of his community, rather than people who dropped in, demanded a certain set of attitudes by resolving to "educate" the ignorant, then left within a few days. What if, for example, Vick and his peers could have taken their dogs to a local recreation center for a free weekly training class? What if they had someone to ask about behavior problems if a dog growled at them, or soiled the house, or chewed up a pair of shoes? What if they learned early, from positive models in their own neighborhood, how life-changing the bond between humans and dogs can be because they had so many ways to experience and celebrate it?

It might have made no difference for Michael Vick or his friends; fate might have pulled them through the cracks anyway. But it would have made a difference to *somebody.*

When I arrived at Hunting Park early the next morning at eight thirty, a small line had already begun to form, an hour and a half before the event was scheduled to begin. "Some of these people arrived here at seven," Mutch said. I took that as a positive sign and began helping the staff set up their tents and card tables. For the next forty minutes, I concentrated on toting boxes of food, sorting bags of treats, matching leashes and collars, and barely looked up.

"Hey," Mutch called to me, "would you mind passing out these intake forms to people in line?"

"Sure thing," I said, grabbing a stack of about two dozen forms. I turned around to a line of PFL clients that had, unbeknownst to me, grown so long that it wrapped around a football-field-sized section of the park. There were elderly pet owners with their walkers, toddlers with their toys, couples leaning on each other, exhausted single mothers, men who had taken off from work, and teenagers frenetically checking their cell phones. And there were so many pit bulls I couldn't count them all. There they were, mixed right in with all the Chihuahuas, Yorkies, Labradors, Akitas, poodles, and mutts. There were energetic ones, sluggish ones, young ones, old ones. None were separated out or chained up or fretted over. No one screamed in terror or recoiled. When the event was finished and the numbers were tallied, I would find out that more than 980 people received services that day. More than 600 signed up their pets for spay/neuter surgery. Enrollment in free training classes filled up. There was so much interest that several dozen more clients had to be turned away when the supplies ran out.

Despite everything that has happened to these dogs over the past two hundred years, I realized, "people" do not hate or fear pit bulls. To believe that "people hate pit bulls," you have to believe that only those who grab the microphone and scream the loudest into it matter. For the hundreds of Philadelphia residents I saw line up that April morning and wait upwards of five hours in the cold just to get their dog something as simple as a rabies shot, there was no debate about whether or not pit bulls made good family pets. They answered that for themselves long ago.

Client with pit bull puppy and kitten at a Pets for Life Community Pet Day

While activists took sides and launched verbal missiles at each other in the press, while shelters and animal controls fretted over the so-called pit bull problem, while breeders debated what counted as a pit bull and whether or not pit bulls have "breed traits," all the people in this line had been peacefully living their lives with their companions, oblivious to the madness going on around them. It could not have been more clear to me that what I had gleaned hints of all along was in fact true: The dogs moved out of the darkness a hundred years ago. We are the ones who are stuck there.

If I felt any tightness in my chest that day, it arose when I looked at all the children. They were as yet completely unaware of the battles raging over pit bulls in boardrooms, courtrooms, and city halls. They took such luminous pride in their pets, and clung to them as though hanging on for dear life. Would they eventually be told by people who did not know them that there was something ugly and flawed about the dogs they grew up with—the dogs they loved, the dogs that made them feel safe? Would they be scrutinized as possible criminals and treated with suspicion? I did not want that for them. I did not want that for anybody.

So much energy has been devoted to this one subject for more than forty years, and the conversation has changed but little. Asking whether or not the dogs are "different" is a bit like asking how many angels can

Pets for Life clients

dance on the head of a pin. The group has grown so large and diverse that generalizations fail. If at some point in the distant future science does establish a link between the genes that determine a dog's head size and a risk for aberrant behavior, that still won't change the fact that all domestic animals are captives in a world of our own making. As the cognitive scientist Steven Pinker has observed, fairness should not depend on sameness. Dogs will do what we ask of them; that is both the blessing and the burden of their loyalty.

The music critic Greil Marcus once wrote "cultural awakening comes not when one learns the contours of the master-narrative, but when one realizes . . . that what one has always been told is incomplete, backward, false, a lie." As I stood on that patch of grass in the middle of so many people and the animals they loved so dearly, I felt the lie crumble and turn to dust. Only one inviolable truth remained: every dog is different. For all its triumph and tragedy, glory and gore, the history of the American pit bull is as messy and complicated as that of its people.

The term "America's dog" took on a whole new meaning for me then. I hoped that, if anything, what pit bulls have taught us is that justice for animals cannot happen at the expense of social justice for humans. The divide between the two that has existed for almost two centuries needs to be bridged. If we fail to do that, the cycle is destined to repeat itself with another type of dog and another group of people. In some ways, it already is.

According to one study conducted by Dr. Emily Weiss at the ASPCA, there are now almost as many Chihuahuas as there are pit bulls in many American animal shelters. A significant portion of those "Chihuahuas" are likely small mixed breeds that have been misidentified, but nevertheless a steady stream of worried headlines has already started to appear. In 2014, *The New York Times* attributed the Chihuahua boom to "geography, pop culture, and immigrant tastes." "Young women put them in their purses to make an impression," the chairwoman of the Chihuahua Club of America said, "kind of like the big macho guys get pit bulls to look tough." While at PFL's Community Pet Day in Philadelphia, Cory Smith, the director of pet protection and policy for HSUS, told me, "Almost every day someone new asks me if we have so many Chihuahuas because of, quote, all the Mexicans." The city of Hollister, California, has added Chihuahuas to its list of "restricted breeds." While their diminutive size may inoculate them from being caught up in a fear narrative, Chihuahuas may likely be the next dogs to face widespread stereotyping and discrimination in other ways.

Pit bulls are not dangerous or safe. Pit bulls aren't saints or sinners. They are no more or less deserving than other dogs of love and compassion, no more or less deserving of good homes. They didn't cause society's ills, nor can their redemption—real or imagined—solve them. There is nothing that needs to be redeemed, anyway; they were never to blame in the first place. To frame anything in such narrow terms is to look at human-animal relationships through the wrong end of the telescope. More important, there never was a "pit bull problem." What happened to these animals was a byproduct of human fears, and what humans feared most was one another.

After all we have put them through, maybe it is time to let pit bulls show us who they are, to let them have a part in writing their own story. Pit bulls are not dogs with an asterisk. Pit bulls are just . . . dogs.

EPILOGUE

> They hate because they fear, and they fear because they feel that the deepest feelings of their lives are being assaulted and outraged. And they do not know why; they are powerless pawns in a blind play of social forces.
>
> —RICHARD WRIGHT, *Native Son*

On August 9, 2014, an unarmed black teenager named Michael Brown was shot and killed by a white police officer in Ferguson, Missouri. Darren Wilson, the officer who killed Brown, claimed that the teen "made like a grunting noise and had the most intense aggressive face I have ever seen on a person." In his interview with ABC News, Wilson likened the experience to that of a "five-year-old hanging on to Hulk Hogan." He shot Brown six times.

The death of Michael Brown inspired a massive Hands Up, Don't Shoot protest movement around the country, which would then become #BlackLivesMatter. Aside from the questionable legality of Brown's shooting, the Ferguson protesters' biggest complaint was that their local government—the city council and their police department, both of which were overwhelmingly white—did not listen to their needs and did not represent their community, which was predominantly African-American.

On the first night of major protests, I watched the news in horror as law enforcement rolled through the streets in armored vehicles, dressed in full riot gear, teargassing the protesters. It did not take long before they brought out the German shepherd and Belgian Malinois police dogs, which were shown snarling and snapping at the people in ways that much too closely resembled the iconic images from Birmingham in 1963. Three local and state officials told a reporter from *Mother Jones* that one of the officers allowed his dog to urinate on the memorial that had been created for Brown on Canfield Drive, the street where the teenager was slain.

When the U.S. Department of Justice released the report from its formal investigation of the Ferguson Police Department in March 2015, it quoted an e-mail written by a Ferguson police officer in June 2011 that referenced a man needing "welfare" for his dogs because they were "mixed in color, unemployed, lazy, can't speak English and have no frigging clue who their Daddies are." One of the report's key points involved the Ferguson PD's use of police dogs:

> FPD engages in a pattern of deploying canines to bite individuals when the articulated facts do not justify this significant use of force. The department's own records demonstrate that, as with other types of force, canine officers use dogs out of proportion to the threat posed by the people they encounter, leaving serious puncture wounds to nonviolent offenders, some of them children. Furthermore, in every canine bite incident for which racial information is available, the subject was African American. This disparity, in combination with the decision to deploy canines in circumstances with a seemingly low objective threat, suggests that race may play an impermissible role in officers' decisions to deploy canines.

Out of curiosity, I looked up Ferguson's animal control ordinances. Sure enough, there it was, written into the municipal code as Section 6-21: "It shall be unlawful to keep, harbor, own, or in any way possess" a pit bull in Ferguson. How did the code define "pit bull"? By the one-drop rule. Purebred APBTs, AmStaffs, and Staffords were banned, but so were "any mixed breed of dog which contains as an element of its breeding" one of those breeds or "any dog which has the appearance and characteristics of being predominately" one of those breeds.

So when I saw a Ferguson protester walk across the frame with a burly block-headed dog alongside him, in defiance of the city council that had ignored the plight of its people so completely in other ways, I almost cheered. Several weeks later, Kim Wolf sent me a Dogs of New York photograph that she had taken of a woman and her dog at a protest honoring another unarmed African-American man, Eric Garner, who had also been killed by white police officers in New York City. The dog was a pit bull, and it wore a T-shirt that read "Justice for Eric Garner."

Ferguson protesters and dogs, 2014

Diane Jessup now lives on a twenty-acre farm in rural Washington that she bought with the money she inherited from her father, who died in early 2015. She keeps goats, sheep, cows, and prairie dogs, in addition to her pit bulls, several of which have passed away since I met them. "I'm so happy," she told me, "I don't even need a quarter of the pain medication." She says that her most recent litter of pups, which were born in late 2014, will be her last.

Randall Lockwood continues his work as senior vice president of anti-cruelty initiatives for the ASPCA. He vows never to return to "the dog bite business." "I'm not as interested in cruelty as I am in empathy," he told me. "That's the fascinating question: Where does empathy come from?" Karen Delise continues her investigations of dog-bite-related fatalities for the National Canine Research Council, where the trainer Janis Bradley has recently signed on as associate director of communications and publications.

Dave Wilson no longer breeds Razor's Edge dogs. He believes the world of dog breeding has gotten far too political. Instead, he prefers to focus on his family and on his duties as president of the American Bully Kennel Club, which is growing rapidly. An offshoot organization, the International Bully Kennel Club, was founded in 2012 and currently registers dogs from 150 countries.

In April 2014, the Maryland legislature passed Senate Bill 247, which abrogated the state court of appeals decision that pit bulls should be considered "inherently dangerous." Now all Maryland dog owners are subject to the same standard of strict liability in the event of a dog bite, except in the cases of victims who were trespassing or committing a crime when they were bitten. Colleen Lynn of DogsBite.org was disappointed but wrote that the 2012 *Solesky* ruling "still accomplished many feats. For instance, every tenant lease in the state was rewritten to prohibit pit bulls during the period the ruling stood. Will this change because of the

Nola, 2010

new state law? Not likely." Tony Solesky responded, "This problem [of pit bulls] is self-evident. When you have a problem that is self-evident, and you bring awareness to this problem only to be met with great resistance and a concerted effort to suppress this awareness, it shows you that the forces behind this suppression are sinister." He maintains a large presence on the Internet, advocating for breed-ban laws.

Though Animal Farm Foundation has not discontinued its grant program for pit-bull-specific shelter promotions entirely, Jane Berkey hopes to see animal welfare move toward community-based work, rather than separating pit bulls out, for good or ill. In 2013, she made a sizable donation to HSUS's Pets for Life program.

Ashley Mutch was promoted to national mentorship and training coordinator for all of PFL's programs. She is proud to report that PFL has spread to almost thirty American cities and has served more than one hundred thousand pets. The ASPCA has also expanded its community-based programs.

Lori and Doris, 2013

I would eventually confirm, with the help of the Mars Wisdom Panel, that my dog, Nola—the one who originally inspired me to write about pit bull history and culture—is not really a pit bull at all. Or, rather, that she is only about half of one. Her test results came back as 25 percent American Staffordshire terrier, 25 percent Staffordshire bull terrier, 25 percent unknown "terrier mix," and 25 percent . . . Australian shepherd.

Lori Hensley continues her outreach work as director of the Coalition to Unchain Dogs. In the spring of 2013, she arrived at Doris's house for one of their regular visits, only to discover that Doris had suffered a serious stroke and fallen out of her wheelchair. Doris's doctors then determined that she would need to move into a nursing home. Once there, she showed little interest in socializing, preferring instead to sleep most days. Her roommates know her only as "the woman looking for her dog."

I last saw Doris, who is now ninety-two and very fragile, in mid-2014, when Lori took me to visit her. Doris asked Lori the same questions she always asks: How is Robert, Lori's husband? What time does he get home from work? What is Lori making for supper? But the most important of these was: Has she seen Pretty Girl?

Lori sat down on Doris's bed, gave her some ginger ale, and told her that Pretty Girl was doing well; Doris's neighbor, George, was taking very good care of her. In fact, the last time she saw the two of them, Pretty Girl was curled up with George in his recliner, sleeping in front of the small space heater he uses when the weather chills.

When we finally turned to go, Doris raised her thin arms and put them around Lori's neck. "I love you," she said.

ACKNOWLEDGMENTS

Anyone who has ever written a book, especially a book that required as much work as this one did, knows that the final product is the combined effort of many people other than the one whose name ends up on the cover. Early in my career I had the tremendous fortune of meeting my agent, Kristine Dahl at ICM, who took a chance on me years before I knew what I wanted to write about. It is because of her support and guidance that this project found a home, and for that I owe her the world.

My editor, Jonathan Segal, pushed me harder than I have ever been pushed to refine and clarify my ideas. All journalists hope for an editor like him, but few are lucky enough to get one. With kindness and patience, he taught me more about writing than I have learned in the past ten years. I am a better reporter, a better writer, and a better human being for knowing him. Also at Knopf, I am indebted to Julia Ringo, Victoria Pearson, Erinn McGrath, Jordan Rodman, and Oliver Munday.

The old journalistic saw that a reporter is only as good as her sources could not have been more appropriate in my case. I was blessed with some of the best. Though there is not room enough to list them all here, I hope you know who you are.

In 2013, Diane Jessup allowed me to interview her for a full week at her home in Washington, despite having vague concerns that I had no idea what I was talking about. Rarely have I met anyone with a deeper love for dogs or a better sense of humor about them. Diane, thank you for telling me to trust my own ideas about this controversial topic, and I hope I haven't caused you too much grief by following that advice.

Jane Berkey sat for multiple weeks' worth of interviews in Amenia, New York, between the spring and fall of 2012. Without her honesty and generosity, not to mention that of her colleagues at Animal Farm Foundation and the National Canine Research Council (specifically Donald Cleary, who introduced me to the work of Harry Parrish and Stanley Cohen, as well as Janis Bradley, Stacey Coleman, Bernice Clifford, Caitlin Quinn, Elizabeth Arps, Tom White, Kate Fraser, Charlotte Blake, Hanna Fushihara, and Stacy Dykema), my research would have taken ten times longer. Jane, your passion for animals and your sense of justice are inspiring to all who know you, especially me.

Karen Delise opened up her complete archive to me and was quick to supply backup documentation for every fact-checking query I ever threw at her. I've often said that she missed her calling as an investigative reporter, but I am so glad that both dogs and the human victims of dog-bite-related fatalities have her in their corner, searching for the truth.

Randall Lockwood gave a full accounting of the media panic of the 1980s over several long meetings and phone calls from 2012 to 2015. He also supplied me with the research equivalent of a time machine: several large boxes of journal articles and news clips from that period. I sincerely appreciate his time, insights, and expertise.

There isn't an acknowledgments section long enough to contain the contributions of Kim Wolf, who sent me hundreds, if not thousands, of primary-source historical documents and photographs from library archives around the country. She was also extraordinarily generous with her experiences in the field of social work and animal welfare. Without her as a thoughtful, skeptical sounding board, this project would inevitably have taken a much different direction.

Thank you, as well, to Tony Solesky, Alan Beck, Gary Wilkes, and the others who shared with me their perspectives from the opposite side of the pit bull wars. It takes courage to stand up for one's beliefs in print, especially when they are controversial, and my understanding of these issues was deepened by their participation.

Elsewhere, I would like to thank the following people:

At the American Bully Kennel Club (Spotsylvania, Virginia): Dave Wilson and Deanna Smith.

At the Animal Protection Society of Durham (North Carolina): Shafonda Davis, Jane Marshall, Alison Baillie, and Kari Linfors.

At the ASPCA (New York, New York): Steve Zawistowski and Terry Mills.

At BAD RAP (Oakland, California): Donna Reynolds and Tim Racer.

At Best Friends Animal Society: Ledy Vankavage.

At the Center for Shelter Dogs (Boston, Massachusetts): Amy Marder and Gary Patronek.

At the Centers for Disease Control and Prevention (Atlanta, Georgia): Jeffrey Sacks.

At Downtown Dog Rescue (Los Angeles, California): Lori Weise and Amanda Arreola.

At HSUS, past and present (Gaithersburg, Maryland): Amanda Arrington, Kenny Lamberti, Ralph Hawthorne, Tim Freeman, Ashley Mutch, Janice Poleon, Stephen Parker, Devell Brookins, Alana Yunez, Robert Sotelo, Chris Schindler, Janette Reever, and Eric Sakach.

At Maddie's Fund: Richard Avanzino.

At the MSPCA in Boston: Kelly D'Agostino.

At the No-Kill Advocacy Center: Nathan Winograd.

At the North Carolina State University School of Veterinary Medicine: Barbara Sherman and Kristine Alpi.

At Orange County Animal Services (North Carolina): Bob Marotto.

At Pit Bull Rescue Central: Marcy Setter.

At SafeHumane Chicago (Chicago, Illinois): Cynthia Bathurst, Kyla Page, Callie Cozzolino, and Josh Feeney.

At the Wake County SPCA (Raleigh, North Carolina): Molly Stone.

Additionally, I would like to thank (in no particular order): Scottie Westfall, Hal Herzog, Ian Dunbar, Jean Donaldson, Stephanie Filer, Michael Rhodes, Selma Mulvey, Fred Kray, Karen Breslin, Sonya Dias, Bonnie Beaver, David Banks, Julie Hecht, Karen London, Pete Colby, Aimee Sadler, Rhea Moriarty, Brian Hare, Brent Toellner, Josh Liddy, Ken Foster, Molly Gibb, Robin Cubbon, Laurie Lambing, Hendey Hostetter, Mindy Oshrain, Russell Colver, Jonathan Nyberg, Joel Halverson, Heidi Tufto, Nichole Hamilton, Jeff Theman, Adrianne Lefkowitz, Joe Stafford, Jim Crosby, Jen Shryock, Annie Pruitt, James Evans, the members of the Coalition to Unchain Dogs, the members of the Western North Carolina American Pit Bull Terrier Club, Cindy Muse/North Paw Animal Hospital, Mike Sager, and the staff at the following library archives: the American Kennel Club, North Carolina State University School of Veterinary Medicine, Edinburgh University, Abraham Lincoln Library and Museum, and the Mark Twain House and Museum.

This book could never have been written had I not met Patricia O'Toole, a mentor and friend who believed in me when I found it very difficult to believe in myself. At Columbia University, I'd also like to thank Lis Harris and Leslie Sharpe. At Duke, Donna Lisker, the late Lawrence Goodwyn, and the late Reynolds Price.

An ocean of gratitude to every editor who has ever given me work and to the other writers in my life: Richard Bausch, Pat Conroy, Sean Flynn, Louise Jarvis Flynn, Jim Keane, Haven Kimmel, Meg Flaherty, Allison Wright, Matthew Shaer, Skip Horack, Ron Rash, George Singleton, John Lane, Dot Jackson, Andrew Park, Sam Stephenson, Daniel Wallace, Jason Zengerle, Paul Reyes, Barry Yeoman, Maggie Messitt, Alan Felsenthal, Christopher Kemp, and Tom Zoellner.

And to my friends: Megan and Joey Chorley, Stephen and Susan Taylor, Heather Lowe,

Bryce Williams, John Svara, Jeff Polish, Julia Bell, Miller Harris, Matt Kramer, Paul Fleschner, Vinny Eng, and Michael Klinger, as well as Jules Szabo and the wonderful women of TDW. I am sure that Buzz Williams and Nicole Hayler had no idea in 2008 that when they introduced me to their pit bull, Angel, they would start me down this long and winding path, but I am profoundly grateful that they did.

And of course my wonderful family: Christopher and Carol Dickey, Madeline Salvatore, Kacey and Gary Eichelberger, Katie Dickey, Molly Dickey, Deirdre Young, Nick Seminoff, Shirley Danner, Douglas Danner, James and Claire Dickey, Dorian Dickey, Johnny Hodgins, Anna and Gordon Kangas, Katie Dickey Marbut, and Deb and John Shea. To my brother Kevin, I love you and hope that you are well.

There is a special place in paradise reserved for Lori and Robert Hensley (not that they believe in it) for their warmth, humor, openness, and dedication to improving the lives of everyone they meet. I hope they know that they have irrevocably changed mine.

The largest debt of all is owed to my father, the late James Dickey, who taught me to ask questions, and to my mother, Deborah Dickey, who encouraged me to answer them.

Sean, you are the best thing that ever happened to me.

NOTES

PROLOGUE

These scenes were reconstructed from the Smoaks' citizen complaint file with the Cookeville police department (No. 2003–01–01–1714–33), which contains the official incident report, witness statements from Pamela Smoak and Brandon Hayden, and the transcripts of Officer Hall's and Officer McWhorter's internal affairs interviews, as well as the 6th Circuit Court's summary and decision in *Smoak v. Hall.*

The first portion of the officers' dashcam recording, which includes the shooting of Patton, is available on YouTube: https://www.youtube.com/watch?v=LvoT2X1dXcI.

3 *The sky had just begun to darken*: "'Stop' Leaves Family Traumatized," *Cookeville (Tenn.) Herald-Citizen* Jan. 2, 2003; "Officer Who Shot Dog 'Protecting Self,'" *Herald-Citizen,* Jan. 3, 2003; "Officials Apologize to Family About Pet," *Knoxville News-Sentinel,* Jan. 9, 2003; "Cops Kill Dog as Cuffed Family Watches," CBS News, February 12, 2003.

4 *One of the officers*: See also "Cops Kill Dog as Cuffed Family Watches," CBS News, Feb. 12, 2003.

4 *Within the hour*: *Smoak et al. v. Hall et al.* (6th Cir., September 2, 2009).

5 *He would also undergo surgery*: Ibid.

5 *For Hall, a professed animal lover*: "Eric and Lisa Hall Tell of the 'Trauma' They Have Also Suffered," *Cookeville (Tenn.) Herald-Citizen,* Feb. 8, 2003.

5 *this was the version of events*: Internal affairs interviews with Eric Hall and Mead McWhorter, as contained in the Smoaks' citizen complaint file, Cookeville Police Department.

6 *"Mace doesn't work on animals"*: "Eric and Lisa Hall Tell of the 'Trauma' They Have Also Suffered."

6 *The judge who presided*: *Smoak v. Hall* (2009).

CHAPTER 1: PARIAH DOGS

8 *In his classic essay*: John Berger, *About Looking* (New York: Vintage International, 1991), 3.

8 *In 2003, Nike ignited a media firestorm*: Rob Walker, "Hating to Love Nike," Slate.com, March 10, 2003.

8 *Likewise, in 2012*: Kent Russell, "Doggy Style," *GQ,* April 2012.

9 *When the Republican vice presidential candidate*: "Transcript: Palin's Speech at the Republican National Convention," *New York Times,* Sept. 3, 2008.

9 *the latest genetic research indicates*: Victoria L. Voith et al., "Comparison of Adoption Agency Breed Identification and DNA Breed Identification of Dogs," *Journal of Applied Animal Welfare Science* 12, no. 3 (2009): 253–62.

10 *Between 1995 and 2005*: Hank Greenwood (ADBA president) to Diane Jessup, personal communication, May 14, 2014. Courtesy of Diane Jessup.

10 *The American Bully Kennel Club*: Deanna Smith, interview with author, Nov. 11, 2014; Dave Wilson, personal communication, July 2015.

10 *In 2011, when the first "mutt census"*: "First Mutt Census Reveals Strong Dog DNA Trends," Today.com, April 4, 2011.

10 *Banfield pet hospitals*: "Banfield Pet Hospital: State of Pet Health 2013 Report," http://www.stateofpethealth.com.

10 *A separate analysis*: "Top Dogs Across America: 10 Most Popular Breeds by State," http://www.vetstreet.com. Confirmed by Allan Halprin (Vetstreet representative), personal communication, March 29, 2012.

11 *Cruelty investigators*: Terry Mills (ASPCA), interview with author, April 2012, Dr. Randall Lockwood (ASPCA), interview with author, Jul. 2012; Chris Schindler and Janette Reever (HSUS), interview with author, Jul. 2012; Eric Sakach (HSUS), interview with author, Sept. 2012; John Goodwin (HSUS), personal communication, Dec. 23, 2014.

11 *For the better part of two hundred years*: For these references, please see notes for chapter 3.

16 *This was greatly exacerbated*: Lori Hensley, interview with author, Durham, N.C., Jan. 2012; Amanda Arrington, interview with author, Durham, N.C., Jan. 2012.

16 *Since 2007, the group has unchained*: Lori Hensley, personal communication, March 15, 2015.

16 *In her youth*: Doris R., interviews with author, Apr. 2013–Dec. 2013.

17 *At a city council meeting*: Robert and Lori Hensley, interview with author, Feb. 2012.

18 *As Dr. Laurel Braitman observes*: Braitman, *Animal Madness*, 3.

18 *"Nature versus nurture only exists"*: Dr. Brian Hare, inverview with author, March 2012.

19 *To date, "pit bulls" have been either banned*: Kim Wolf, BSL spreadsheet by city and state; "U.S. Cities Increasingly Ditching Pit Bull Bans," *USA Today*, Nov. 18, 2014.

19 *at least one small organization*: "About Us," DogsBite.org.

19 *In nearly every municipality*: See notes for chapter 14.

19 *In 2013, even the White House*: "Breed-Specific Legislation Is a Bad Idea," We the People, Aug. 12, 2013, https://petitions.whitehouse.gov.

20 *People for the Ethical Treatment of Animals*: "PETA's Position on Pit Bulls," Peta.org, July 21, 2009.

20 *At some facilities*: Sonya Dias, interview with author, Jul. 2012; Jared Jacang Maher, "Leaked: Photos of Pit Bulls Killed Due to Denver Ban," *Westword*, Oct. 7, 2009.

20 *It calls to mind what the historian*: Coleman, *Vicious*, 2.

20 *On Facebook, the interest page*: See https://www.facebook.com/pages/Labradors/111733895511768; https://www.facebook.com/pages/Pit-Bulls/639531392723629; https://www.facebook.com/pages/Dogs/114197241930754.

20 *In February 2014*: "Mauling of Phoenix Boy Keys Debate Over Fate of Pit Bull," *Arizona Republic*, March 11, 2014.

21 *the dog's supporters vowed*: "Watch Mickey Beat Cancer," http://www.facebook.com.

21 *complete with a $1,000 Webcam*: "Mickey the Pit Bull Soon to Be a 24-Hour Jail Webcam Star," Associated Press, Dec. 3, 2014.

21 *raised less than $2,500*: "Which Matters More: A Boy's Life or a Pit Bull's?" *Arizona Republic*, March 13, 2014. In 2015, the boy's supporters renewed their efforts and managed to raise almost $26,000 on the fund-raising Web site IndieGoGo.com. Only 242 people donated.

22 *In the thirty-five thousand years*: Nikolai Ovodov et al., "A 33,000-Year-Old Incipient Dog from the Altai Mountains of Siberia: Evidence of the Earliest Domestication Disrupted by the Last Glacial Maximum," *PLoS ONE* 6, no. 7 (2011): e22821.

22 *"The magic that gleams"*: Loren C. Eiseley, *The Unexpected Universe* (New York: Harcourt Brace Jovanovich, 1985), 23.

23 *the French anthropologist Claude Lévi-Strauss*: Claude Lévi-Strauss, "The Structural Study of Myth," *Journal of American Folklore* 68, no. 270 (1955), 428–44.

CHAPTER 2: THE KEEP

Tearing myself away from this rich and fascinating history proved to be an incredibly difficult task. Fortunately, the staff at the American Kennel Club's library in Manhattan and the veterinary sciences library at North Carolina State University helped me find my way through some of the early material. Entire books can, and have, been written on the daily struggle for survival in New York's poorest neighborhoods during the nineteenth century. The most helpful to me were Herbert Asbury's colorful (if not always entirely accurate) portrait, *The Gangs of New York;* Luc Sante's *Low Life;* and Tyler Anbinder's meticulous account, *Five Points.* Anyone interested in this time period should also read *The Jungle* by Upton Sinclair, the novel that forced America to confront the suffering of the urban poor, and at least one volume on the culture of bare-knuckle boxing and its significance in the lives of early Irish-American immigrants. I recommend Elliot Gorn's *The Manly Art: Bare-Knuckle Prize Fighting in America* for a good overview, and *John L. Sullivan and His America* by Michael T. Isenberg for a more in-depth look.

Harriet Ritvo's influence on my work was profound. I turned to *The Animal Estate* so many times that my copy is now falling apart. The same goes for James Turner's *Reckoning with the Beast.* Carson Ritchie's *The British Dog* is the most comprehensive account of canine history and culture in the U.K. that I have yet come across.

25 *third-oldest building in Manhattan*: "Historic House May Get New Life at Last," *New York Times,* Nov. 28, 1989; "A Scruffy Old Tavern Is Now Luxury Apartments," *New York Times,* Oct. 27, 1998.

25 *mahogany trader named Joseph Rose*: "Historic American Buildings Survey," National Park Service (Department of the Interior), 1983.

25 *From at least 1863 to 1870*: Asbury, *Gangs of New York;* Sante, *Low Life.*

25 *"the Cruelest Man in New York"*: "New York and Vicinity," *Independent,* Oct. 1, 1868, 4.

25 *sometime around 1845*: Chris Pomorski, "The Brutal Honesty of a Bloodsport Baron," *Narratively,* Aug. 8, 2013.

25 *When he died in 1870*: "Funeral of 'Kit Burns,'" *New York Times,* December 24, 1870.

26 *The Five Points at mid-century*: See Asbury, *Gangs of New York;* Sante, *Low Life;* Anbinder, *Five Points.*

26 *"Lower than the Five Points"*: As quoted in Pomorski, "Brutal Honesty of a Bloodsport Baron."

26 *In 1857, one newspaper correspondent*: "Discouraging," *New York Times,* Feb. 6, 1857.

27 *"The facial and cranial appearance"*: As quoted in Pomorski, "Brutal Honesty of a Bloodsport Baron."

27 *he gravitated toward a saloon called the Sawdust House*: Sullivan, *Rats.* Also M. Kaufman and H. J. Kaufman, "Henry Bergh, Kit Burns, and the Sportsmen of New York," *New York Folklore Quarterly* 28, no. 1 (1972): 15.

27 *clandestine bare-knuckle matches*: Gorn, *Manly Art,* 75–76; also Redmond, *Irish and the Making of American Sport;* and Michael T. Isenberg, *John L. Sullivan and His America* (Chicago: University of Illinois Press, 1988).

28 *"Dogs will fight," he liked to say*: As quoted in Kaufman and Kaufman, "Henry Bergh, Kit Burns, and the Sportsmen of New York"; and Pomorski, "Brutal Honesty of a Bloodsport Baron."

28 *the industrial moonscape of the English Midlands*: Homan, *Staffordshire Bull Terrier in History and Sport*, 49.

28 *Both Miltiades*: Kalof, *A Cultural History of Animals in Antiquity*, 104.

28 *Roman rulers*: Auguet, *Cruelty and Civilization*; George Dennison, *Animals for Show and Pleasure in Ancient Rome* (1937; Philadelphia: University of Pennsylvania Press, 2007).

29 *the English medieval spectacle of bullbaiting*: Ritchie, *British Dog*, 56–180; Turner, *Reckoning with the Beast*, 20–29; Leslie Hotson, "Bear Gardens and Bear-Baiting During the Commonwealth," *Publications of the Modern Language Association of America* 40, no. 2 (1925): 276–88; Emma Griffin, "Popular Culture in Industrializing England," *Historical Journal* 45, no. 3 (2002): 619–35; Giles E. Dawson, "London's Bull-Baiting and Bear-Baiting Arena in 1562," *Shakespeare Quarterly* 15, no. 1 (1964): 97–101.

29 *Queen Elizabeth, in particular*: Marc Bekoff, ed., *Encyclopedia of Animal Rights and Animal Welfare, Second Edition* (ABC-CLIO, 2009), 93.

29 *Shakespeare's Globe Theatre*: Jason Scott-Warren, "When Theaters Were Bear Gardens; or, What's at Stake in the Comedy of Humors," *Shakespeare Quarterly* 54, no. 1 (2003): 63–82.

30 *a representative of the Crown*: Ritchie, *British Dog*, 14, 110.

30 *It would not be until 1789*: Jeremy Bentham, *An Introduction to the Principles of Morals and Legislation* (London: Clarendon Press, 1789); Ritvo, *Animal Estate;* Turner, *Reckoning with the Beast.*

31 *"spread amongst the lower orders"*: Ritvo, *Animal Estate*, 135.

31 *a symbolic assault on what few freedoms they had left*: Turner, *Reckoning with the Beast*, 23–30.

32 *"Of all dogs, it stands confess'd"*: George Richard Jesse, *Researches into the History of the British Dog: From Ancient Laws, Charters, and Historical Records, with Original Anecdotes and Illustrations of the Nature and Attributes of the Dog, from the Poets and Prose Writers of Ancient, Mediaeval, and Modern Times*, vol. 2 (London: Robert Hardwicke, 1866).

32 *slow, stupid, and "savage"*: William Youatt, *The Dog* (Philadelphia: Blanchard and Lea, 1855).

32 *eerily impervious to pain*: "Interesting Account of Dogs," *Edinburgh Magazine, or Literary Miscellany*, July 1802, 20.

32 *"In many cases"*: H. G. Hitchcock, "Dogs," as cited in *Report of the Rugby School Natural History Society* (1874): 59.

32 *Dickens never explicitly stated*: Charles Dickens, *Oliver Twist* (London: Richard Bentley, 1838); Beryl Gray, *The Dog in the Dickensian Imagination* (Farnham: Ashgate, 2014), 112–13.

32 *"There is no dog"*: John Meyrick, "The Bull-Dog," in *House Dogs and Sporting Dogs* (London: John van Voorst, 1861), 61.

33 *a more agile dog for badger hunting*: William Taplin, as quoted in William D. Drury, *British Dogs: Their Points, Selection, and Show Preparation* (London: L. Upcott Gill, 1903), 408.

33 *"the handsomest and best of all terriers"*: James Page Stinson, "The Bull-Terrier," *Century Magazine*, May 1885, 34.

33 *"Such is the fancy for this dog"*: Thomas Brown, *Biographical Sketches and Authentic Anecdotes of Dogs* (Edinburgh: Oliver & Boyd, 1829), 405.

33 *"Of all the dogs"*: "Life Among Bulldogs," *Spirit of the Times*, Nov. 26, 1868, 236.

33 *"I hate . . . seeing a bulldog ill-treated"*: "Letter 48 (December 1874)," in *The Works of John Ruskin* (London: George Allen, 1907), 28:218.

33 *"He was of great strength"*: James Hay, *Sir Walter Scott* (London: James Clarke, 1899), 163.

34 *"gentle as a lamb among the children"*: Percy Stevenson, "Sir Walter Scott and His Dogs," *Living Age* 304 (1920): 45.

34 *Never did Scott's daughter*: John Gibson Lockhart, *Memoirs of the Life of Sir Walter Scott* (Boston: Houghton Mifflin, 1901), 2:128–29.

34 *"[Camp] has made a sort of blank"*: Sir Walter Scott, *The Letters of Sir Walter Scott: 1831–1832 and Appendices of Early Letters* (New York: AMS Press, 1971), 311.

35 *During the 1890s, opticians*: "Glass Eyes for Animals," Delphos (Ohio) *Daily Herald*, Nov. 13, 1897, 7.

35 *George Leese, better known as Snatchem*: Isenberg, *John L. Sullivan and His America*, 85.

35 *The dogs lived downstairs*: "Kit Burn's [*sic*] Dog-Pit," *Daily Evening Telegraph*, Sept. 23, 1868, 7.

36 *"Two huge bull-dogs"*: James Dabney McCabe (as Edward Winslow Martin), *The Secrets of the Great City* (Philadelphia: National Publishing Company, 1868), 389.

36 *"nothing was too good"*: John Henry Walsh, "The Bull Terrier," in *The Dogs of the British Islands* (London: H. Cox, 1872), 129.

37 *In the spring of 1861*: Nigel Cawthorne, *Canine Commandos: The Heroism, Devotion, and Sacrifice of Dogs in War* (Berkeley, Calif.: Ulysses Press, 2012), 23–24; A. M. Stewart, *Camp, March, and Battle-Field; or, Three Years and a Half with the Army of the Potomac* (Philadelphia: Jas. B. Rodgers, 1865).

38 *A bulldog named Old Harvey*: Timothy R. Brookes, "Harvey: Civil War Dog" (private collection). Photographs of Harvey with the 104th Ohio Volunteers and other related documents can be found in the archives of the Massillon Museum in Massillon, Ohio.

39 *Sallie, a "brindle bulldog"*: Marilyn Seguin, *Dogs of War: And Stories of Other Beasts of Battle in the Civil War* (Boston: Brandon Publishing, 1998), 24.

39 *The state of New York formally outlawed*: "New York Revised Statutes 1867: Chapter 375: Sections 1–10," Michigan State University, Animal Legal and Historical Center, https://www.animallaw.info/statute/new-york-revised-statutes-1867-chapter-375-sections-1–10.

39 *In 1868, he publicly announced*: "The Revival in Water Street," *New York Herald*, Sept. 20, 1868, 8; "More About the Wickedest Man in New York," *Baltimore Sun*, Sept. 21, 1868, 1; "Religion in a Dog Pit," *Chicago Tribune*, Sept. 25, 1868, 2; "The Church of the Holy Dog Pit," *New York Herald*, Feb. 9, 1870. See also *Galaxy: A Magazine of Entertaining Reading*, Nov. 1868, 723.

40 *"unlimited sympathy and substantial aid"*: Turner, *Reckoning with the Beast*, 47–48. More on Bergh and the founding of the ASPCA can be found in Ritvo, *Animal Estate*; Beers, *For the Prevention of Cruelty;* and Lane and Stephen Zawistowski, *Heritage of Care*.

40 *"The history of the world"*: "The Lash for Criminals," *New York Times*, July 29, 1877.

41 *"the only argument that appeals directly"*: "Treatment of Criminals," *New York Times*, Dec. 26, 1880.

41 *"glared down upon the Irish"*: Turner, *Reckoning with the Beast*, 55.

41 *tinges of respect, even admiration*: "Revival in Water Street," 8.

41 *The odds-on favorite*: "A Night at Kit Burns'," *New York Herald*, Nov. 22, 1870.

42 *drowned them in the East River*: "The Dog Days and the New Dog Law," *New York Times*, June 2, 1867; "Night at Kit Burns'."

CHAPTER 3: IN THE BLOOD

44 *99.9 percent of its DNA*: David Grimm, *Citizen Canine* (New York: PublicAffairs, 2014), 23.

44 *formally bred by humans*: Ritchie, *British Dog*, 96–103.

44 *only sixteen broad types*: Ibid.

44 *Robert Bakewell*: Martin Wallen, "Foxhounds, Curs, and the Dawn of Breeding: The Discourse of Modern Human-Canine Relations," *Cultural Critique* 79, no. 1 (2011): 125–51.

44 *Bakewell's breeding methods also played a significant role*: Charles Darwin and James T. Costa, *The Annotated "Origin": Facsimile of the First Edition of "On the Origin of Species"* (Cambridge, Mass.: Harvard University Press, 2009), 35. More on the ways in which dog breeding influenced Darwin's work can be found in Townshend, *Darwin's Dogs.*

44 *Hugo Meynell*: Tim Blanning, *The Pursuit of Glory: The Five Revolutions That Made Modern Europe, 1648–1815* (New York: Penguin, 2007), 424.

45 *Pape never intended to invert*: Ritchie, *British Dog;* Ritvo, *Animal Estate;* Turner, *Reckoning with the Beast.*

46 *perceived purity of its bloodline*: Ritvo, *Animal Estate.*

46 *Dogs with larger eyes are also more likely*: Bridget M. Waller et al., "Paedomorphic Facial Expressions Give Dogs a Selective Advantage," *PLoS ONE* 8, no. 12 (2013): e82686; Julie Hecht and Alexandra Horowitz, "Seeing Dogs: Human Preferences for Dog Physical Attributes," *Anthrozoös* 28, no. 1 (2015): 153–63; John Archer and Soraya Monton, "Preferences for Infant Facial Features in Pet Dogs and Cats," *Ethology* 117, no. 3 (2011): 217–26; Marta Borgi et al., "Baby Schema in Human and Animal Faces Induces Cuteness Perception and Gaze Allocation in Children," *Frontiers in Psychology* 5 (2014): 411; Angelo Gazzano et al., "Dogs' Features Strongly Affect People's Feelings and Behavior Toward Them," *Journal of Veterinary Behavior: Clinical Applications and Research* 8, no. 4 (2013): 213–20.

46 *Bill George*: Joan McDonald Brearley, *The Book of the Bulldog* (Neptune City, N.J.: TFH, 1992).

47 *"They are ferocious in aspect"*: Charles Dickens, *All the Year Round* 15 (1876): 375.

47 *James Hinks*: *Outing* 60 (1912): 751.

47 *Robert Hooper*: Ethel Braunstein, *The Complete Boston Terrier* (New York: Howell Book House, 1968), 8–18.

47 *Ota Benga*: David Samuels, "Wild Things," *Harper's,* June 2012.

47 *"a race that scientists do not rate high"*: "Bushman Shares a Cage with Bronx Park Apes," *New York Times,* Sept. 9, 1906.

47 *Sir Francis Galton*: Alexandra Stern, *Eugenic Nation: Faults and Frontiers of Better Breeding in Modern America* (Berkeley: University of California Press, 2005), 11.

48 *Madison Grant*: Jonathan Peter Spiro, "Patrician Racist: The Evolution of Madison Grant" (Ph.D. diss., University of California at Berkeley, 2000).

48 *Leon Fradley Whitney*: Welling Savo, "The Master Race," *Boston,* Dec. 2002.

48 *"Fitter Families for Future Firesides"*: Laura L. Lovett, *Conceiving the Future: Pronatalism, Reproduction, and the Family in the United States, 1890–1938* (Chapel Hill: University of North Carolina Press, 2009).

48 *Max von Stephanitz*: Max von Stephanitz, Joseph Schwabacher, and Carrington Charke, *The German Shepherd Dog in Word and Picture* (Jena, Germany: Anton Kämpfe, 1923).

49 *The very word "race"*: Charles de Miramon, "Noble Dogs, Noble Blood: The Invention of the Concept of Race in the Late Middle Ages," in *The Origins of Racism in the West,* ed. Miriam Eliav-Feldon, Benjamin H. Isaac, and Joseph Ziegler (Cambridge, U.K.: Cambridge University Press, 2009), 200–16.

49 *"I somehow never feel"*: W. Gordon Stables, *The Practical Kennel Guide: With Plain Instructions on How to Rear and Breed Dogs for Pleasure, Show, and Profit* (London: Cassell Petter & Galpin, 1877), 19.

49 *"The commercial value of canine monstrosities"*: Thorstein Veblen, *The Theory of the Leisure Class: An Economic Study of Institutions* (New York: Macmillan, 1912), 142.

50 *"there are so many sizes"*: Ritvo, *Animal Estate,* 106.

50 *Between 1945 and 1972*: Harold Herzog, "Forty-Two Thousand and One Dalmatians: Fads, Social Contagion, and Dog Breed Popularity," *Society and Animals* 14, no. 4 (2006): 383–97.

50 *Marketing research has confirmed*: Paul Clark and Jay Page, "Examining Role Model and Information Source Influence on Breed Loyalty: Implications in Four Important Product Categories," *Journal of Management and Marketing Research* 2, no. 1 (2008).

50 *While media culture plays a part*: Stefano Ghirlanda, Alberto Acerbi, and Harold Herzog, "Dog Movie Stars and Dog Breed Popularity: A Case Study in Media Influence on Choice," *PLoS ONE* (2014): e106565. Also Harold Herzog, Alexander Bentley, and Matthew W. Hahn, "Random Drift and Large Shifts in Popularity of Dog Breeds," *Proceedings of the Royal Society of London, Series B: Biological Sciences* 271, supplement 5 (2004): S353–S356.

51 *the demand for giant Tibetan mastiffs*: "Once-Prized Tibetan Mastiffs Are Discarded as Fad Ends in China," *New York Times,* April 17, 2015.

51 *Dr. Kristopher Irizarry*: Phone interview with author, May 2012.

52 *"based on looks alone"*: As quoted in Karyn Grey, "Breed-Specific Legislation Revisited: Canine Racism or the Answer to Florida's Dog Control Problems?" *Nova Law Review* 27 (2002): 415.

52 *more than half of America's seventy-seven million dogs*: Voith et al., "Comparison of Adoption Agency Breed Identification and DNA Breed Identification of Dogs," *Journal of Applied Animal Welfare Science* 12, no. 3 (2009): 253–62; "Results of Nation's First Ever Mutt Census Reveal Paw Print of Nation's Mixed Breed Population," Mars Veterinary, April 5, 2011.

52 *one million new dogs a year*: In 2010, the country's largest registry, the AKC, registered 563,611 dogs (per Hal Herzog, AKC data, personal communication). The UKC is roughly half the size of the AKC, and the other clubs are much smaller still. Most veterinary professionals place the ballpark estimate of new registrations of purebred dogs each year at about 1 million. Because a number of clubs (like the UKC) do not release their registration data, an exact number is impossible to determine.

52 *co-authored several papers*: Voith et al.,"Comparison of Adoption Agency Breed Identification and DNA Breed Identification of Dogs"; Voith et al., "Comparison of Visual and DNA Breed Identification of Dogs and Inter-Observer Reliability," *American Journal of Sociological Research* 3, no. 2 (2013): 17–29.

53 *nineteen thousand mapped genes*: Heidi Parker and Elaine Ostrander, "Canine Genomics and Genetics: Running with the Pack," *PLoS Genetics* 1, no. 5 (2005) e58, doi:10.1371/journal.pgen.0010058.

54 *researchers at Stanford University*: Adam Boyko et al., "A Simple Genetic Architecture Underlies Morphological Variation in Dogs," *PLoS Biology* 8, no. 8 (2010): 1948.

54 *Almost sixty thousand phenotypes*: The Mammalian Phenotype Browser, which contains all the mouse phenotypes mapped to date, can be searched here: http://www.informatics.jax.org/searches/MP_form.shtml.

55 *Angela Hughes*: Edie Lau, "Dog Breed Genetic Tests Put to the Test," Veterinary Information News Service, July 10, 2012, http://news.vin.com.

55 *"Imagine if the situation"*: Scottie Westfall, personal communication, July 10, 2015.

57 *The Mars Wisdom Panel is now able to match*: "FAQ," Mars Wisdom Panel/Mars Veterinary, http://www.wisdompanel.com/why_test_your_dog/faqs/.

58 *87.5 percent of the time*: Voith et al., "Comparison of Adoption Agency Breed Identification and DNA Breed Identification of Dogs."

58 *In a follow-up study*: Voith et al., "Comparison of Visual and DNA Breed Identification of Dogs and Inter-observer Reliability." See also Christy L. Hoffman et al., "Is That Dog a Pit Bull? A Cross-Country Comparison of Perceptions of Shelter Workers

Regarding Breed Identification," *Journal of Applied Animal Welfare Science* 17, no. 4 (2014): 322–39.

58 *Subsequent research confirmed*: K. R. Olson et al., "Inconsistent Identification of Pit Bull-Type Dogs by Shelter Staff," *Veterinary Journal* (2015).

58 *One woman I know*: Kim Wolf, personal communication, July 20, 2015.

58 *some veterinarians and behaviorists*: Robert John Simpson, Kathyrn Jo Simpson, and Ledy VanKavage, "Rethinking Dog Breed Identification in Veterinary Practice," *Journal of the American Veterinary Medical Association* 241, no. 9 (2012): 1163–66; Amy Marder and Victoria Voith, "The American Shelter Dog: Identification of Dogs by Personality," *Journal of Veterinary Behavior: Clinical Applications and Research* 5, no. 1 (2010): 26.

59 *Shelters that have abandoned*: "Orange Shelter No Longer Will Identify Dogs by Breed to Avoid Pit-Bull Stigma," *Orlando Sentinel*, March 8, 2014; Kristen Auerbach, Interim Director of Fairfax County Animal Shelter, "Removing Breed Labels: Easier Than You Think," https://animalfarmfoundation.wordpress.com.

59 *"If we don't know for sure"*: Victoria Voith, interview with author, Sept. 2012.

CHAPTER 4: AMERICA'S DOG

60 *the nattily dressed John P. Colby*: Pete Colby, interview with author, April 2012; Louis B. Colby and Diane Jessup, *Colby's Book of the American Pit Bull Terrier.*

61 *he bred an estimated five thousand*: Ibid.

61 *a roster of clients that included*: Pete Colby, interview with author, Newburyport, Mass., April 2012.

61 *In 1898, Bennett established his own registry*: Ted Kerasote, *Pukka's Promise: The Quest for Longer-Lived Dogs* (Boston: Houghton Mifflin Harcourt, 2013), 63.

61 *Historians of the breed*: Diane Jessup, *The Working Pit Bull;* Richard F. Stratton, *The Book of the American Pit Bull Terrier* (Neptune City, N.J.: TFH, 1981).

61 *considered him a "puppy peddler"*: Diane Jessup, interview with author, Nov. 2013; "Choosing Your Bloodline," *Sporting Dog News*, June 29, 2015.

61 *Burt Leadbetter*: "Boston Terrier Kills Child," *Harrisburg (Pa.) Daily Independent*, Feb. 3, 1909; "Shaken to Death by a Bulldog," *Middlebury Register*, Feb. 5, 1909.

62 *since 1900, there have been fewer than ten*: Karen Delise, interview with author, July 2012.

62 *"very secretive about the affair"*: "Bull Terrier Kills Child," *Boston Daily Globe*, Feb. 3, 1909.

62 *In 1911, a separate class*: "Pet Pit Bulls Will Form New Dog Show Class," *New York Evening World*, Feb. 11, 1911, 4.

63 *"There is no limit to the merits"*: *Country Life Illustrated* 5 (1899): 815.

63 *"the Bulldog, like all other noble animals"*: George O. Shields, *The American Book of the Dog: The Origin, Development, Special Characteristics, Utility, Breeding, Training, Points of Judging, Diseases, and Kennel Management of All Breeds of Dogs* (Chicago: Rand, McNally, 1891), 603.

63 *In the summer of 1917*: "Pet Bulldog Keeps Girl Aged 4 from Death When Lost High in Mountains," *Fitchburg* (Mass.) *Sentinel*, June 26, 1917, 12.

64 *"Her father's gun"*: Laura Ingalls Wilder, *Little House in the Big Woods* (New York: HarperTrophy, 2004), 3.

64 *until the 1970s*: "A Breed That Came Up the Hard Way," *New York Times*, Sept. 19, 1971.

65 *during the 1960s, John Steinbeck*: Steinbeck to Howard Gossage, May 6, 1965, auctioned by Sotheby's in 2011.

65 *the humorist*: James Thurber, "Snapshot of a Dog," *New Yorker*, March 9, 1935, 15.

66 *"When Thurber was writing"*: Adam Gopnik, "A Note on Thurber's Dogs," *New Yorker,* Nov. 1, 2012.

66 *"I wish to preach"*: Kathleen Dalton, *Theodore Roosevelt: A Strenuous Life* (New York: Knopf Doubleday, 2007), 186.

67 *fifteen of the University of Michigan's twenty mascot dogs*: "Frat Dogs of the U. of M.," *Detroit Free Press,* March 13, 1910.

67 *Votes*: "Driver of Suffrage Automobile," *Marble Hill* (Mo.) *Press,* May 13, 1915.

67 *"little sense of humor"*: "Pete, the Bulldog, Guards Roosevelt," *New York Times,* May 1, 1907, 1.

67 *Roosevelt then promptly banished the dog*: "Pete," *Cincinnati Inquirer,* May 2, 1907, 6.

67 *They called the dog's last dustup*: "Pete Meets Waterloo," *Harrisburg Telegraph,* May 21, 1907, 2.

67 *"nature faker"*: "Liars, Says President," *Washington Post,* Aug. 21, 1907, 9.

68 *"I would like to match"*: "Offers to Bet President," *Washington Post,* June 10, 1907, 2.

68 *"damnable feminization"*: Henry James, as quoted in Michael Bellesiles, *1877: America's Year of Living Violently* (New York: New Press, 2010), 275–76.

68 *nicknamed the Crackers*: "Georgia Badly Beaten," *Red and Black,* 1900. Trilby's story can be found in "History of Our Mascot," University of Georgia Athletics, http://www.georgiadogs.com/genrel/102208aab.html.

68 *Pig Bellmont*: Jim Nicar, "A Dog Named Pig," UT History Central, http://www.texas-exes.org.

70 *by far the most common animal mascot*: Arluke and Bogdan, *Beauty and the Beast.*

70 *"People who haven't been at the front"*: Bausum, *Sergeant Stubby,* 43.

70 *As America readied its troops for war*: Ibid.

72 *"Sergeant" Helen Kaiser*: "D.C. War Hero, First American Dog to Enter German Territory," *Washington Times,* March 10, 1919, 1.

73 *Willie had a reputation*: Alan Axelrod, *Patton* (New York: Palgrave Macmillan, 2009), 124.

74 *the American military police used the dogs*: Associated Press, August 20, 1944.

CHAPTER 5: DOGS OF CHARACTER

Material for this chapter came primarily from my interviews with Diane Jessup conducted in Olympia, Washington (Nov. 2013), and Seattle, Washington (March 2014).

77 *tight inbreeding puts dogs at higher risk*: Jane M. Reid, Peter Arcese, and Lukas F. Keller, "Inbreeding Depresses Immune Response in Song Sparrows (*Melospiza melodia*): Direct and Inter-Generational Effects," *Proceedings of the Royal Society of London B: Biological Sciences* 270, no. 1529 (2003): 2151–57; D. E. Wildt et al., "Influence of Inbreeding on Reproductive Performance, Ejaculate Quality, and Testicular Volume in the Dog," *Theriogenology* 17, no. 4 (1982): 445–52; Federico Calboli et al., "Population Structure and Inbreeding from Pedigree Analysis of Purebred Dogs," *Genetics* 179, no. 1 (2008): 593–601; Mija Jansson and Linda Laikre, "Recent Breeding History of Dog Breeds in Sweden: Modest Rates of Inbreeding, Extensive Loss of Genetic Diversity, and Lack of Correlation Between Inbreeding and Health," *Journal of Animal Breeding and Genetics* 131, no. 2 (2014): 153–62.

81 *William Koehler*: William Koehler, *The Koehler Method of Dog Training* (New York: Howell Book House, 1962); "William Koehler, Dog Trainer, 82," *New York Times,* Nov. 19, 1993; Martin Kihn, "Good Dog, Bad Dog," Slate.com, Dec. 22, 2010.

83 *social referencing*: Brian Hare, Josep Call, and Michael Tomasello, "Communication of Food Location Between Human and Dog (*Canis familiaris*)," *Evolution of Commu-*

nication 2, no. 1 (1998): 137–59; Isabella Merola, Emanuela Prato-Previde, and Sarah Marshall Pescini, "Dogs' Social Referencing Towards Owners and Strangers," *PLoS ONE* (2012): e47653; Wolfgang M. Schleidt and Michael D. Shalter, "Co-Evolution of Humans and Canids," *Evolution and Cognition* 9 (2003): 57–72. For more on the co-evolution between humans and dogs, see Derr, *How the Dog Became the Dog;* and Shipman, *Animal Connection.*

84 *Lucas promised Durfee*: Judith Watt and Peter Dyer, *Men and Dogs: A Personal History from Bogart to Bowie* (New York: Atria Books, 2006), 56.

84 *Mack Sennett*: Stuart Oderman, *Roscoe "Fatty" Arbuckle: A Biography of the Silent Film Comedian, 1887–1933* (Jefferson, N.C.: McFarland, 2005), 70.

84 *Arbuckle signed a contract*: "Fatty Arbuckle's Dog Signs for Fifty Bones," *Los Angeles Times*, June 2, 1918.

84 *more films than any other actor*: "In 224 Films," *Baltimore Sun*, Oct. 3, 1926, MF4.

84 *a barn outside Bordeaux*: "Canine Star Is War Puppy," *San Francisco Chronicle*, July 1, 1923, 3.

84 *heavyweight wrestler*: "Harry Lucenay Is Killed in Hollywood," *Corsicana (Tex.) Daily Sun*, May 29, 1944, 8.

85 *a famous Oklahoma fighting dog*: The most complete pedigree available for Pal (http://pedigrees.co/node/895/public) actually lists him as "Peter." Pal and Pete are often confused because both belonged to Lucenay and both were famous Hollywood stars, but Pete was most definitely sired by Pal. Pal was smaller, more brindle than white, and his ears were cropped. Lucenay often dressed him in a studded Victorian show harness. Pete, on the other hand, was larger and mostly white, with natural ears. The dogs were extensively photographed by the press and easily distinguishable from each other. See "Our Gang Dog in 'Personal' Show at State," *Altoona (Pa.) Tribune*, Feb. 1, 1935, 4.

85 *He was insured for $10,000*: "Q&A," *El Paso Herald*, May 6, 1922, 19.

85 *"The dogs we were using had no personality"*: Maltin and Bann, *The Little Rascals*, 281–83.

86 *including the young Fred Rogers*: Ibid.

86 *Pete made Lucenay more than $21,000 a year*: "From Films to Fourth Estate," *Logansport (Ind.) Pharos-Tribune*, Oct. 1, 1932, 8.

86 *"Doubtless"*: Robert Louis Stevenson, "The Character of Dogs," *The English Illustrated Magazine* 1 (1884): 300–305.

90 *"an authentic and intelligent admirer"*: Hearne, *Adam's Task*, 202.

90 *"No one wants [dogfights]"*: Ibid.

CHAPTER 6: TOOTH AND CLAW

Interview sources for this chapter included Dr. Ian Dunbar, Dr. Jeffrey Sacks, Dr. James Serpell, Dr. Karen London, Dr. Randall Lockwood, Dr. Bonnie Beaver, Dr. Stephen Zawistowski, Dr. Brian Hare, Dr. Gary Patronek, Dr. Amy Marder, Dr. Barbara Sherman, Jean Donaldson, Janis Bradley, Jennifer Shryock, Julie Hecht, Bernice Clifford, and Aimee Sadler.

96 *"Among the earliest forms"*: Quammen, *Monster of God*, 3.

97 *A study conducted by Dr. Jeffrey Sacks*: Jeffrey J. Sacks, Marcie-Jo Kresnow, and Barbara Houston, "Dog Bites: How Big a Problem?," *Injury Prevention* 2, no. 1 (1996): 52–54.

97 *Only about 316,200 bites*: Laurel Holmquist and Anne Elixhauser, "Emergency Department Visits and Inpatient Stays Involving Dog Bites, 2008," Statistical brief, Healthcare Cost and Utilization Project (2010).

97 *Compare dog bite deaths with those*: Centers for Disease Control and Prevention, Web-Based Injury Statistics Query and Reporting System (WISQARS), http://www.cdc.gov/injury/wisqars/.

98 *equestrian sports result in approximately twelve thousand cases*: Karen E. Thomas et al., "Non-Fatal Horse Related Injuries Treated in Emergency Departments in the United States, 2001–2003," *British Journal of Sports Medicine* 40, no. 7 (2006): 619–26.

98 *According to one Canadian study*: Chad G. Ball et al., "Equestrian Injuries: Incidence, Injury Patterns, and Risk Factors for 10 Years of Major Traumatic Injuries," *American Journal of Surgery* 193, no. 5 (2007): 636–40.

98 *Deer cause roughly two hundred fatal motor vehicle crashes*: Marcel P. Huijser et al., *Wildlife-Vehicle Collision Reduction Study: Report to Congress* (U.S. Department of Transportation, No. FHWA-HRT-08–034, 2007).

98 *a 2014 study conducted by the Mayo Clinic*: Nikola Babovic, Cenk Cayci, and Brian T. Carlsen, "Cat Bite Infections of the Hand: Assessment of Morbidity and Predictors of Severe Infection," *Journal of Hand Surgery* 39, no. 2 (2014): 286–90.

98 *prompted Dr. Henry Parrish*: Henry M. Parrish et al., "Epidemiology of Dog Bites," *Public Health Reports (1896–1970)* 74, no. 10 (1959): 891–903.

100 *Most dog bite studies*: Harold Weiss, Deborah I. Friedman, and Jeffrey H. Coben, "Incidence of Dog Bite Injuries Treated in Emergency Departments," *JAMA* 279, no. 1 (1998): 51–53; Karen Overall and Molly Love, "Dog Bites to Humans: Demography, Epidemiology, Injury, and Risk," *JAMA* 218, no. 12 (2001): 1923–35; Thomas V. Brogan et al., "Severe Dog Bites in Children," *Pediatrics* 96, no. 5 (1995): 947–50; Douglas Boenning, Gary R. Fleisher, and Joseph M. Campos, "Dog Bites in Children: Epidemiology, Microbiology, and Penicillin Prophylactic Therapy," *American Journal of Emergency Medicine* 1, no. 1 (1983): 17–21; Yoon-Taek Chun, Jay E. Berkelhamer, and Terry E. Herold, "Dog Bites in Children Less Than 4 Years Old," *Pediatrics* 69, no. 1 (1982): 119–20; Norma C. Guy et al., "Risk Factors for Dog Bites to Owners in a General Veterinary Caseload," *Applied Animal Behaviour Science* 74, no. 1 (2001): 29–42; Loren J. Borud and David W. Friedman, "Dog Bites in New York City," *Plastic and Reconstructive Surgery* 106, no. 5 (2000): 987–90; Norma C. Guy et al., "A Case Series of Biting Dogs: Characteristics of the Dogs, Their Behaviour, and Their Victims," *Applied Animal Behaviour Science* 74, no. 1 (2001): 43–57; "The Role of Breed in Dog Bite Risk and Prevention," American Veterinary Medical Association, April 17, 2012.

100 *an inverse relationship between household income and dog bites*: Carrie M. Shuler et al., "Canine and Human Factors Related to Dog Bite Injuries," *Journal of the American Veterinary Medical Association* 232, no. 4 (2008): 542–46; John A. Ndon, Gregory J. Jach, and William B. Wehrenberg, "Incidence of Dog Bites in Milwaukee, Wis.," *Wisconsin Medical Journal* 95, no. 4 (1996): 237–41; Malathi Raghavan et al., "Exploring the Relationship Between Socioeconomic Status and Dog-Bite Injuries Through Spatial Analysis," *Rural and Remote Health* 14, no. 3 (2014): 2846; Laura A. Reese, "The Dog Days of Detroit: Urban Stray and Feral Animals," *City and Community* 14, no. 2 (2015): 167–82; "Dog Bites Highest in Deprived Areas," BBC News, April 24, 2014.

100 *children who have been exposed to domestic violence*: Elizabeth DeViney, Jeffery Dickert, and Randall Lockwood, "The Care of Pets Within Child Abusing Families," *Cruelty to Animals and Interpersonal Violence: Readings in Research and Application* (West Lafayette, Ind.: Purdue University Press, 1998), 305–13; see also Arnold Arluke et al., "The Relationship of Animal Abuse to Violence and Other Forms of Antisocial Behavior," *Journal of Interpersonal Violence* 14, no. 9 (1999): 963–75; and Arluke, *Just a Dog.*

101 *Growing up in the United Kingdom*: Dr. Ian Dunbar, interview with author, Berkeley, Calif., Sept. 2012.

102 *"dark figure" of crime*: Jock Young, *The Exclusive Society: Social Exclusion, Crime, and Difference in Late Modernity* (Thousand Oaks, Calif.: Sage, 1999), 37.

103 *In the Chicago suburb of Palatine, Illinois*: "Palatine Code of Ordinances, Chapter Five: Animals," https://www.palatine.il.us/assets/1/code_of_ordinances/Chapt_051.pdf.

103 *psychological trauma of a dog bite injury*: Kenneth A. Deffenbacher et al., "A

Meta-Analytic Review of the Effects of High Stress on Eyewitness Memory," *Law and Human Behavior* 28, no. 6 (2004): 687.

103 *"Dog bite statistics"*: Bonnie V. Beaver et al., "A Community Approach to Dog Bite Prevention," *Journal of the American Veterinary Medical Association* 218 (2001): 1732–49.

103 *Animal behaviorists divide aggression*: Bonnie V. Beaver, "Clinical Classification of Canine Aggression," *Applied Animal Ethology* 10, no. 1 (1983): 35–43; Randall Lockwood, "The Ethology and Epidemiology of Canine Aggression," as collected in Serpell, *The Domestic Dog*, 131–38.

104 *"For every anecdotal report"*: Donaldson, *Culture Clash*, 54.

104 *"Just a generation ago"*: Jean Donaldson, interview with author, Oakland, Calif., Sept. 2012.

105 *"Dogs may view children"*: Molly Love and Karen Overall, "How Anticipating Relationships Between Dogs and Children Can Help Prevent Disasters," *Journal of the American Veterinary Medical Association* 219, no. 4 (2001): 446–53.

106 *Janis Bradley, a former educator*: Janis Bradley, interview with author, Oakland, Calif., Sept. 2012.

107 *researchers from the Institute for Animal Welfare and Behaviour*: Esther Schalke et al., "Is Breed-Specific Legislation Justified? Study of the Results of the Temperament Test of Lower Saxony," *Journal of Veterinary Behavior: Clinical Applications and Research* 3, no. 3 (2008): 97–103; Stefanie A. Ott et al., "Is There a Difference? Comparison of Golden Retrievers and Dogs Affected by Breed-Specific Legislation Regarding Aggressive Behavior," *Journal of Veterinary Behavior: Clinical Applications and Research* 3, no. 3 (2008): 134–40.

107 *A 2011 Canadian study*: A. MacNeil-Allcock et al., "Aggression, Behaviour, and Animal Care Among Pit Bulls and Other Dogs Adopted from an Animal Shelter," *Animal Welfare* 20, no. 4 (2011): 463–68.

108 *Canine Behavioral Assessment and Research Questionnaire*: Deborah L. Duffy, Yuying Hsu, and James A. Serpell, "Breed Differences in Canine Aggression," *Applied Animal Behaviour Science* 114, no. 3 (2008): 441–60. See also Lindsay R. Mehrkam and Clive D. L. Wynne, "Behavioral Differences Among Breeds of Domestic Dogs (*Canis lupus familiaris*): Current Status of the Science," *Applied Animal Behaviour Science* 155 (2014): 12–27.

108 *American Temperament Test Society*: "ATTS Breed Statistics," http://atts.org.

108 *which the biologist Ray Coppinger defines*: Coppinger and Coppinger, *Dogs*, 199 (N.B.: The wolf biologist L. David Mech refers to these behaviors as "fixed action patterns.")

109 *"Dog breeding is a well-established art"*: "Study of Canine Genes Seeks Hints on Behavior," *New York Times*, Dec. 3, 1991.

109 *A 2015 meta-analysis*: Lenka Hradecká et al., "Heritability of Behavioural Traits in Domestic Dogs: A Meta-Analysis," *Applied Animal Behaviour Science* 170 (Sept. 2015): 1–13.

109 *Pointing, for example, has been observed*: Mark Neff, "A Fetching Model for Understanding the Mammalian Mind: Genetics and the Natural History of Pointing," TGen and the Van Andel Institute, 2011.

110 *"Dogs are artifacts"*: Scottie Westfall, "Bulldogs, Newfoundlands, and Britishness," *Canis Lupus Hominis: The Retriever, Dog, and Wildlife Blog*, July 28, 2010, http://retrieverman.net.

110 *In 2005, Kenth Svartberg*: Kenth Svartberg, "Breed-Typical Behaviour in Dogs—Historical Remnants or Recent Constructs?," *Applied Animal Behaviour Science* 96, no. 3 (2006): 293–313.

111 *While conducting his research at Bar Harbor*: John Paul Scott, "Agonistic Behavior of Mice and Rats: A Review," *American Zoologist* 6, no. 4 (1966): 683–701.

111 *certain lines of the English springer spaniel*: Ilana R. Reisner, Katherine A. Houpt, and Frances S. Shofer, "National Survey of Owner-Directed Aggression in English Springer

Spaniels," *Journal of the American Veterinary Medical Association* 227, no. 10 (2005): 1594–603; Anthony L. Podberscek and James A. Serpell, "The English Cocker Spaniel: Preliminary Findings on Aggressive Behaviour," *Applied Animal Behaviour Science* 47, no. 1 (1996): 75–89.

111 *disruption of the 5-hydroxytryptamine (5-HT) receptors*: Kathelijne Peremans et al., "Estimates of Regional Cerebral Blood Flow and 5-HT2A Receptor Density in Impulsive, Aggressive Dogs with 99mTc-ECD and 123I-5-I-R91150," *European Journal of Nuclear Medicine and Molecular Imaging* 30, no. 11 (2003): 1538–46; Kathelijne Peremans et al., "The Effect of Citalopram Hydrobromide on 5-HT2A Receptors in the Impulsive–Aggressive Dog, as Measured with 123I-5-I-R91150 SPECT," *European Journal of Nuclear Medicine and Molecular Imaging* 32, no. 6 (2005): 708–16.

112 *"Dog breeds develop reputations"*: Coppinger and Coppinger, *Dogs*, 262.

CHAPTER 7: A FEAR IS BORN

114 *By 1969, owners of guard-dog kennels:* "A Reply to Rising Crime: Guard Dogs," *New York Times*, Nov. 23, 1969; "Rising Fear of Crime Brings More Jobs for Private Guards," *New York Times*, July 20, 1969; "Attack Dogs Newest Weapon in City Life," *Mansfield (Ohio) News Journal*, Aug. 20, 1970; "Move One Muscle and He'll Kill You," *Chicago Tribune*, Nov. 1, 1970.

114 *some companies rented out "attack dogs"*: "Attack Dogs for Rent," *Palm Beach Post Times*, May 17, 1970.

115 *American Kennel Club registrations*: American Kennel Club registration data, 1923–2010; "Poodles Replaced by German Shepherds," *Beaver (Pa.) County Times*, May 6, 1970; "Bigger Barkers," *Washington Post*, April 4, 1976.

115 *"violent Negro crime"*: "City of Fear and Crime," *New York Times*, Jan. 22, 1969.

115 *"up to 90 percent"*: "Crime Rise Brings Guard Dog Boom," *Chicago Tribune*, Sept. 3, 1972.

115 *"standard fixtures"*: "Dogs Used Here to Combat Crime," *New York Times*, Dec. 26, 1971; "Fear Is Steady Companion of Many Harlem Residents," *New York Times*, June 3, 1971; "Inflation and Crime Fuel Public Housing Crisis," *New York Times*, June 4, 1970; "'You Must Be Out of Your Mind to Go Out After Dark in a Neighborhood Like This,'" *New York Times*, March 22, 1970; "Crime Puts Hollis in a State of Alarm," *New York Times*, March 5, 1973; "Armed Citizens Patrol Bushwick, Saying the Police Do Not Do Job," *New York Times*, July 21, 1977.

116 *As early as 1966*: "The Watchdog Fraud," *St. Petersburg (Fla.) Times*, Sept. 18, 1966; "Beware of Fierce Breeders!" *Sports Illustrated*, Nov. 10, 1969.

116 *thousands of hastily acquired dogs*: "Large, Vicious Guard Dogs Put Bite on City Life," *Fremont (Calif.) Argus*, June 26, 1974; "Guard Dogs Creating Public Health Woes," *Ocala (Fla.) Star-Banner*, Dec. 16, 1975; "Owners Give Up Dogs as Fido Takes Bigger Bite Out of the Dollar," *Lewiston (Maine) Daily Sun*, May 20, 1975; "Stray Dogs Peril Jersey City Area," *New York Times*, July 27, 1975.

116 *the city's canine population topped 700,000*: David Harris, P. J. Imperato, and B. Oken, "Dog Bites: An Unrecognized Epidemic," *Bulletin of the New York Academy of Medicine* 50, no. 9 (1974): 981.

116 *"man's worst friend"*: "Dog, 'Man's Worst Friend,' Called Beast in Town," *Long Beach (Calif.) Independent*, July 13, 1970.

116 *There is substantial evidence*: Harris, Imperato, and Oken, "Dog Bites."

117 *Today, the visual image of the Doberman*: Isabelle Viaud-Delmon et al., "Auditory-Visual Virtual Environments to Treat Dog Phobia," *7th International Conference on Disability, Virtual Reality, and Associated Technologies with Art Abilitation* (Reading, U.K.: ICDVRAT and the University of Reading, 2008), 119–24.

117 *police having to shoot rampaging Saint Bernard "attack dogs"*: "Attack Dog Killed by Police Officer," *Los Angeles Times,* Jan. 4, 1973.

117 *"more vicious and unpredictable"*: Robert Goldwyn, "Editorial," *Archives of Surgery,* March 1976.

117 *In 1975, officials in Georgia reported*: "Fierce Packs of 'Devil Dogs' Run Wild and Mean in South," *Milwaukee Journal,* Oct. 16, 1975.

118 *He was one of the first activists*: "Inhumane 'Puppy Mills' a Growing Blight," *New York Times,* Oct. 23, 1972.

118 *"A society that encourages"*: "Cruelty to Animals Saddens SPCA Head," *Sarasota (Fla.) Herald-Tribune,* Aug. 7, 1977.

118 *"you can be fined $500"*: "Organized Dog Fight Rings Spring Up All Over U.S.," *Oakland Tribune,* Oct. 13, 1974.

119 *a Dalmatian named Pepper*: Daniel Engber, "Pepper Goes to Washington," *Slate,* June 3, 2009.

119 *The following year,* Life *magazine published*: "Concentration Camps for Dogs," *Life,* Feb. 4, 1966; "Legislative History of the Animal Welfare Act: An Introduction," U.S. Department of Agriculture, https://awic.nal.usda.gov; Unti, *Protecting All Animals,* 64–72, 161–81.

120 *"Numbers are created and repeated"*: Joel Best, *Damned Lies and Statistics,* 13.

120 *"Stranger danger"*: Joel Best, "Missing Children, Misleading Statistics," *Public Interest* 92 (1988): 84–92.

121 *Jack Kelly*: Jack Kelly, "The Law and the Pit Bull Terrier," *Sporting Dog Journal,* July 2012, 29; "Sparks Just Breeds the Dogs, Doesn't Make Them Fight," *Ocala (Fla.) Star-Banner,* Feb. 18, 1979; "Magazine on Breeding and Matching of Pit Dogs Under Inquiry," *New York Times,* Oct. 2, 1974.

121 *Match reports from the early 1970s*: The publications consulted were *Pit Dogs, Pit Dog Report, Your Friend and Mine, The Sporting Dog Journal,* and *The American Gamedog Times,* as well as Ed Faron, *The Complete Gamedog* (Charlotte, N.C.: Walsworth, 1995); Bob Stevens, *Dogs of Velvet and Steel: A Manual for Owners* (Charlotte, N.C.: Walsworth, 1983); George C. Armitage, *Thirty Years with Fighting Dogs*; Bobby Hall, *Bullyson and His Sons* (Charlotte, N.C.: Walsworth, 1986); Carl Semencic, *The World of Fighting Dogs* (Neptune, N.J.: TFH, 1992); and Dieter Fleig, *History of Fighting Dogs.*

121 *Pat Bodzianowski and Sonny Sykes*: "Dog Fighting: Illegal, Brutal, Growing," *New York Times,* Aug. 15, 1974; Kelly, "Law and the Pit Bull Terrier."

122 *"kills one hundred dogs"*: "Dog Fighting a Nationwide Clandestine 'Sport' for 4,000," *Chicago Tribune,* Aug. 24, 1974.

122 *Nowhere in the underground fighting literature*: In the publications listed above, the only references to "bait" animals referred to the use of live chickens or rabbits, which were put in cages in front of treadmills to encourage dogs in keep to run longer. The dogs were not allowed to "have" the animal after running.

122 *Chris Schindler*: Chris Schindler and Janette Reever, interview with author, Gaithersburg, Md., July 2012.

122 *Terry Mills*: interview with author, Leesburg, Va., April 2012.

123 *When the Senate hearings*: "Hearings Before the Subcommittee on Livestock and Grains of the Committee on Agriculture, House of Representatives, Ninety-Third Congress, Second Session on H.R. 15843, H.R. 16738, and Related Bills," Sept. 30 and Oct. 2, 1974.

124 *a 1982* High Times *article*: Ike Abbott, "Dogfight!," *High Times,* July 1982.

124 *three-page profile of Pete Sparks*: "A Breeder of Fighting Dogs Defends an Illegal Sport," *People,* Nov. 11, 1974.

124 *"very moral individuals"*: Abbott, "Dogfight!"

124 *"I can no longer say"*: David Epstein, "Pit Bull Fighting: The Bloodiest Sport," *Hustler,* March 1977.

125 *"An American Pastime"*: Edward Meadows, "An American Pastime," *Harper's*, March 1, 1976.

125 *Benno Kroll*: Benno Kroll, "The Savage Pit," *GEO*, Nov. 1979.

125 *"For two thousand years"*: Gary Cartwright, "Leroy's Revenge," *Texas Monthly*, Aug. 1975.

125 *"an efficient, nearly inexhaustible"*: "Dogfighting: Cashing In on Blood and Gore," *Capital Times*, Aug. 15, 1975.

125 *The true bite force of any living animal*: Dean Dessem, "Interactions Between Jaw Muscle Recruitment and Jaw-Joint Forces in *Canis familiaris*," *Journal of Anatomy* 164 (1989): 101; J. J. Thomason, "Cranial Strength in Relation to Estimated Biting Forces in Some Mammals," *Canadian Journal of Zoology* 69, no. 9 (1991): 2326–33; Per Christiansen and Stephen Wroe, "Bite Forces and Evolutionary Adaptations to Feeding Ecology in Carnivores," *Ecology* 88, no. 2 (2007): 347–58; Stephen Wroe, Colin McHenry, and Jeffrey Thomason, "Bite Club: Comparative Bite Force in Big Biting Mammals and the Prediction of Predatory Behaviour in Fossil Taxa," *Proceedings of the Royal Society of London: Biological Sciences* 272, no. 1563 (2005): 619–25; Stephen Wroe, Michael B. Lowry, and Mauricio Anton, "How to Build a Mammalian Super-Predator," *Zoology* 111, no. 3 (2008): 196–203; G. J. Slater, E. R. Dumont, and B. Van Valkenburgh, "Implications of Predatory Specialization for Cranial Form and Function in Canids," *Journal of Zoology* 278, no. 3 (2009): 181–88; J. L. Davis et al., "Predicting Bite Force in Mammals: Two-Dimensional Versus Three-Dimensional Lever Models," *Journal of Experimental Biology* 213, no. 11 (2010): 1844–51; Jennifer Lynn Ellis et al., "Calibration of Estimated Biting Forces in Domestic Canids: Comparison of Post-Mortem and In Vivo Measurements," *Journal of Anatomy* 212, no. 6 (2008): 769–80; Jennifer Lynn Ellis et al., "Cranial Dimensions and Forces of Biting in the Domestic Dog," *Journal of Anatomy* 214, no. 3 (2009): 362–73.

126 *it seems to have been carried over*: G. H. Chambers and J. F. Payne, "Treatment of Dog-Bite Wounds," *Minnesota Medicine* 52, no. 3 (1969): 427–30.

126 *"It looks really savage"*: Conan Le Cilaire, interview, *The Original Faces of Death: 30th Anniversary Edition* (Blu-ray), Gorgon Video, 2008.

126 *Teenagers who have been exposed*: Robert Hornik et al., "Effects of the National Youth Anti-Drug Media Campaign on Youths," *American Journal of Public Health* 98, no. 12 (2008): 2229.

127 *sociologists have also mapped the clusters of suicides*: Madelyn Gould, Patrick Jamieson, and Daniel Romer, "Media Contagion and Suicide Among the Young," *American Behavioral Scientist* 46, no. 9 (2003): 1269–84.

127 *mass shootings:* Sherry Towers et al., "Contagion in Mass Killings and School Shootings," *PLoS ONE* 10, no. 7 (2015): e0117259–e0117259.

127 *"It's excessive media attention"*: "School Shootings, Mass Killings Are 'Contagious,' Study Finds," CNN.com, Oct. 2, 2015.

127 *"You can't push around"*: Randall Lockwood, interview with author, Falls Church, Va., July 2012.

CHAPTER 8: THE SLEEP OF REASON

128 *"You would think they would have a way"*: "Brian Kilmeade of Fox News: Why Can't They Clear the Ocean of Sharks?," *Kansas City Star*, July 21, 2015.

128 *In Massachusetts during the 1690s*: Marilynne K. Roach, *The Salem Witch Trials: A Day-by-Day Chronicle of a Community Under Siege* (Lanham, Md.: Taylor Trade Publications, 2004).

129 *In 1738, a bulldog owned by a Scottish butcher*: Homan, *Staffordshire Bull Terrier in History and Sport*, 115.

130 *a number of American physicians theorized*: "Selections," *Cincinnati Lancet and Observer* 19, no. 37 (E. B. Stevens, 1876), 1125; "Medical Notes," *Boston Medical and Surgical Journal* 96 (1877); O. R. Ford, "Atlanta Academy of Medicine," *Atlanta Medical and Surgical Journal* 15 (1878), 170.

131 *some 200 out of 240 bites in New York City*: "Destroying the Dogs," *New York Times,* July 6, 1877.

131 *"There are but four venomous beasts"*: "A Venomous Beast," *New York Times,* Nov. 17, 1876.

131 *One of the only people to defend the spitz*: "Beware of the Dog!," *New York Herald,* March 16, 1877.

131 *"No trust or dependence"*: "Hydrophobia," *Manufacturer and Builder* 9 (1877), 171.

132 *the town of Long Branch banned the spitz*: "An Edict Against Spitz Dogs," *New York Times,* April 7, 1878.

132 *Originally imported to the United States*: John Campbell, "The Seminoles, the 'Bloodhound War,' and Abolitionism, 1796–1865," *Journal of Southern History* 72, no. 2 (2006): 259–302.

132 *Captain Henry Wirz*: "Bloodhounds," *Oliver Optic's Magazine,* March 12, 1870, 169; John M. Kistler, *Animals in the Military: From Hannibal's Elephants to the Dolphins of the U.S. Navy* (Santa Barbara, Calif.: ABC-CLIO, 2011), 14.

132 *It is unclear from the testimony*: Government Printing Office, *The Trial of Henry Wirz: Letter from the Secretary of War Ad Interim, in Answer to a Resolution of the House of April 16, 1866, Transmitting a Summary of the Trial of Henry Wirz,* pt. 1 (1868), 97.

133 *a number of famous stage dogs*: "The Dogs of Dutchtown," *Wallace's Monthly,* Sept. 1883, 597; John W. Frick, *Uncle Tom's Cabin on the American Stage and Screen* (New York: Palgrave Macmillan, 2012), 124–26.

133 *the state of Massachusetts banned the Cuban bloodhound*: *The Revised Laws of the Commonwealth of Massachusetts* (Boston: Wright and Potter, 1902), 886.

134 *In 1920, a resident of Navarre, Ohio, reported*: "News of Soldiers and Sailors," *Flagstaff (Ariz.) Coconino Sun,* Feb. 6, 1920, 5.

134 *breeders scrambled to rename them*: "The Dachshund Triumphs over Prejudice," *Saturday Evening Post,* Jan. 12, 1946; "Badgered Dog," *Sports Illustrated,* Dec. 12, 1955.

134 *the German "shepherd" was intended not to herd*: Max von Stephanitz, Joseph Schwabacher, and Carrington Charke, *The German Shepherd Dog in Word and Picture* (1923); Aaron Skabelund, "Breeding Racism: The Imperial Battlefields of the 'German' Shepherd Dog," *Society and Animals* 16, no. 4 (2008): 354–71.

134 *"If a man or his wife"*: "Lardner Says You Must Have Police Dog to Belong," *Temple (Tex.) Daily Telegram,* Dec. 31, 1922.

134 *a New York newspaper proudly reported*: "Ravenous Police Dogs Are Escorting New York Girls to Guard Against Mashers," *Star,* Jan. 23, 1914.

135 *The dean of the Veterinary College at New York University*: "German Police Dog Gains Favor Here," *New York Times,* Feb. 13, 1921.

135 *"Being the fashionable dog"*: "Not Mad, Only Natural," *New York Times,* July 7, 1924.

135 *"The police dog situation"*: "Ban on Police Dogs in Queens Urged by Magistrate Conway," *New York Times,* Jan. 7, 1925.

136 *German shepherds owned by bootleggers*: "Kill Dog in Dry Raid," *New York Times,* May 5, 1927; "Raiders Kill Police Dog in Liquor Haul," *Danville (Va.) Bee,* April 4, 1934; "Police Dogs Used to Protect Stills," *Harrisburg (Pa.) Telegraph,* April 3, 1925; "Police Dogs Guarded Brewers' Operations," *Ironwood (Mich.) Daily Globe,* Dec. 22, 1925; "7 Chicago Gangsters Slain by Rivals," *New York Times,* Feb. 15, 1929.

136 *In New Jersey in 1930*: "Wild Dog Resists Bullets and Tear Gas," *New York Times,* Dec. 29, 1930.

136 *"the highly developed watch-dog instincts"*: "It Sprang Like a Wolf at My Throat," *Galveston (Tex.) Daily News*, Aug. 3, 1924.

136 *a canine "aristocrat"*: "This East Side Landlord Encourages Big Families," *New York Times*, Aug. 12, 1923.

137 *panicked news articles*: See notes for chapter 14.

137 *"had time to digest before disseminating"*: Howard Rosenberg and Charles S. Feldman, *No Time to Think: The Menace of Media Speed and the 24-Hour News Cycle* (New York: Continuum, 2008), 17.

138 *33 percent of Americans reject the theory of evolution*: "Public's Views on Human Evolution," Pew Research Center, Dec. 30, 2013, http://www.pewforum.org.

138 *almost a third of Americans don't believe there is solid evidence*: "Climate Change in the American Mind," Yale Project on Climate Change Communication, Nov. 2013, environment.yale.edu.

138 *more than a third believe extraterrestrials*: "The Fear That Drives Our Alien Belief," *Washington Post*, May 24, 2013.

138 *23 percent have not read a book in the last year*: Jordan Weissmann, "The Decline of the American Book Lover," *Atlantic*, Jan. 21, 2014.

138 *One 2015 poll revealed*: "Why Many Americans Hold False Beliefs About WMDs in Iraq and Obama's Birth Place," *Christian Science Monitor*, Jan. 7, 2015.

138 *"Morality binds and blinds"*: Haidt, *Righteous Mind*, 366.

138 *"backfire effect"*: Brendan Nyhan and Jason Reifler, "When Corrections Fail: The Persistence of Political Misperceptions," *Political Behavior* 32, no. 2 (2010): 303–30; Brendan Nyhan et al., "Effective Messages in Vaccine Promotion: A Randomized Trial," *Pediatrics* 133, no. 4 (2014): e835–e842; Brendan Nyhan, "Why the 'Death Panel' Myth Wouldn't Die: Misinformation in the Health Care Reform Debate," *Forum* 8, no. 1 (2010); Brendan Nyhan, Jason Reifler, and Peter A. Ubel, "The Hazards of Correcting Myths About Health Care Reform," *Medical Care* 51, no. 2 (2013): 127–32.

CHAPTER 9: THE PHANTOM MENACE

140 *Stanley Cohen first laid out his theory*: Cohen, *Folk Devils and Moral Panics*.

140 *Anger and anxiety are highly contagious feelings*: Rui Fan et al., "Anger Is More Influential Than Joy: Sentiment Correlation in Weibo," *PLoS ONE* 9, no. 10 (2014): e110184.

141 *not one case of a child*: Victor, *Satanic Panic*.

141 *polls indicate that Americans fear crime*: Glassner, *The Culture of Fear*.

141 *Dozens of case-controlled medical studies*: Mnookin, *Panic Virus*.

141 *Reports of violent crime in Miami*: James Kelly and Michael Wallis, "Absolute War in Our Streets: Southern Florida, Riots, Refugees, and Now a Crime Wave," *Time*, Nov. 24, 1980.

142 *A higher tolerance*: "Problems with Vicious Dogs Increase in County," *Sarasota (Fla.) Journal*, Feb. 20, 1980.

142 *Frankie Scarbrough and his neighbor*: "Youth in Serious Condition After Dog Attack," *Boca Raton (Fla.) News*, Dec. 3, 1979.

143 *The Miami Dolphins football team*: "Benefit for Pit Bull Victim," *Miami News*, Jan. 29, 1980.

143 *Several months later, Scarbrough was readmitted*: "Doc Says Bitten Boy Was Neglected," *Daytona Beach Morning Journal*, Aug. 23, 1980.

143 *Soon after, Scarbrough's parents abandoned him*: "Victim in Dog Attack Recovering," *Gainesville (Fla.) Sun*, Dec. 10, 1989.

143 *Ethel Tiggs*: "The Attack of the Pit Bulls," *Miami News*, Jan. 16, 1980.

143 *local pet shops reported*: "Pet Shop Owners Getting Requests for 'the Dog That Bit the Old Lady,'" *Miami News*, Jan. 17, 1980.

144 *The town of Hollywood then passed a formal ordinance*: "Hollywood Officials Battle Pro-dog Crowd," *Miami News*, Jan. 17, 1980.

144 *"the [people] who don't fight"*: "Pit Bulls: Part Terrier, Part Terror," *Chicago Tribune*, May 15, 1985.

145 *And, of course, they owned "killer dogs"*: "Killer Dog Attacks Rack Bay Area," *Los Angeles Times*, Feb. 10, 1982.

145 *"A German shepherd or Doberman"*: "Dogfighting: Illegal but Proliferating," *Los Angeles Times*, March 3, 1980.

145 *a pit bull named Frog*: "The Region," *Los Angeles Times*, Nov. 16, 1986.

146 *a trend story about California pit bull "attacks"*: "Fighting Dogs' Attacks Raise Alarm on Coast," *New York Times*, Feb.12, 1982.

146 *"Our industry contributed to the pit bull hysteria"*: Richard Avanzino, interview with author, Aug. 2015.

147 *African-Americans made up only 4 percent*: Sue-Ellen Brown, "The Under-Representation of African Americans in Animal Welfare Fields in the United States," *Anthrozoös* 18, no. 2 (2005): 98–121.

147 *when the U.S. Bureau of Labor Statistics*: Bureau of Labor Statistics, U.S. Department of Labor, *Occupational Outlook Handbook* (2013–14); Derek Thompson, "The 33 Whitest Jobs in America," *Atlantic*, Nov. 6, 2013.

148 *"is aggressive to anything"*: "Director Discusses Pit Bull Issue," *Arizona Humane Society Magazine*, Summer/Fall 1987, 2–4.

148 *"[Pit bulls] chew"*: Dale Dunning and John G. Reynolds, "Responsible Pit Bull Ownership," Arizona Humane Society, 1987.

148 *Donald Clifford*: Donald Clifford to the William and Charlotte Parks Foundation, April 23, 1986; Donald Clifford, Mary Pat Boatfield, and Judy Rubright, "Observations on Fighting Dogs," *Journal of the American Veterinary Medical Association* 183, no. 6 (1983): 654–57; Pratik S. Multani and Donald Clifford, "Are Pit Bulls Giving Good Dogs a Bad Name?" *Community Animal Control*, May/June 1985; Merrilyn Segrest and Donald Clifford, "Are Pit Bulls Different? Part I," *Community Animal Control*, July/Aug. 1986; Donald Clifford, Kay Ann Green, and R. M. Watterson, *The Pit Bull Dilemma: The Gathering Storm: 1,000 Annotated Abstracts from Books, Journals, Magazines, Newspapers, and Reports* (Philadelphia: Charles Press, 1990).

149 *"You are not going to change my opinion"*: Donald Clifford to Diane Jessup, April 5, 1991, and May 30, 1990.

149 *"the most dangerous dog in America"*: Michael Satchell, "The Most Dangerous Dog in America," *U.S. News and World Report*, April 20, 1987.

149 *"the macho dog to have"*: Mark D. Uehling and Sue Hutchinson, "The Macho Dog to Have," *Newsweek*, July 14, 1986.

149 *"This is* a dog in name only": "For Outlawing Pit Bulls," *Los Angeles Times*, Sept. 4, 1983.

149 *"Every pit bull in the country"*: "Pit Bull Deserves World's Worst Fate," *Seattle Post-Intelligencer*, May 19, 1986.

149 *"availability cascade"*: Timur Kuran and Cass R. Sunstein, "Availability Cascades and Risk Regulation," *Stanford Law Review* (1999): 683–768; Cass R. Sunstein, *Laws of Fear: Beyond the Precautionary Principle* (Cambridge, U.K.: Cambridge University Press, 2005); Cass R. Sunstein, "The Law of Group Polarization," University of Chicago Law School, John M. Olin Law and Economics Working Paper 91, 1999.

CHAPTER 10: KNOWN UNKNOWNS

Interviews for this chapter include Dr. Randall Lockwood, Dr. Jeffrey Sacks, Dr. Amy Marder, Dr. Bonnie Beaver, Dr. Victoria Voith, Dr. Gary Patronek, Karen Delise, Donald Cleary, Richard Avanzino, and Michael Mountain.

150 *The event's organizers expected*: Andrew Rowan, ed., *Dog Aggression and the Pit Bull Terrier: Proceedings of a Workshop Organized by the Tufts Center for Animals on July 17, 1986* (North Grafton, Mass.: Tufts University School of Veterinary Medicine, 1986), 1.

151 *Drs. Lee Pinckney and Leslie Kennedy*: Lee E. Pinckney and Leslie A. Kennedy, "Traumatic Deaths from Dog Attacks in the United States," *Pediatrics* 69, no. 2 (1982): 193–96.

151 *in 1976 the UKC registered 23,500 APBTs*: I. Lehr Brisbin, "Letter to the Editor," Community Animal Control, July/Aug. 1984.

152 *Dr. Randall Lockwood*: Randall Lockwood, interviews with author, Falls Church, Va., July 2012, and Myrtle Beach, S.C., Nov. 2013.

152 *"Are 'Pit Bulls' Different?"*: Randall Lockwood and Kate Rindy, "Are 'Pit Bulls' Different? An Analysis of the Pit Bull Terrier Controversy," *Anthrozoös* 1, no. 1 (1987): 2–8.

153 *A lawsuit in Milwaukee*: "Milwaukee Police Shoot, Kill Dozens of Dogs Each Year," *Minneapolis Star Tribune*, June 4, 2014.

153 *Between 2009 and 2012*: "Collateral Damage: Police Shooting Dogs in the Line of Duty," WGRZ News (Buffalo), Nov. 11, 2014.

153 *Police in Buffalo, New York*: Ibid.

153 *"Dog shootings are part"*: Balko, *Rise of the Warrior Cop*, 292.

153 *In 2014, a police officer shot and killed*: " 'Vicious' Pit Bull Fatally Shot by Police Officer Was Actually a Black Lab," Gawker.com, July 7, 2014.

154 *Angela Hands*: "Debate Widens on Plans to Restrict Pit Bull Dogs," *New York Times*, Dec. 30, 1985.

154 *Loose dogs of all types had been killing livestock*: "Domestic Dogs Start Spring Slaughter," *Taos (N. M.) News*, April 3, 1980.

154 *Even small children*: "Some Cities Ban Fighting Dogs," *Weekly Reader*, Nov. 21, 1986.

155 *"the poor man's guard dog"*: "Rash of Attacks: Pit Bulls Taking Rap for Owners?," *Los Angeles Times*, June 26, 1987.

155 *the number of Americans* killed *by dogs actually dropped*: Dr. Jeffrey Sacks and Karen Delise, combined American DBRF data (1979–2001).

156 *Hayward Turnipseed*: "Friends, Neighbors Start Drive for Boy's Funeral," *Atlanta Daily World*, Nov. 27, 1986; "Owner of Three Pit Bulldogs Indicted," *Atlanta Daily World*, Dec. 7, 1986; "Bulldog Owner Gets Five Years Sentence," *Atlanta Daily World*, March 1, 1987.

156 *Lockwood later evaluated Turnipseed's dogs*: Randall Lockwood and John C. Wright, "Behavioral Testing of Dogs Implicated in a Fatal Attack on a Young Child" (paper presented at a meeting of the Animal Behavior Society, Williamstown, Mass., June 1987).

156 *DeKalb County, Georgia*: DeKalb County Animal Control, "An Ordinance for Dogs," report presented to the DeKalb County Chief Executive Officer and Board of Commissioners, April 1987.

157 *subject of many neighborhood complaints*: "Fatal Mauling Spurs Review of Dog Complaints," *Atlanta Journal-Constitution*, Nov. 23, 1986.

157 *"a perfect storm"*: Malcolm Gladwell, "Troublemakers: What Pit Bulls Can Teach Us About Profiling," *New Yorker*, Feb. 6, 2006.

157 *Dr. William Eckman*: "2 Pit Bulldogs Kill Retired Surgeon and Attack Owner," *Atlanta Daily World*, April 10, 1987.

157 *agreed that Eckman had come to her house*: "Couple Indicted in Fatal Pit Bull Attack," *Lakeland Ledger,* May 27, 1987; "Dismissal of Charges Sought in Pit Bull Attack," *Toledo Blade,* Oct. 13, 1987; "Pit Bull Owners Found Innocent," *The Bryan (Ohio) Times* Oct. 16, 1987.

157 *ran a full-page feature*: Michael Satchell, "The Most Dangerous Dog in America," *U.S. News and World Report,* April 20, 1987.

158 *prompted Ohio legislators to pass a bill*: "Celeste Gets Bills to Control Vicious Dogs, Regulate Thrifts," *Toledo Blade,* July 1, 1987.

158 *"We can't get a veterinarian"*: "Pit Bull Attack Has Left Public in the Jaws of Hysteria," *San Jose Mercury-News,* June 18, 1987.

158 *laws on exotic animal ownership were notoriously weak*: "Police Kill Dozens of Animals Freed on Ohio Reserve," *New York Times,* Oct. 19, 2011.

158 *Ohio overturned its statewide pit bull law*: "House Axes 25-Year-Old 'Pit Bull' Law," *Toledo Blade,* Feb. 9, 2012.

158 *James Soto*: "Pit Bull Kills California Boy," *New York Times,* June 15, 1987; "Pit Bull's Owner Blames Parents," *Santa Cruz Sentinel,* June 16, 1987; "Pit Bull's Owner Pleads Innocent in Boy's Death," *Santa Cruz Sentinel,* July 1, 1987.

158 *Timothy Nicolai*: "Dog Kills Tot," *Daily Sitka (Alas.) Sentinel,* April 17, 1987; "Police Report," *Anchorage Daily News,* April 18, 1987.

158 *Jennifer Evon*: "Dog Kills Child," *Daily Sitka (Alas.) Sentinel,* April 1, 1988; "Police Report," *Anchorage Daily News,* April 1, 1988.

159 *"A headline with a question mark"*: Andrew Marr, *My Trade: A Short History of British Journalism* (London: Pan Macmillan, 2009), 253.

159 *Ullrich Ecker*: Stephan Lewandowsky et al., "Misinformation and Its Correction: Continued Influence and Successful Debiasing," *Psychological Science in the Public Interest* 13, no. 3 (2012): 106–31; Ullrich Ecker, Stephan Lewandowsky, and Joe Apai, "Terrorists Brought Down the Plane!—No, Actually It Was a Technical Fault: Processing Corrections of Emotive Information," *Quarterly Journal of Experimental Psychology* 64, no. 2 (2011): 283–310.

159 *"There's a kind of mob psychology"*: "Some Pit Bull Owners Giving Up Their Dogs," *Ukiah (Calif.) Daily Journal,* June 21, 1987.

159 *her adult female dog had been beaten to death*: "It's the Fad: Angry Citizens Turn on Pit Bulls," *USA Today,* July 20, 1987.

160 *More than three hundred pit bulls were surrendered*: "Pit Bull Fear Spreads to Owners, Neighbors Alike," *Los Angeles Times,* July 2, 1987.

160 *"People come in here with their dogs"*: "'Hysteria' Condemns Pit Bulls," *San Jose Mercury-News,* Aug. 12, 1987.

160 *"It was really just heartbreaking"*: "Pit Bull Attack Has Left Public in the Jaws of Hysteria."

160 *80 to 90 percent of the city's dogs*: "Pit Bulls: Months of Hysteria Leads to Distorted Response," *Detroit Free Press,* Aug. 10, 1987.

160 *a camera crew*: "Pit Bull Bloodies Animal Control Officer," *Los Angeles Times,* June 23, 1987.

161 *Angelenos began seeing loose "pit bulls"*: "Woman Charged in Pet's Attack as Other Reports Flood County," *Los Angeles Times,* June 26, 1987.

161 *dogs from twenty-seven different breeds*: "Outlawing Pit Bulls Won't Solve Problem," *USA Today,* Jan. 22, 1990.

161 *"weapon of choice"*: "Pit Bulls Becoming Weapon of Choice Among Drug Dealers," *Tyrone (Pa.) Daily Herald,* June 30, 1987.

161 *"ghetto youths often keep pit bulls"*: "Pit Bulls Popular with Underworld," *Seattle Times,* July 12, 1987.

162 *news coverage of crack*: The best scholarship on this issue (which is truly enlightening)

can be found in Reinarman and Levine's *Crack in America* and Reeves and Campbell's *Cracked Coverage.* Other important works are Michelle Alexander's *New Jim Crow: Mass Incarceration in the Age of Colorblindness* (New York: New Press, 2012); and the documentary film *The House I Live In,* directed by Eugene Jarecki.

162 *Robert Stutman*: Katherine Beckett, *Making Crime Pay: Law and Order in Contemporary American Politics* (New York: Oxford University Press, 2000), 56.

163 *"Much of the drug-related news"*: Ibid., 57.

165 *a combined readership*: "Time, Inc., Warner to Merge," *New York Times,* March 5, 1989.

165 *In 2006, a news station*: "Elderly Man Shaken by Pit Bull," KBCD.com, July 26, 2006.

165 *"thousands of urban trees mauled"*: "Thousands of Urban Trees Mauled and Destroyed as 'Weapon Dog' Owners Train Animals for Fighting," *Guardian,* Aug. 11, 2009.

165 *A pit bull "knocked down" a five-year-old child*: "Pit Bull Knocks Down 5-Year-Old," *Tyler (Tex.) Morning Telegraph,* n.d., http://www.kiiitv.com/story/14640510/pit-bull-child-minor-injuries.

165 *a woman reportedly "fended off"*: "Elderly Woman Uses Hand Sanitizer to Fight Off Pit Bulls," KSDK.com, May 1, 2012.

165 *from Nanaimo, British Columbia*: "Pit Bull in Life or Death Struggle with Giant Beaver," *Alberni Valley Times* (Nanaimo, B.C.), July 19, 2012.

165 *Wendy Bergen*: "Televised Dogfight Stirs Furor," *Chicago Tribune,* May 9, 1990; Alan Predergast, "Wendy Bergen's Exclusive," *American Journalism Review,* Oct. 1991; "TV Reporter Guilty of Staging Dogfight but Not of Perjury," *Deseret News* (Salt Lake City), Aug. 8, 1991.

166 *"This says some sad things"*: "Pit Bulls Are Latest Fad Scare," *San Francisco Chronicle,* July 30, 1987.

168 *Lockwood partnered with a public health epidemiologist*: Jeffrey J. Sacks et al., "Fatal Dog Attacks, 1989–1994," *Pediatrics* 97, no. 6 (1996): 891–95.

169 *a retrospective analysis of all American DBRFs*: Jeffrey J. Sacks et al., "Breeds of Dogs Involved in Fatal Human Attacks in the United States Between 1979 and 1998," *Journal of the American Veterinary Medical Association* 217, no. 6 (2000): 836–40.

CHAPTER 11: LOOKING WHERE THE LIGHT IS

In addition to the sources already mentioned, information for this chapter comes from my interviews with Karen Delise in May 2012, July 2012, and October 2012. I consulted official coroners' reports, animal control records, and legal depositions for extra verification. To ensure that my understanding of the scientific literature was correct, I also dissected over two hundred studies on dog bites and fatalities. Delise's book *The Pit Bull Placebo* explores the issue of DBRFs and media reporting at greater length.

171 *in 1993, a two-year-old boy*: "Neighbor's 121-Pound Dog Kills 2-Year-Old Boy in Hicksville," *New York Times,* May 15, 1993.

173 *"Often we get our news tips"*: Jeremy Jojola, Facebook.com comment, Nov. 11, 2014.

175 *Diane Whipple*: Aphrodite Jones, *Red Zone: The Behind-the-Scenes Story of the San Francisco Dog Mauling* (New York: Morrow, 2003).

176 *Julia Mazziotto*: "Pet Pit Bulls Kill Woman, 80, in Her Home," *New York Times,* Dec. 11, 2002; "Killer Pit Bulls Rip Granny to Shreds," *New York Post,* Dec. 11, 2002.

177 *A neighbor of Mazziotto's told* Newsday: "Grandmother Mauled to Death by Family's Pit Bulls," *Newsday,* Dec. 10, 2002.

178 *Ethel Horton*: Autopsy and necropsy report filed by Dr. Janice Ross of Newberry Pathology (Lee County, South Carolina), March 5, 2010; toxicology report, South Carolina Law Enforcement Division Forensic Services Toxicology Department, filed by Toni M. Broome, March 16, 2010.

178 *James Chapple*: Dr. Lisa Funte, "Report of Autopsy Examination," Office of Shelby County Medical Examiner Regional Forensic Center, May 17, 2007; "Report of Investigation by County Medical Examiner," Tennessee Department of Health and Environment, Oct. 7, 2007.

179 *"There's just something about that breed"*: "In Defense of Pit Bulls," SeacoastOnline.com, Sept. 29, 2007.

180 *after almost two decades of independent research*: Gary J. Patronek et al., "Co-Occurrence of Potentially Preventable Factors in 256 Dog Bite–Related Fatalities in the United States (2000–2009)," *Journal of the American Veterinary Medical Association* 243, no. 12 (2013): 1726–36.

181 *A 1994 case-control study of biting dogs*: Kenneth A. Gershman, Jeffrey J. Sacks, and John C. Wright, "Which Dogs Bite? A Case-Control Study of Risk Factors," *Pediatrics* 93, no. 6 (1994): 913–17.

181 *One series of studies*: Tyrone J. Burrows and William J. Fielding, "Views of College Students on Pit Bull 'Ownership': New Providence, The Bahamas," *Society and Animals* 13, no. 2 (2005): 139–52; William J. Fielding, "Determinants of the Level of Care Provided for Various Types and Sizes of Dogs in New Providence, The Bahamas," *International Journal of Bahamian Studies* 16 (2009): 63–77; William J. Fielding and Susan J. Plumridge, "The Association Between Pet Care and Deviant Household Behaviors in an Afro-Caribbean, College Student Community in New Providence, the Bahamas," *Anthrozoös* 23, no. 1 (2010): 69–78.

182 *the next two most cited medical journal articles*: Steven F. Viegas, J. H. Calhoun, and J. Mader, "Pit Bull Attack: Case Report and Literature Review," *Texas Medicine* 84, no. 11 (1988): 40; B. R. Baack et al., "Mauling by Pit Bull Terriers: Case Report," *Journal of Trauma* 29, no. 4 (1989): 517–20.

183 *dog bites in Los Angeles*: Peter C. Meade, "Police and Domestic Dog Bite Injuries: What Are the Differences? What Are the Implications About Police Dog Use?," *Injury Extra* 37, no. 11 (2006): 395–401.

183 *a 1980 paper on prophylactic antibiotics*: Michael Callaham, "Prophylactic Antibiotics in Common Dog Bite Wounds: A Controlled Study," *Annals of Emergency Medicine* 9, no. 8 (1980): 410–14.

184 *a Siberian husky lifted a two-day-old infant*: "McKeesport Mother of Infant Killed by Dog Charged with Harboring a Dangerous Dog, Other Violations," WXPI.com, Feb. 21, 2012.

184 *A 2014 study of facial dog bites*: Raffi Gurunluoglu et al., "Retrospective Analysis of Facial Dog Bite Injuries at a Level I Trauma Center in the Denver Metro Area," *Journal of Trauma and Acute Care Surgery* 76, no. 5 (2014): 1294–300.

184 *"The differentiation of dog races is problematic"*: R. Lessig, M. Weber, and V. Wenzel, "Bite Mark Analysis in Forensic Routine Case Work," *EXCLI Journal* 5 (2006): 93–102.

184 *Dr. Kenneth Cohrn*: Personal communication, Nov. 27, 2012.

184 *In 1989, Dr. Bret Baack*: Bret R. Baack et al., "Mauling by Pit Bull Terriers: Case Report," *Journal of Trauma* 29, no. 4 (1989): 517–20.

184 *a 1983 paper on dog bites and penicillin*: Douglas A. Boenning, Gary R. Fleisher, and Joseph M. Campos, "Dog Bites in Children: Epidemiology, Microbiology, and Penicillin Prophylactic Therapy," *American Journal of Emergency Medicine* 1, no. 1 (1983): 17–21.

185 *Cheryl Loewe*: Cheryl L. Loewe, Francisco J. Diaz, and John Bechinski, "Pitbull Mauling Deaths in Detroit," *American Journal of Forensic Medicine and Pathology* 28, no. 4 (2007): 356–60.

185 *The cited source is a 2001 paper*: Casey M. Calkins et al., "Life-Threatening Dog Attacks: A Devastating Combination of Penetrating and Blunt Injuries," *Journal of Pediatric Surgery* 36, no. 8 (2001): 1115–17.

185 *a paper written by doctors at the University of Texas*: John K. Bini et al., "Mortality, Mauling, and Maiming by Vicious Dogs," *Annals of Surgery* 253, no. 4 (2011): 791–97.

186 *Colleen Lynn*: "About the Founder," Dogsbite.org.

186 *During the incident*: Officer Julie Maenhout, "Seattle Animal Shelter Investigation Report (SAS# 07–8404)," July 5, 2007.

186 *Lynn maintained the fortune-telling Web site*: "About" and "Divine Lady's Guide to the Runes: The General and Relationship Meanings," RuneCast.com, accessed Aug. 24, 2015.

187 *The site is also littered with childish ad hominems*: "Maul Talk Manual," Dogsbite.org.

187 *"Pit Bulls should be used for target practice"*: Posted by an anonymous user, July 31, 2008, under the site's comment policy: http://blog.dogsbite.org/2008/06/dogsbiteorg-comment-policy-no-pit.html.

187 *headings such as*: "Fuck Pit Bulls and the Faggots That Own Pit Bulls," *Craven Desires* (blog), accessed Aug. 24, 2015, http://cravendesires.blogspot.com.

187 *Merritt Clifton*: Charlotte Alter, "The Problem with Pit Bulls," Time.com, June 20, 2014.

187 *his research methods are limited to scanning*: Merritt Clifton, "Dog Attack Deaths and Maimings, U.S. & Canada: September 1982 to December 31, 2014," DogsBite.org.

187 *The "award" he received*: "ProMED-mail Anniversary Award for Excellence in Outbreak Reporting on the Internet," ProMED-mail.org.

188 *"clip newspaper reports"*: Rory Coker, "Distinguishing Science and Pseudoscience," QuackWatch.com, accessed Aug. 24, 2015, http://www.quackwatch.com.

188 *The Agency for Healthcare Research and Quality estimates*: Holmquist and Elixhauser, "Emergency Department Visits and Inpatient Stays Involving Dog Bites, 2008."

188 *the American Society of Plastic Surgeons reported*: "2012 Plastic Surgery Report" and "2013 Reconstructive Plastic Surgery Statistics," American Society of Plastic Surgeons, National Clearinghouse of Plastic Surgery Procedural Statistics.

189 *"I have more than a hundred"*: Clifton's impromptu interview at the 2014 Animal Rights Conference in Los Angeles is available at: https://www.youtube.com/watch?v=Cv9-2kC-hHw.

189 *Both were papers on rabies*: Merritt Clifton, "How Not to Fight a Rabies Epidemic: A History in Bali," *Asian Biomedicine* 4, no. 4 (2011): 663; Merritt Clifton, "How to Eradicate Canine Rabies: A Perspective of Historical Efforts," *Asian Biomedicine (Research Reviews and News)* 5, no. 4 (2011): 559.

189 *Jeffrey Sacks likened dog bite research*: Jeffrey Sacks, interview with author, June 2012.

CHAPTER 12: "DON'T BELIEVE THE HYPE"

Interview sources for this chapter include Dave Wilson, Deanna Smith, Shawn Mullins, Hector Lopez, and Kim Wolf, as well as several dozen other American bully owners I met at the American Bully Kennel Club Nationals 2014 (Secaucus, New Jersey) and 2015 (Atlanta, Georgia). In Atlanta, I volunteered for a daylong community outreach event through the Humane Society of the United States, where Ralph Hawthorne and Tim Freeman were critical to my understanding of pet ownership in underserved communities. In Compton and Watts, California, I spoke with Lori Weise and Amanda Arreola of Downtown Dog Rescue, Kevon Gulley, and Larry Hill. In East Los Angeles, Alana Yañez and Robert Sotelo of HSUS allowed me to observe their daily outreach work. Stephanie Shain of the Washington Humane Society provided valuable insight into shelter policy at the 2013 HSUS Animal Sheltering Expo in Nashville.

The best scholarship I have found on hip-hop's transition to the mainstream has been Jeff Chang's *Can't Stop Won't Stop* and Tricia Rose's *The Hip-Hop Wars: What We Talk About When We Talk About Hip-Hop—and Why It Matters* (New York: BasicCivitas, 2008).

Few people have written about dogfighting in poor neighborhoods with the clarity, insight, and unflinching honesty that Jesmyn Ward employs in both her National Book Award–winning novel, *Salvage the Bones,* and her memoir, *Men We Reaped.* I highly recommend these books for anyone interested in the emotional complexities of that subject.

To learn more about life in America's urban neighborhoods during the mid-1980s and the 1990s, I leaned heavily on Elijah Anderson's *Streetwise: Race, Class, and Change in an Urban Community* (Chicago: University of Chicago Press, 1990); Douglas Massey and Nancy Denton's *American Apartheid;* William Julius Wilson's *When Work Disappears: The World of the New Urban Poor* (New York: Knopf, 1996); and Paul Jargowsky's *Poverty and Place: Ghettos, Barrios, and the American City* (New York: Russell Sage Foundation, 1997).

192 *Tone Lōc bragged to reporters*: "Tone Loc Fits Right Into the Groove," *Chicago Sun-Times,* July 2, 1988.

192 *"built codes, rules, and vocabularies"*: Jeff Chang, "I Gotta Be Able to Counterattack: Rap and the Los Angeles Riots," *Los Angeles Review of Books,* May 3, 2012.

193 *Among the most iconic images*: "Dogs and Hoses Repulse Negroes at Birmingham," *New York Times,* May 4, 1963. Susan Orlean discusses these photographs and their impact at length in *Rin Tin Tin.*

193 *"Black males, we are America's pit bull"*: "'Fruitvale Station' Star Michael B. Jordan: 'Black Males, We Are America's Pit Bull,'" *Huffington Post,* Dec. 18, 2013.

193 *"the compiled interest"*: Ice-T and Heidi Seigmund, *The Ice Opinion* (New York: St. Martin's, 1994), 15.

193 *almost twenty-five thousand Americans were murdered in 1993*: Alexia Cooper and Erica L. Smith, "Homicide Trends in the United States, 1980–2008," Office of Justice Programs, U.S. Department of Justice, 2011.

193 *"Many of the dogfighters"*: Joel Halverson, interview with author, Nov. 2013.

194 *The new brand of urban street fighter developed a reputation*: Terry Mills, interview with author, April 2012. For more on the world of street dogfighting, see the undercover documentary *Off the Chain* (2005) and Simon Harding's *Unleashed.*

194 *In his 2002 memoir*: DMX and Smokey Fontaine, *E.A.R.L.: The Autobiography of DMX* (New York: HarperCollins, 2003).

194 *"I think Boomer"*: "Dressing Up to the (Ca)nines," *Billboard,* Sept. 20, 2003.

195 *He was arrested twice*: "New Details Are Released in DMX Animal Cruelty Case," *Arizona Republic,* Oct. 6, 2007; "Rapper DMX Arrested for Second Time in a Week on Animal Cruelty, Drug Charges," FoxNews.com, May 9, 2008.

197 *In a 2013 Web post*: Dave Wilson, Facebook post, July 16, 2013.

197 *In 1999 alone, the Pennsylvania SPCA*: "SPCA: Bulls Are the Pits," *Philadelphia Daily News,* Aug. 2, 2000.

197 *"very high percentage of inner-city kids"*: "Despite Popularity Among City Kids, Some Say It's Time to Ban the Dogs of War," *Philadelphia Inquirer,* Aug. 8, 1997.

197 *"There's a definite racial element"*: As quoted in Daryl Gale, "Mortal Combat," *Philadelphia City Paper,* June 1, 2000.

197 *more difficult to adopt a pit bull*: Lisa Davis, "What's Best: Pistol or Pit Bull?," *SF Weekly,* April 9, 1997.

198 *"The people interested"*: "Borough Struggling with Pit Bull Problem," *Norristown (Pa.) Times-Herald,* May 28, 2000.

198 *"You should see the people"*: Kim Wolf, Stephanie Shain, and Mark Kumpf, "Moving Beyond Breed" (presentation at 2013 HSUS Animal Sheltering Expo, Nashville).

198 *held stray pit bulls at gunpoint*: "Pit Bull Problem Emerges in Harrisburg," *Harrisburg (Ill.) Daily Register,* July 21, 2000.

198 *Shelter staff at a number of facilities*: Rhea Moriarty, interview with author, July 2012.

199 *"Pit bulls are perhaps the most abused dogs"*: Ingrid Newkirk, "Controlling an Animal as Deadly as a Weapon," *San Francisco Chronicle,* June 8, 2005.

199 *to eradicate pit bulls from American communities*: "PETA's Position on Pit Bulls."

199 *"The UKC told us"*: Hector Lopez, interview with author, 2014 ABKC Nationals, Atlanta.

199 *"Since the UKC didn't seem"*: Shawn Mullins, interview with author, 2013 ABKC Nationals, Secaucus, N. J.

200 *Antwan Patton*: "About Us," and "Testimonials," PitfallKennels.com.

201 *A 2006 editorial*: "Bully for the Bulls?," *Fredericksburg (Va.) Free Lance-Star,* Oct. 3, 2006.

201 *"Not everything primitive"*: "Canine Killers," *Fredericksburg (Va.) Free Lance-Star,* March 10, 2005.

201 *"It's outrageous"*: "Dog Events Have Foes Howling in Oklahoma," *McClatchy-Tribune Regional News,* March 24, 2009.

201 *"Anything that portrays pit bulls"*: "Bullies Buck Bad Reputation," *Riverside (Calif.) Press-Enterprise,* June 3, 2008.

202 *preparing a show dog for Westminster*: "Cost of Getting a Dog to Westminster," *CNN Money,* Feb. 17, 2015.

204 *Dr. Emily Weiss*: Emily Weiss et al., "Why Did You Choose This Pet? Adopters and Pet Selection Preferences in Five Animal Shelters in the United States," *Animals* 2, no. 2 (2012): 144–59.

204 *In 2013, researchers in California*: Lisa Gunter, "Breed Stereotype and Effects of Handler Appearance on Perceptions of Pit Bulls," Interdisciplinary Forum for Applied Animal Behavior (2013).

204 *Two studies conducted in 2006 and 2009*: Jaclyn E. Barnes et al., "Ownership of High-Risk ('Vicious') Dogs as a Marker for Deviant Behaviors: Implications for Risk Assessment," *Journal of Interpersonal Violence* 21, no. 12 (2006): 1616–34; Laurie Ragatz et al., "Vicious Dogs: The Antisocial Behaviors and Psychological Characteristics of Owners," *Journal of Forensic Sciences* 54, no. 3 (2009): 699–703; Allison M. Schenk, Laurie L. Ragatz, and William J. Fremouw, "Vicious Dogs Part 2: Criminal Thinking, Callousness, and Personality Styles of Their Owners," *Journal of Forensic Sciences* 57, no. 1 (2012): 152–59.

205 *an anonymous online survey of 758 college students*: Laurie Ragatz et al., "Vicious Dogs: The Antisocial Behaviors and Psychological Characteristics of Owners," *Journal of Forensic Science* 54, no. 2 (2009): 699–703.

206 *many people in prison*: Jerome R. Koch et al., "Body Art, Deviance, and American College Students," *Social Science Journal* 47, no. 1 (2010): 151–61.

206 *researchers in the U.K. interviewed*: Gordon Hughes, Jenny Maher, and Claire Lawson, "Status Dogs, Young People, and Criminalisation: Towards a Preventative Strategy" (report presented to the Royal Society for the Prevention of Cruelty to Animals, April 2011).

206 *"People [in the U.K.] talk about 'weapon dogs'"*: "Who Do We Fear More: The Dogs or Their Owners?," *Guardian,* Jan. 25, 2012.

207 *"When vented indirectly"*: Arnold Arluke, "Our Animals, Ourselves," *Contexts* (Summer 2010): 39.

207 *"Members of this task force"*: Arnold Arluke, "Ethnozoology and the Future of Sociology," *International Journal of Sociology and Social Policy* 23, no. 3 (2003): 26–45.

207 *a Republican representative*: "Dangerous Dog Bill Would Restrict Pit Bulls," *Jackson (Miss.) Clarion-Ledger,* Jan. 28, 2015.

207 *"the animals who own the animals"*: "Dazed and Confused," *LaPlace (La.) L'Observateur,* July 5, 2000.

208 *"the other kind of pit bull owner"*: "No Excuse for Pit Bulls," *Schenectady (N.Y.) Daily Gazette,* Jan. 6, 2003.

208 *"open letter to pit bull defenders"*: "An Open Letter to Pit Bull Defenders," *Shawnee (Okla.) News-Star,* Aug. 18, 2007.

208 *The most telling statement*: "Sterling Heights Officials to Examine Vicious Dog Laws," *Advisor & Source Newspapers,* June 17, 2010.

CHAPTER 13: TRAINING THE DOG

Interviews for this chapter included Tony Solesky, Dr. Alan Beck, Gary Wilkes, Karen Breslin, Sonya Dias, Bob Rohde, Kristina Vourax, Rhea Moriarty, Aimee Sadler, Fred Kray, Brent Toellner, Selma Mulvey, Chris Mitchell, and Dr. David Banks. I consulted the available legal documents (summaries, briefs, expert reports, depositions, and opinions) for the following court cases: *Tracey v. Solesky* (2012); *Dias v. City and County of Denver* (2009); *City of Toledo v. Tellings* (2007); *Matthews v. Amberwood Associates Limited* (1998); *Cardelle v. Miami-Dade County Code Enforcement* (2010).

209 *Tara FitzGerald had lived in New York City*: Tara FitzGerald, "Face," n.d., *Vela,* http://www.velamag.

210 *Ten-year-old Dominic*: "Pit Bull Attack Injures Boy, 10," *Towson (Md.) Times,* May 2, 2007; "Pit Bull Owner Facing Charges," *Towson Times,* June 26, 2007.

211 *Her insurance company, State Farm*: Ron Cassie, "Tough Love," *Baltimore,* Feb. 2013.

211 *"Because of it's [sic] aggressive"*: J. Cathell, *Tracey v. Solesky* No. 53, Sept. term 2012, Maryland Court of Appeals.

212 *tenants across Maryland*: "Pit Bull Owners Face Decision to Abandon Their Dogs or Be Evicted from Their Homes," CBS Baltimore, Aug. 29, 2012.

212 *the president of Armistead Homes Corporation*: Sharon F. Vick to leaseholders, Aug. 10, 2012.

212 *According to Charles Edwards*: "Baltimore Housing Community Residents Seek Restraining Order to Keep Their Pit Bulls," CBS Baltimore, Sept. 13, 2012.

212 *"It's been heartbreaking at times"*: Cassie, "Tough Love."

212 *Lawrence Grandpre*: "You Can't Separate Pit Bull Prejudice from Racial Prejudice," *Baltimore Sun,* May 1, 2012.

213 *Prince George's County*: "Report of the Vicious Animal Legislation Task Force," Prince George's County, Md., 2003.

213 *Taylor spoke out against the* Solesky *ruling*: Testimony of Rodney Taylor, "Pit Bull Task Force Hearing," Maryland Judicial Proceedings Committee, June 19, 2012.

215 *In 1981, he called for unsupervised German shepherds*: "Guard Dog Puts Some Meaning into Production," *New York Times,* Oct. 2, 1981.

215 *"I perceive the pit bull"*: "Koch Unleashes Pit Bull Bill," *Newsday,* Aug. 18, 1987.

215 *Pit bulls were banned*: "New York Will Ban Pit Bulls After Oct. 1 and Restrict Others," *New York Times,* March 28, 1989.

215 *a dog attack that badly scarred*: "Mauling Rekindles Dog-Control Effort," *St. Petersburg (Fla.) Times,* May 25, 1989; "Miami Lawmakers: Extend Pit Bull Ban Statewide," *Palm Beach Post,* April 22, 1989; "Law Bans Pit Bulls from Entering Dade, Confines Existing Ones," *Palm Beach Post,* April 5, 1989; "Florida City Sets Restrictions on Pit Bulls," *Miami Herald,* April 2, 1989.

215 *In Denver, a similarly high-profile attack*: Jared Jacang Maher, "For Two Decades, Pit Bulls Have Been Enemy #1 in Denver. But Maybe It's Time for a Recount," *Westword,* Sept. 23, 2009.

216 *Heidi Tufto*: Personal communication, Oct. 8, 2015; Maher, "For Two Decades."

216 *"There appears to be a racial end of this"*: "Johnson: Pit Bull Ban May Reveal Unwarranted Prejudice," *Rocky Mountain News,* May 11, 2005.

217 *Denver impounded almost 5,300 "pit bulls"*: Maher, "For Two Decades."

218 *during the twelfth-century reign of Henry II*: Ritchie, *British Dog*, 61–66.

218 *"It is not for any good purpose"*: George Washington, as quoted in John Campbell, "'My Constant Companion': Slaves and Their Dogs in the Antebellum South," in *Working Toward Freedom: Slave Society and Domestic Economy in the American South*, Larry E. Hudson, ed. (Rochester, N.Y.: University of Rochester Press, 1994); see also Grier, *Pets in America*, 35.

218 *For Charles Ball*: Campbell, "'My Constant Companion.'"

219 *"a colored man accused"*: Carl Sandburg, *The Chicago Race Riots, July 1919* (New York: Harcourt, Brace, and World, 1969), 63.

219 *James K. Vardaman*: Ta-Nehisi Coates, "The Case for Reparations," *Atlantic*, June 2014.

220 *"You will find just as much difference"*: "The Southern Negro's Dog," *Anderson (S.C.) Intelligencer*, July 10, 1884.

220 *"The nigger dog"*: "Essay on the Dog," *Frankfort (Ky.) Roundabout*, Sept. 19, 1891.

220 *"Before the war no legislation was needed"*: "The Dog Evil Has Increased," *Anderson (S.C.) Intelligencer*, Feb. 16, 1901.

220 *the German Nazis passed a battery*: Boria Sax, *Animals in the Third Reich;* Arnold Arluke and Boria Sax, "Understanding Nazi Animal Protection and the Holocaust," *Anthrozoös* 5, no. 1 (1992): 6–31; Boria Sax, "What Is a 'Jewish Dog'? Konrad Lorenz and the Cult of Wildness," *Society and Animals* 5, no. 1 (1997): 3–21.

220 *In 1989, panic swept across the U.K.*: "How Fear Made Man Shun His Best Friend," *Sunday Times* (London), June 4, 1989; "Grandparents Find Boy Savaged by Rottweiler," *Glasgow Herald*, June 3, 1989.

220 *"although the German shepherd"*: Anthony L. Podberscek, "Dog on a Tightrope: The Position of the Dog in British Society as Influenced by Press Reports on Dog Attacks (1988 to 1992)," *Anthrozoös* 7, no. 4 (1994): 232–41.

221 *Animal wardens in the U.K.*: "An Unlikely Battle Pits Englishmen Against Mad Dogs," *Wall Street Journal*, June 12, 1990.

221 *"being bought by criminals"*: Kenneth Baker, *The Turbulent Years* (London: Faber & Faber, 1993).

222 *Dangerous Dogs Act*: "Bad Dogs and Englishmen: What's to Be Done?" *New York Times*, May 29, 1991.

222 *Dog bite injuries and hospitalizations are higher than ever*: "Dog Bites Data: How Likely Are You to Get Bitten," *Guardian*, March 27, 2013; "Dog Bites in Wales Up 81% in Last 10 Years," BBC News, March 18, 2014; "Hospitals Swamped with Victims of Dog Attacks," *Sunday Post*, May 31, 2015.

223 *Aiden McGrew . . . Isabelle Dinoire*: "Infant Boy Killed When Mauled by Family Dog in Ridgeville as Father Slept," *Charleston (S.C.) Post and Courier*, April 21, 2012; "Chesterfield Woman Hospitalized After Golden Retriever Attack," NBC 12 News (Chesterfield County, Virginia), March 20, 2013; "Dog Shot, Killed by Police After Attack," *Barrie (Ont.) Examiner*, Aug. 27, 2014; "French Face-Transplant Patient Tells of Her Ordeal," *New York Times*, February 7, 2006.

224 *the research he conducted on the ecology of stray dogs*: Alan M. Beck, *The Ecology of Stray Dogs: A Study of Free-Ranging Urban Animals* (Baltimore: York Press, 1973).

225 *"Imagine a dog breed"*: Alan Beck, personal communication, Dec. 2012.

225 *According to a 2009 report*: "Dog Bites in Colorado: A Report of Dog Bite Incidents Reported to Animal Control July 2007–June 2008" (prepared by Corona Research, Inc., 2009).

226 *Austin Cussins*: "Dog Attacks, Kills Toddler," *Rocky Mountain News*, June 15, 1998.

226 *Drs. Gary Patronek*: Gary J. Patronek, Margaret Slater, and Amy Marder, "Use of a Number-Needed-to-Ban Calculation to Illustrate Limitations of Breed-Specific Legislation in Decreasing the Risk of Dog Bite–Related Injury," *Journal of the American Veterinary Medical Association* 237, no. 7 (2010): 788.

226 *Between January 2012 and May 2014*: "Breed-Specific Legislation on the Decline," National Canine Research Council, June 16, 2014.

227 *Studies conducted in the U.K.*: B. Klaassen, J. R. Buckley, and Aziz Esmail, "Does the Dangerous Dogs Act Protect Against Animal Attacks: A Prospective Study of Mammalian Bites in the Accident and Emergency Department," *Injury* 27, no. 2 (1996): 89–91; Tiny De Keuster, Jean Lamoureux, and André Kahn, "Epidemiology of Dog Bites: A Belgian Experience of Canine Behaviour and Public Health Concerns," *Veterinary Journal* 172, no. 3 (2006): 482–87; Jessica Cornelissen and Hans Hopster, "Dog Bites in The Netherlands: A Study of Victims, Injuries, Circumstances and Aggressors to Support Evaluation of Breed-Specific Legislation," *Veterinary Journal* 186, no. 3 (2010): 292–98; Belén Rosado et al., "A Comprehensive Study of Dog Bites in Spain, 1995–2004," *Veterinary Journal* 179, no. 3 (2009): 383–91; Nancy M. Clarke and David Fraser, "Animal Control Measures and Their Relationship to the Reported Incidence of Dog Bites in Urban Canadian Municipalities," *Canadian Veterinary Journal* 54, no. 2 (2013): 145; Chiara Mariti, Carlo Ciceroni, and Claudio Sighieri, "Italian Breed-Specific Legislation on Potentially Dangerous Dogs (2003): Assessment of Its Effects in the City of Florence (Italy)," *Dog Behavior* 1, no. 2 (2015): 25–31.

227 *"[BSL]'s not designed"*: "U.S. Communities Increasingly Ditching Pit Bull Bans," *USA Today,* Nov. 18, 2014.

228 *Liam Perk*: "Toddler Dies After Being Bitten by Dog," *Cape Coral (Fla.) Daily Breeze,* Dec. 22, 2009; "About," Liam J. Perk Foundation, http://www.liamjperkfoundation.org.

228 *Chris Mitchell*: Chris Mitchell, interview with author, Dec. 2012; "DeKalb Woman Killed in Home By Pet Dog," WSB TV (Atlanta, Georgia), August 16, 2012.

230 *"The problem isn't that the white woman"*: "Bruce A. Jacobs on Bigotry," available at https://www.youtube.com/watch?v 3kUqMwPC1Y.

CHAPTER 14: DIFFERENT IS DEAD

Interview sources include Jane Berkey, Stacey Coleman, Donald Cleary, Bernice Clifford, Donna Reynolds, Tim Racer, Ken Foster, Marcy Setter, Ledy VanKavage, Kenny Lamberti, Jim Crosby, Cynthia Bathurst, Michael Mountain, Dr. James Serpell, Kelly D'Agostino, Josh Liddy, Brent Toellner, Bob Marotto, Shafonda Davis, Jane Marshall, and Molly Gibb.

The Great Deluge (New York: Morrow, 2006) by Douglas Brinkley; and *Five Days at Memorial* (New York: Crown, 2013) by Sheri Fink gave me a good overview of Katrina's wrath in New Orleans, as did several documentaries: *When the Levees Broke,* directed by Spike Lee; *Dark Water Rising,* directed by Mike Shiley; and *Mine,* directed by Geralyn Pezanoski. Jim Gorant's *Lost Dogs* was critical for understanding the complex rescue and rehabilitation of the dogs seized from Michael Vick's fighting operation.

231 *Some residents had to be forced at gunpoint*: Douglas Brinkley, *The Great Deluge,* 516.

232 *"There is a class issue involved here"*: "Best Friends Need Shelter, Too," *Washington Post,* Sept. 10, 2005.

232 *Workers made their way*: Kenny Lamberti, interview with author, July 2015.

232 *"nasty, despicable animals"*: Daryl Gale, "Mortal Combat," *Philadelphia City Paper,* June 1, 2000.

233 *But even Gulf Coast residents*: "Pet Addendum to FEMA Occupant Lease," Federal Emergency Management Agency, Feb. 12, 2007.

234 *"The message is, 'You're poor'"*: Megan McNabb, "Pets in the Eye of the Storm: Hurricane Katrina Floods the Courts with Pet Custody Disputes," *Animal Law* 14 (2007): 71.

234 *Kara Keyes*: "There's a Dog in This Hunt," *Los Angeles Times,* Aug. 27, 2007.

237 *"When I show [clients]"*: Kelly D'Agostino, interview with author, April 2012.

241 *"My philosophy and my reason"*: "Newspaper Publishes Addresses of Pit Bull Owners," CBS 4 Denver, Jan. 25, 2006.

241 *Dr. Melinda Merck*: Personal communication, March 8, 2015.

241 *a two-year-old female pit bull was doused with gasoline*: Charles Siebert, "The Animal Cruelty Syndrome," *New York Times Magazine,* June 11, 2010.

241 *"Pit bulls should be boiled alive"*: Dan Savage, "Pit Bulls Should Be Boiled Alive Like Lobsters and Fed to Their Idiot Owners," *Stranger,* Aug. 23, 2007.

244 *"With other breeds"*: Marcy Setter, interview with author, May 2012.

244 *"I've worked with hundreds"*: Aimee Sadler, interview with author, July 2012.

246 *"You might say, 'come on Mike'"*: "Michael Vick Blogs: 'My Heart Hurts . . . to Think of What I've Done," *Global Grind,* Aug. 16, 2009.

247 *"Officials from our organization"*: "Government Makes a Case, and Holds Dogs as Evidence," *New York Times,* Aug. 1, 2007.

248 *"What the hell did Michael Vick do, man?"*: "Dog Lovers Cry Foul over Comments from Jay Leno and Chris Rock About Michael Vick," *Los Angeles Times Blogs,* Nov. 11, 2009: http://latimesblogs.latimes.com.

248 *"I'm a Christian"*: "Tucker Carlson: Michael Vick Deserves to Die," CBS News, Dec. 29, 2010.

248 *"Not only have animals been used"*: "Michael Vick, Racial History, and Animal Rights," *Nation,* Dec. 30, 2010.

249 *"I am not saying dogfighting"*: "Michael Vick, Dog Fighting and Race," *Duke Today,* Aug. 29, 2007.

249 *"Matched him with a Doberman"*: Tex Maule, "When the Saints Go Stumbling Out," *Sports Illustrated,* Oct. 27, 1969.

249 *Ed Faron*: "Dogfighting God Father Given Prison," *Winston-Salem (N.C.) Journal,* Feb. 13, 2009.

249 *When presented with fictional "news" items*: Hal Herzog, "Why People Care More About Pets Than Other Humans," *Wired,* April 13, 2015; "Empathy with Dogs Stronger Than with Humans: Study," *Huffington Post Canada,* Aug. 18, 2013.

250 *what seems to be the most powerful factor*: Tehila Kogut and Ilana Ritov, "The 'Identified Victim' Effect: An Identified Group, or Just a Single Individual?" *Journal of Behavioral Decision Making* 18, no. 3 (2005): 157.

250 *"Vick showed the worst of us"*: Gorant, *Lost Dogs,* 3.

251 *"If pit bulls are so difficult"*: Dr. James Serpell, interview with author, Aug. 2012.

252 *One widely circulated figure*: John Goodwin, personal communication, Dec. 23, 2014.

252 *"Overstating the problem of dog fighting"*: Donna Reynolds, "The Color of Dog Fighting," BAD RAP (blog), May 10, 2011: http://badrap-blog.blogspot.com.

252 *This was a keen observation*: "The Top Ten Reasons for Pet Relinquishment in the United States," National Council on Pet Population Study & Policy, 2010; M. D. Salman et al., "Human and Animal Factors Related to Relinquishment of Dogs and Cats in 12 Selected Animal Shelters in the United States," *Journal of Applied Animal Welfare Science* 1, no. 3 (1998): 207–26; John New, M. D. Salman et al., "Moving: Characteristics of Dogs and Cats and Those Relinquishing Them to 12 U.S. Animal Shelters," *Journal of Applied Animal Welfare Science* 2, no. 2 (1999): 83–96; Elsie R. Shore, Connie L. Petersen, and Deanna K. Douglas, "Moving as a Reason for Pet Relinquishment: A Closer Look," *Journal of Applied Animal Welfare Science* 6, no. 1 (2003): 39–52; Natalie DiGiacomo, Arnold Arluke, and Gary Patronek, "Surrendering Pets to Shelters: The Relinquisher's Perspective," *Anthrozoös* 11, no. 1 (1998): 41–51.

253 *costs related to dog bite injuries made up one-third*: Brad Tuttle, "Dog Bites Insurance Companies: Man's Best Friend Behind One Third of All Homeowner Claims," *Time,* May 28, 2013.

253 *only 4.8 percent of insured homeowners*: "Homeowners and Renters Insurance Facts," Insurance Information Institute, 2013.

253 *health expenditures in the United States*: "Accounting for the Cost of Healthcare in the United States," McKinsey Global Institute, 2011.

253 *estimated insurance payouts for dog bite claims*: "Claims Over Dog Bites Are Ripping $1 Billion Hole in Insurers' Pockets," *Wall Street Journal*, Mar. 11, 1996.

253 *American insurance companies paid out $4 billion*: "Car and Deer Collisions Cause 200 Deaths, Cost $4 Billion a Year," *Insurance Journal*, October 24, 2012.

253 *Brian J. Glenn*: Brian J. Glenn, "The Shifting Rhetoric of Insurance Denial," *Law and Society Review* 34, no. 3 (2000): 779–808. See also Tom Baker, "Insuring Morality," *Economy and Society* 29, no. 4 (2000): 559–77; and Jonathan Simon, "The Ideological Effects of Actuarial Practices," *Law and Society Review* 22, no. 4 (1988): 771–800.

254 *insurance companies often invoked actuarial risk*: Coates, "Case for Reparations."

CHAPTER 15: FOR LIFE

Almost everything in this chapter comes from the week I spent in North Philadelphia in April 2013, shadowing Ashley Mutch, Janice Poleon, Stephen Parker, Devell Brookins, and Jonathan Bricker as they took to the streets to engage their community. I spoke at length with Amanda Arrington and Kenny Lamberti of HSUS, as well as with Cory Smith and Betsy McFarland. As previously noted, I also spent time with PFL outreach teams in Atlanta and Los Angeles.

258 *It didn't work*: "Anti-dogfighting Program Fails to Catch On in Philadelphia," *Newsworks Tonight*, July 11, 2012, http://www.newsworks.org.

259 *Eighty-eight percent of PFL's clients*: "Pets for Life Annual Report, 2013/2014," Humane Society of the United States. Courtesy of Amanda Arrington.

266 *fairness should not depend on sameness*: Pinker, *Blank Slate*.

266 *"cultural awakening"*: Greil Marcus, *The Dustbin of History* (Cambridge, Mass.: Harvard University Press, 1998), 28.

266 *there are now almost as many Chihuahuas*: Emily Weiss, "Climbing Out of the Pit," ASPCA Professional Blog, May 28, 2015: http://www.aspcapro.org.

267 *"geography, pop culture"*: "Arizona's All Points Bulletin: Who Can Take in a Chihuahua?" *New York Times*, May 17, 2014.

EPILOGUE

269 *an unarmed black teenager named Michael Brown*: "Grief and Protests Follow Shooting of a Teenager," *New York Times*, Aug. 10, 2014.

269 *Three local and state officials told a reporter from*: Mark Follman, "Michael Brown's Mom Laid Flowers Where He Was Shot—and Police Crushed Them," *Mother Jones*, Aug. 27, 2014.

270 *When the U.S. Department of Justice released the report*: "Investigation of the Ferguson Police Department," U.S. Department of Justice, Civil Rights Division, March 4, 2015.

271 *the Maryland legislature passed Senate Bill 247*: "O'Malley Signs Pre-K, Dog-Bite Bills," *Baltimore Sun*, April 8, 2014.

271 *"still accomplished many feats"*: "Maryland Legislature Mutes Landmark Solesky Ruling, Tracey v. Solesky, During 2014 Legislative Session," DogsBite.org, June 2, 2014.

SELECTED BIBLIOGRAPHY

Allport, Gordon W. *The Nature of Prejudice.* Cambridge, Mass.: Addison-Wesley, 1954.
Altheide, David. *Creating Fear: News and the Construction of Crisis.* New York: Aldine de Gruyter, 2002.
Anbinder, Tyler. *Five Points: The 19th-Century New York City Neighborhood That Invented Tap Dance, Stole Elections, and Became the World's Most Notorious Slum.* New York: Free Press, 2001.
Anderson, Elijah. *Streetwise: Race, Class, and Change in an Urban Community.* Chicago: University of Chicago Press, 1992.
Arluke, Arnold. *Just a Dog: Understanding Animal Cruelty and Ourselves.* Philadelphia: Temple University Press, 2006.
Arluke, Arnold, and Robert Bogdan. *Beauty and the Beast: Human-Animal Relations as Revealed in Real Photo Postcards, 1905–1935.* Syracuse, N.Y.: Syracuse University Press, 2010.
Arluke, Arnold, and Clinton Sanders. *Regarding Animals.* Philadelphia: Temple University Press, 1996.
Armitage, George C. *Thirty Years with Fighting Dogs.* Read Country, 2004.
Asbury, Herbert. *The Gangs of New York.* Leicester: W. F. Howes, 2003.
Asma, Stephen T. *On Monsters: An Unnatural History of Our Worst Fears.* Oxford: Oxford University Press, 2009.
Auguet, Roland. *Cruelty and Civilization: The Roman Games.* London: Routledge, 1994.
Balko, Radley. *Rise of the Warrior Cop: The Militarization of America's Police Forces.* New York: PublicAffairs, 2014.
Bausum, Ann. *Sergeant Stubby: How a Stray Dog and His Best Friend Helped Win World War I and Stole the Heart of a Nation.* Washington, D.C.: National Geographic, 2015.
Becker, Howard. *Outsiders: Studies in the Sociology of Deviance.* New York: Free Press, 1963.
Beers, Diane L. *For the Prevention of Cruelty: The History and Legacy of Animal Rights Activism in the United States.* Athens: Swallow/Ohio University Press, 2006.
Bekoff, Marc. *The Emotional Lives of Animals.* Novato, Calif.: New World Library, 2008.
Berns, Gregory. *How Dogs Love Us: A Neuroscientist and His Adopted Dog Decode the Canine Brain.* New York: New Harvest/Houghton Mifflin Harcourt, 2013.
Best, Joel. *Damned Lies and Statistics: Untangling Numbers from the Media, Politicians, and Activists.* Berkeley: University of California Press, 2001.
Bradley, Janis. *Dogs Bite . . . But Balloons and Slippers Are More Dangerous.* Berkeley, Calif.: James & Kenneth, 2005.
Bradshaw, John. *Dog Sense: How the New Science of Dog Behavior Can Make You a Better Friend to Your Pet.* New York: Basic Books, 2011.
Braitman, Laurel. *Animal Madness: How Anxious Dogs, Compulsive Parrots, and Elephants in Recovery Help Us Understand Ourselves.* New York: Simon & Schuster, 2014.
Chang, Jeff. *Can't Stop, Won't Stop: A History of the Hip-Hop Generation.* New York: St. Martin's, 2005.

Cohen, Stanley. *Folk Devils and Moral Panics: The Creation of the Mods and Rockers.* London: MacGibbon and Kee, 1972.

Colby, Louis B., and Diane Jessup. *Colby's Book of the American Pit Bull Terrier.* Neptune City, N. J.: T.F.H., 1997.

Coleman, Jon T. *Vicious: Wolves and Men in America.* New Haven, Conn.: Yale University Press, 2004.

Coppinger, Raymond, and Lorna Coppinger. *Dogs: A Startling New Understanding of Canine Origin, Behavior, and Evolution.* New York: Scribner, 2001.

Csányi, Vilmos. *If Dogs Could Talk: Exploring the Canine Mind.* New York: North Point, 2005.

Delise, Karen. *The Pit Bull Placebo: The Media, Myths, and Politics of Canine Aggression.* New York: Anubis, 2007.

Derr, Mark. *A Dog's History of America: How Our Best Friend Explored, Conquered, and Settled a Continent.* New York: North Point, 2004.

———. *How the Dog Became the Dog: From Wolves to Our Best Friends.* New York: Overlook Duckworth, 2011.

Donaldson, Jean. *The Culture Clash.* Berkeley, Calif.: James & Kenneth, 1996.

Dozier, Rush W. *Why We Hate: Understanding, Curbing, and Eliminating Hate in Ourselves and Our World.* Chicago: Contemporary, 2002.

Dunbar, Ian. *Before and After Getting Your Puppy: The Positive Approach to Raising a Happy, Healthy, and Well-Behaved Dog.* Novato, Calif.: New World Library, 2004.

Eleftheriou, Basil E., and John Paul Scott. *The Physiology of Aggression and Defeat.* New York: Plenum Press, 1971.

Fleig, Dieter. *History of Fighting Dogs.* Neptune City, N.J.: T.F.H., 1996.

Fromm, Erich. *The Anatomy of Human Destructiveness.* New York: Holt, 1992.

Gardner, Dan. *The Science of Fear: How the Culture of Fear Manipulates Your Brain.* New York: Plume, 2009.

Garland, David. *The Culture of Control: Crime and Social Order in Contemporary Society.* Chicago: University of Chicago Press, 2001.

Glassner, Barry. *The Culture of Fear: Why Americans Are Afraid of the Wrong Things.* New York: Basic Books, 1999.

Goode, Erich, and Nachman Ben-Yehuda. *Moral Panics: The Social Construction of Deviance.* Oxford: Blackwell, 1994.

Gorant, Jim. *The Lost Dogs: Michael Vick's Dogs and Their Tale of Rescue and Redemption.* New York: Gotham, 2010.

Gorn, Elliott J. *The Manly Art: Bare-Knuckle Prize Fighting in America.* Ithaca, N. Y.: Cornell University Press, 1986.

Gould, Stephen Jay. *The Mismeasure of Man.* New York: Norton, 1981.

———. *The Panda's Thumb: More Reflections in Natural History.* New York: Norton, 1992.

Grandin, Temple, and Catherine Johnson. *Animals Make Us Human: Creating the Best Life for Animals.* Boston: Houghton Mifflin Harcourt, 2009.

Grier, Katherine C. *Pets in America: A History.* Chapel Hill: University of North Carolina Press, 2006.

Haidt, Jonathan. *The Righteous Mind: Why Good People Are Divided by Politics and Religion.* New York: Pantheon, 2012.

Harding, Simon. *Unleashed: The Phenomena of Status Dogs and Weapon Dogs.* Bristol, U.K.: Policy Press, 2012.

Hare, Brian, and Vanessa Woods. *The Genius of Dogs: How Dogs Are Smarter Than You Think.* New York: Plume, 2013.

Hari, Johann. *Chasing the Scream: The First and Last Days of the War on Drugs.* New York: Bloomsbury, 2015.

Hearne, Vicki. *Adam's Task: Calling Animals by Name.* New York: Knopf, 1986.

———. *Bandit: Dossier of a Dangerous Dog.* New York: HarperPerennial, 1992.
Herzog, Hal. *Some We Love, Some We Hate, Some We Eat: Why It's So Hard to Think Straight About Animals.* New York: Harper, 2010.
Homan, Mike. *The Staffordshire Bull Terrier in History and Sport.* Hampshire, U.K.: Nimrod Press, 1986.
Homans, John. *What's a Dog For? The Surprising History, Science, Philosophy, and Politics of Man's Best Friend.* New York: Penguin, 2012.
Horowitz, Alexandra. *Inside of a Dog: What Dogs See, Smell, and Know.* New York: Scribner, 2009.
Howell, Philip. *At Home and Astray: The Domestic Dog in Victorian Britain.* Charlottesville: University of Virginia Press, 2015.
Huck, Schuyler W., and Howard M. Sandler. *Rival Hypotheses: Alternative Interpretations of Data Based Conclusions.* New York: Harper & Row, 1979.
Huff, Darrell, and Irving Geis. *How to Lie with Statistics.* New York: Norton, 1954.
Hunter, James Davison. *Culture Wars: The Struggle to Define America.* New York: Basic Books, 1991.
Ingebretsen, Edward J. *At Stake: Monsters and the Rhetoric of Fear in Public Culture.* Chicago: University of Chicago Press, 2001.
Irvine, Leslie. *My Dog Always Eats First: Homeless People and Their Animals.* Boulder, Colo.: Lynne Rienner, 2013.
Jablonka, Eva, and Marion J. Lamb. *Evolution in Four Dimensions: Genetic, Epigenetic, Behavioral, and Symbolic Variation in the History of Life.* Cambridge, Mass.: MIT Press, 2005.
Jargowsky, Paul. *Poverty and Place: Ghettos, Barrios, and the American City.* New York: Russell Sage Foundation, 1997.
Jessup, Diane. *The Dog Who Spoke with Gods.* New York: St. Martin's, 2001.
———. *The Working Pit Bull.* Neptune City, N.J.: T.F.H. Publications, 1995.
Kahneman, Daniel. *Thinking, Fast and Slow.* New York: Farrar, Straus and Giroux, 2011.
Kalof, Linda, and Brigitte Pohl-Resl. *A Cultural History of Animals in Antiquity.* New York: Berg, 2007.
Katz, Jon. *The New Work of Dogs: Tending to Life, Love, and Family.* New York: Villard, 2003.
Kerasote, Ted. *Merle's Door: Lessons from a Freethinking Dog.* New York: Mariner Books, 2008.
Knapp, Caroline. *Pack of Two: The Intricate Bond Between People and Dogs.* New York: Dial, 1998.
Lane, Marion, and Stephen Zawistowski. *Heritage of Care: The American Society for the Prevention of Cruelty to Animals.* Westport, Conn.: Praeger, 2008.
Lemish, Michael G. *War Dogs: A History of Loyalty and Heroism.* Washington, D.C.: Brassey's, 1999.
Lindsay, Steven R. *Handbook of Applied Dog Behavior and Training,* Vols. 1–3. Ames, Iowa: Iowa State University Press, 2000.
Lopez, Barry Holstun. *Of Wolves and Men.* New York: Scribner, 1978.
Lorenz, Konrad. *Man Meets Dog.* London: Methuen, 1954.
———. *On Aggression.* New York: Harcourt, Brace & World, 1966.
Maltin, Leonard, and Richard W. Bann. *The Little Rascals: The Life and Times of Our Gang.* New York: Crown, 1992.
Massey, Douglas S., and Nancy A. Denton. *American Apartheid: Segregation and the Making of the Underclass.* Cambridge, Mass.: Harvard University Press, 1993.
Matz, K. S. *The Pit Bull: Fact and Fable.* Sacramento: De Mortmain Books, 1984.
McCaig, Donald. *The Dog Wars: How the Border Collie Battled the American Kennel Club.* Hillsborough, N.J.: Outrun, 2007.
McConnell, Patricia B. *The Other End of the Leash: Why We Do What We Do Around Dogs.* New York: Ballantine, 2002.

Mech, L. David, and Luigi Boitani. *Wolves: Behavior, Ecology, and Conservation.* Chicago: University of Chicago Press, 2003.

Miklósi, Ádám. *Dog Behaviour, Evolution, and Cognition.* Oxford: Oxford University Press, 2007.

Miller, Pat. *The Power of Positive Dog Training.* New York: Howell Book House, 2001.

Mnookin, Seth. *The Panic Virus: A True Story of Medicine, Science, and Fear.* New York: Simon & Schuster, 2011.

Murray, David, Joel Schwartz, and S. Robert Lichter. *It Ain't Necessarily So: How the Media Remake Our Picture of Reality.* New York: Penguin, 2002.

Orlean, Susan. *Rin Tin Tin: The Life and the Legend.* New York: Simon & Schuster, 2011.

Ostrander, Elaine A., Urs Giger, and Kerstin Lindblad-Toh. *The Dog and Its Genome.* Cold Spring Harbor, N.Y.: Cold Spring Harbor Laboratory, 2006.

Pacelle, Wayne. *The Bond: Our Kinship with Animals, Our Call to Defend Them.* New York: William Morrow, 2011.

Pinker, Steven. *The Blank Slate: The Modern Denial of Human Nature.* New York: Viking, 2002.

Pollack, Adam J. *John L. Sullivan: The Career of the First Gloved Heavyweight Champion.* Jefferson, N.C.: McFarland, 2006.

Pryor, Karen. *Don't Shoot the Dog! How to Improve Yourself and Others Through Behavioral Training.* New York: Simon & Schuster, 1984.

Quammen, David. *Monster of God: The Man-Eating Predator in the Jungles of History and the Mind.* New York: Norton, 2003.

Redmond, Patrick R. *The Irish and the Making of American Sport, 1835–1920.* Jefferson, N.C.: McFarland, 2014.

Reeves, Jimmie Lynn, and Richard Campbell. *Cracked Coverage: Television News, the Anti-Cocaine Crusade, and the Reagan Legacy.* Durham, N.C.: Duke University Press, 1994.

Regan, Tom. *The Case for Animal Rights.* Berkeley: University of California Press, 1983.

Reid, Pamela J. *Excel-Erated Learning: Explaining in Plain English How Dogs Learn and How Best to Teach Them.* Berkeley, Calif.: James & Kenneth, 1996.

Reinarman, Craig, and Harry Levine. *Crack in America: Demon Drugs and Social Justice.* Berkeley: University of California Press, 1997.

Ridley, Matt. *The Agile Gene: How Nature Turns on Nurture.* New York: Perennial, 2004.

Ritchie, Carson I. A. *The British Dog: Its History from Earliest Times.* London: Robert Hale, 1981.

Ritvo, Harriet. *The Animal Estate: The English and Other Creatures in the Victorian Age.* Cambridge, Mass.: Harvard University Press, 1987.

Rose, Tricia. *Black Noise: Rap Music and Black Culture in Contemporary America.* Hanover, N.H.: University of New England Press, 1994.

———. *The Hip-Hop Wars: What We Talk About When We Talk About Hip-Hop—and Why It Matters.* New York: Basic Civitas Books, 2008.

Sagan, Carl, and Ann Druyan. *The Demon-Haunted World: Science as a Candle in the Dark.* New York: Random House, 1996.

Sante, Luc. *Low Life: Lures and Snares of Old New York.* New York: Farrar, Straus and Giroux, 1991.

Sax, Boria. *Animals in the Third Reich: Pets, Scapegoats, and the Holocaust.* New York: Continuum, 2000.

Schaffer, Michael. *One Nation Under Dog: Adventures in the New World of Prozac-Popping Puppies, Dog-Park Politics, and Organic Pet Food.* New York: Henry Holt, 2009.

Scott, John Paul. *Aggression.* Chicago: University of Chicago Press, 1975.

Scott, John Paul, and John L. Fuller. *Genetics and Social Behavior of the Dog.* Chicago: University of Chicago Press, 1966.

Serpell, James, ed. *The Domestic Dog: Its Evolution, Behaviour, and Interactions with People.* Cambridge, U.K.: Cambridge University Press, 1995.

Shermer, Michael. *Why People Believe Weird Things: Pseudoscience, Superstition, and Other Confusions of Our Time.* New York: W. H. Freeman, 1997.

Shipman, Pat. *The Animal Connection: A New Perspective on What Makes Us Human.* New York: Norton, 2011.

Siebert, Charles. *The Wauchula Woods Accord: Toward a New Understanding of Animals.* New York: Scribner, 2009.

Sinclair, Leslie, Melinda Merck, and Randall Lockwood. *Forensic Investigation of Animal Cruelty: A Guide for Veterinary and Law Enforcement Professionals.* Washington, D.C.: Humane Society, 2006.

Singer, Peter. *Animal Liberation: A New Ethics for Our Treatment of Animals.* New York: New York Review, 1975.

Sittert, Lance Van, and Sandra Swart. *Canis Africanis: A Dog History of Southern Africa.* Leiden, Netherlands: Brill, 2008.

Skabelund, Aaron Herald. *Empire of Dogs: Canines, Japan, and the Making of the Modern Imperial World.* Ithaca, N.Y.: Cornell University Press, 2011.

Slovic, Paul. *The Perception of Risk.* London: Earthscan, 2000.

Smith, David Livingstone. *Less Than Human: Solving the Puzzle of Dehumanization.* New York: St. Martin's, 2011.

Steele, Claude. *Whistling Vivaldi: How Stereotypes Affect Us and What We Can Do.* New York: Norton, 2011.

Stratton, Richard F. *This Is the American Pit Bull Terrier.* Neptune City, N.J.: T.F.H., 1976.

Sullivan, Robert. *Rats: Observations on the History and Habitat of the City's Most Unwanted Inhabitants.* New York: Bloomsbury, 2004.

Sunstein, Cass R. *Risk and Reason: Safety, Law, and the Environment.* Cambridge, U.K.: Cambridge University Press, 2002.

Sunstein, Cass R., and Martha Craven Nussbaum. *Animal Rights: Current Debates and New Directions.* Oxford: Oxford University Press, 2004.

Townshend, Emma. *Darwin's Dogs: How Darwin's Pets Helped Form a World-Changing Theory of Evolution.* London: Frances Lincoln, 2009.

Tucker, William H. *The Science and Politics of Racial Research.* Urbana: University of Illinois Press, 1994.

Turner, James. *Reckoning with the Beast: Animals, Pain, and Humanity in the Victorian Mind.* Baltimore: Johns Hopkins University Press, 1980.

Unti, Bernard Oreste. *Protecting All Animals: A Fifty-Year History of the Humane Society of the United States.* Washington, D.C.: Humane Society, 2004.

Varner, John Grier, and Jeannette Johnson Varner. *Dogs of the Conquest.* Norman: University of Oklahoma Press, 1983.

Victor, Jeffrey S. *Satanic Panic: The Creation of a Contemporary Legend.* Chicago: Open Court, 1993.

Ward, Jesmyn. *Men We Reaped: A Memoir.* New York: Bloomsbury, 2013.

———. *Salvage the Bones: A Novel.* New York: Bloomsbury, 2011.

Warren, Cat. *What the Dog Knows: The Science and Wonder of Working Dogs.* New York: Touchstone, 2013.

Wasik, Bill, and Monica Murphy. *Rabid: A Cultural History of the World's Most Diabolical Virus.* New York: Viking, 2012.

Wilson, William Julius. *When Work Disappears: The World of the New Urban Poor.* New York: Vintage, 1997.

Woestendiek, John. *Dog, Inc.: The Uncanny Inside Story of Cloning Man's Best Friend.* New York: Avery, 2010.

INDEX

Page numbers in *italics* refer to illustrations.

ILLUSTRATION CREDITS

ABOUT THE AUTHOR

Bronwen Dickey is a contributing editor at the *Oxford American*. Her work has also appeared in *The New York Times, Newsweek, Outside, Slate, Garden & Gun, The Best American Travel Writing 2009,* the *San Francisco Chronicle,* and *The Independent Weekly,* among other publications. She lives in North Carolina.

A NOTE ON THE TYPE

This book was set in Minion, a typeface produced by the Adobe Corporation specifically for the Macintosh personal computer, and released in 1990. Designed by Robert Slimbach, Minion combines the classic characteristics of old-style faces with the full complement of weights required for modern typesetting.

Composed by North Market Street Graphics, Lancaster, Pennsylvania
Printed and bound by Berryville Graphics, Berryville, Virginia
Designed by Maggie Hinders